HMMWV "Hummer" M1113 Series
Operators Manual
TM 9-2320-387-10

Also applies to
M1114
M1151
M1151A1
M1152
M1152A1
M1165
M1165A1
M1167

edited by
Brian Greul

The HMMWV or High Mobility Multipurpose Wheeled Vehicle, commonly called the Hummer was produce by AM General to replace the venerable Jeep or GPW vehicle that the US Military had used for decades. GM later released a similiarly styled, but unrelated Hummer civilian vehicle. This manual is the operator's manual for the HMMWV military vehicle. It is printed as a convenience to enthusiasts, civilian and government operators and others who have need for a professionally printed version of the manual.

Removals: The publisher has removed the following extraneous/excess content from this manual; DOD order signature pages, blank pages next to the same, feedback forms for Army use.

Fonts: A significant effort has been invested to standardize and modernize the fonts used in this manual. The publisher believes the content to be useable but acknowledges that minor layout inconsistencies may exist. This is a regrettable inconvenience associated with reprinting manuals.

An 8.5x11 3 hole punched loose leaf copy may be purchased for your 3 ring binder. Email books@ocotillopress.com for current information.

Should you have suggestions or feedback on ways to improve this book please send email to Books@OcotilloPress.com

---

Edited 2021 Ocotillo Press
ISBN 978-1-954285-32-3

Printed in the United States of America

Ocotillo Press
Houston, TX 77017
Books@OcotilloPress.com

ARMY TM 9-2320-387-10
AIR FORCE TO 36A12-1A-3061-1
MARINE CORPS TM 11033-OR

OPERATOR'S MANUAL

FOR

TRUCK, UTILITY: S250 SHELTER CARRIER,
4X4, M1113
(2320-01-412-0143) (EIC: B6B);

TRUCK, UTILITY: UP-ARMORED
CARRIER, 4X4, M1114
(2320-01-413-3739) (EIC: B6C);

TRUCK, UTILITY: EXPANDED
CAPACITY, ARMAMENT   CARRIER, M1151
(2320-01-518-7330) (EIC: BA5);

TRUCK, UTILITY: EXPANDED
CAPACITY, ARMAMENT CARRIER,
IAP/ARMOR READY, M1151A1
(2320-01-540-2038) (EIC: BEG);

TRUCK, UTILITY: EXPANDED
CAPACITY, ENHANCED,   M1152
(2320-01-518-7332) (EIC: BA6);

TRUCK, UTILITY: EXPANDED
CAPACITY, ENHANCED,
IAP/ARMOR READY, M1152A1
(2320-01-540-2007) (EIC: BEH);

TRUCK, UTILITY: COMMAND AND CON-
TROL/GENERAL PURPOSE VEHICLE, M1165
(2320-01-540-1993) (EIC: BEK);

TRUCK, UTILITY: COMMAND AND CON-
TROL/GENERAL PURPOSE VEHICLE,
IAP/ARMOR READY, M1165A1
(2320-01-540-2017) (EIC: BEJ);

TRUCK, UTILITY: EXPANDED CAPACITY,
TOW ITAS CARRIER, M1167
(2320-01-544-9638).

DISTRIBUTION STATEMENT A: Approved for public release; distribution is unlimited.

HEADQUARTERS, DEPARTMENTS OF THE ARMY,
THE AIRFORCE, AND MARINE CORPS

OCTOBER 1997
PCN 184 110330 00

WARNING

EXHAUST GASES CAN KILL

Brain damage or death can result from heavy exposure. Precautions must be followed to ensure crew safety when the personnel heater or engine of any vehicle is operated for any purpose.

1. Do not operate your vehicle engine in enclosed areas.

2. Do not idle vehicle engine with windows closed.

3. Be alert at all times for exhaust odors.

4. Be alert for exhaust poisoning symptoms. They are:

    - Headache
    - Dizziness
    - Sleepiness
    - Loss of muscular control

5. If you see another person with exhaust poisoning symptoms:

    - Remove person from area
    - Expose to open air
    - Keep person warm
    - Do not permit physical exercise
    - Administer artificial respiration, if necessary*
    - Notify a medic

    *For artificial respiration, refer to FM 21-11.

6. BE AWARE, the field protective mask for Nuclear, Biological, or Chemical (NBC) protection will not protect you from carbon monoxide poisoning.

    THE BEST DEFENSE AGAINST EXHAUST POISONING IS ADEQUATE VENTILATION.

WARNING SUMMARY

- Do not attempt to operate cargo shell door forward latch. The cargo shell door is not to be opened from inside the vehicle. Opening cargo shell door from inside the vehicle may cause damage to equipment or injury to personnel.

- Drycleaning solvent is flammable and will not be used near an open flame. A fire extinguisher will be kept nearby when the solvent is used. Use only in well-ventilated places. Failure to do this may result in injury to personnel and/or damage to equipment.

- Protective gloves, clothing, and/or respiratory equipment must be worn whenever caustic, toxic, or flammable cleaning solutions are used. Failure to do this may result in injury to personnel.

- Do not perform fuel or battery system checks, inspections, or maintenance while smoking or near sparks. Fuel may ignite and batteries may explode, causing damage to vehicle and injury or death to personnel.

- Never use transmission shift lever in place of parking brake. Set parking brake. Ensure transmission shift lever is in P (park) position and transfer case shift lever is NOT in N (neutral) position. Damage to equipment and injury to personnel may occur if these instructions are not followed.

- Ensure all slack from the three-point seatbelt or Improved Personal Restraint System is removed. The three-point seatbelt and Improved Personal Restraint System will lock only during sudden stops or impact. Injury and/or death to personnel may result if an accident occurs and seatbelts or lap and shoulder straps are not in use or adjusted properly.

- This vehicle has been designed to operate safely and efficiently within the limits specified in this TM. Operations beyond these limits are prohibited IAW AR 750-1 without written approval from the Commander, U.S. Army Tank-automotive and Armaments Command, ATTN:AMSTA-CM-S, Warren, MI 48397-5000.

- Use extreme caution when transporting personnel. Although certain design characteristics of the vehicle, such as vehicle width, ground clearance, independent suspension, etc., provide improved capabilities, accidents can still happen.

- Operators are reminded to observe basic safe driving techniques/skills when operating the vehicle, especially when transporting personnel. Vehicle speed must be reduced consistent with weather and road/terrain conditions. Obstacles such as stumps and boulders must be avoided. Failure to use basic safe driving techniques/skills may result in loss of control and an accident or rollover resulting in injury or death to personnel and damage to equipment. Since the troop/cargo area has minimal overhead protections and does not have seatbelts, personnel seated here are at greater risk of serious injury.

- Prior to towing vehicle with rear wheels up, secure steering wheel to prevent front wheels from turning.

- Wear leather gloves when handling winch cable. Do not handle cable with bare hands. When fully extending winch cable, ensure that four wraps of winch cable remain on drum at all times. Direct all personnel to stand clear of winch cable during winch operation. Failure to do this may cause damage to vehicle and injury or death to personnel.

- Protective eye equipment (goggles/shield) must be worn when removing snaprings or springs. Failure to comply may result in injury to personnel.

WARNING SUMMARY (Cont'd)

- Avoid using fire extinguisher in unventilated areas. Prolonged inhalation exposure to extinguishing agent or fumes from burning material may cause injury to personnel. Using fire extinguisher in windy area will cause rapid dispersal of extinguishing agent and reduce effectiveness in fire control.

- Vehicle operation in snow is a hazardous condition. Operator must travel at reduced speeds and be prepared to meet sudden changes in road conditions. Pump brakes gradually when stopping vehicle on ice or snow. Sudden braking will cause wheels to lock and vehicle to slide out of control, causing damage to vehicle and injury or death to personnel.

- Do not rely on wet service brakes. Keep applying brakes until they dry out and uneven braking ceases. Failure to do this may cause damage to vehicle and injury or death to personnel.

- Exhaust components are hot after prolonged vehicle use. Ensure exhaust system components are cool before removing/installing exhaust assembly. Failure to do this may result in injury to personnel.

- Do not use hand throttle as an automatic speed or cruise control. The hand throttle does not automatically disengage when brake is applied, resulting in increased stopping distances and possible hazardous and unsafe operation. Injury to personnel or damage to equipment may result.

- To ensure opening of hood is accomplished safely and effectively, always maintain the proper lifting posture, with legs bent and back straight. When raising and securing hood, ensure hood prop rod is secured to hood support bracket. When releasing hood prop rod, do not pull rod at hook end. Damage to equipment or injury to personnel may occur if hood is not raised, secured, and lowered properly.

- If NBC exposure is suspected, all air filter media should be handled by personnel wearing protective equipment. Consult your unit NBC officer or NBC NCO for appropriate handling or disposal instructions.

- Due to tire load ratings and vehicle load carrying capabilities, only the load range D radial tire can be used.

- Hearing protection is required for driver and passengers when engine is running. Noise levels produced by these vehicles exceed 85 dBA, which may cause injury to personnel.

- Do not operate the vehicle without windshield assembly positioned upright and the B-pillar securely attached. Operation of vehicle without these structures in place may result in injury to personnel and damage to equipment.

- Chock blocks shall be used when parking a vehicle with inoperative parking brakes, operating in extreme cold conditions, parking on inclines, or whenever and wherever maintenance is being performed. Failure to do so may result in injury to personnel or damage to equipment.

- Do not operate the vehicular heater in a closed area without proper exhaust evacuations. Damage to heater or injury to personnel may result.

- Do not operate the vehicular heater when refueling. Damage to heater or injury to personnel may result.

- Use care when opening and closing doors. Do not rest fingers in door opening. Personnel injury may result.

WARNING SUMMARY (Cont'd)

· Do not operate heater when ventilation system is on. Damage to heater or injury to personnel may result.

· Communication shelters AN/GRC-122 and AN/GRC-142 RATT may overload vehicle by up to 500 pounds. Use caution when driving to avoid damage to equipment or injury to personnel.

· Do not use tow pintle as a step when entering or exiting vehicle cargo area. Failure to do so may result in injury to personnel.

· For TOW ITAS vehicles, always utilize the manual crank handle to rotate the turret. Failure to do this may result in injury to personnel.

· For TOW ITAS vehicles, always secure the locking mechanism while on an incline to prevent the T-GPK from unexpectedly rotating, which may result in injury to the crew or damage to the ITAS.

· For TOW ITAS vehicles, always maximize personal space by adjusting storage and placement of personal gear. Failure to do this may result in injury to personnel.

· For TOW ITAS vehicles, the left side and rear T-GPK panels must be lowered to the down position prior to firing IAW with the procedures in FM 3-22.32.

· Frag 5 doors are extremely heavy and difficult to open and close. For vehicles equipped with Frag 5 doors, avoid parking on inclines or sloped terrain if possible. If vehicle must be parked on an incline or sloped terrain, use extreme caution when opening and closing Frag 5 doors. Failure to comply may result in damage to equipment or injury to personnel.

ARMY TM 9-2320-387-10
AIR FORCE TO 36A12-1A-3061-1
MARINE CORPS TM 11033-OR

CHANGE
NO. 6

HEADQUARTERS
DEPARTMENTS OF THE ARMY, AIR FORCE,
AND MARINE CORPS
WASHINGTON, D.CO, *JUNE 2009*

OPERATOR'S MANUAL
FOR

TRUCK, UTILITY: S250 SHELTER CARRIER,
4X4, M1113
(2320-01-412-0143) (EIC: B6B);

TRUCK, UTILITY: UP-ARMORED
CARRIER, 4X4, M1114
(2320-01-413-3739) (EIC: B6C);

TRUCK, UTILITY: EXPANDED
CAPACITY, ARMAMENT   CARRIER, M1151
(2320-01-518-7330) (EIC: BA5);

TRUCK, UTILITY: EXPANDED
CAPACITY, ARMAMENT CARRIER,
IAP/ARMOR READY, M1151A1
(2320-01-540-2038) (EIC: BEG);

TRUCK, UTILITY: EXPANDED
CAPACITY, ENHANCED,   M1152
(2320-01-518-7332) (EIC: BA6);

TRUCK, UTILITY: EXPANDED
CAPACITY, ENHANCED,
IAP/ARMOR READY, M1152A1
(2320-01-540-2007) (EIC: BEH);

TRUCK, UTILITY: COMMAND AND CONTROL/GENERAL
PURPOSE VEHICLE, M1165
(2320-01-540-1993) (EIC: BEK);

TRUCK, UTILITY: COMMAND AND CONTROL/GENERAL
PURPOSE VEHICLE, IAP/ARMOR READY, M1165A1
(2320-01-540-2017) (EIC: BEJ);

TRUCK, UTILITY: EXPANDED CAPACITY,
TOW ITAS CARRIER, M1167
(2320-01-544-9638).

TM 9-2320-387-10, 17 October 1997, is changed as follows:

1. One new model has been added to the front cover. The new cover, located at the end of the change package, replaces the existing cover.
2. Remove old pages and insert new pages as indicated below.
3. New or changed material is indicated by a vertical bar in the margin of the page.
4. File this change sheet in front of the publication for reference purposes.

Marine Corps PCN 184 110330 06

Remove pages	Insert pages
warning c and warning d	warning c and warning d
A and **B**	A through C/(D blank)
i through iv	i through iv
1-1 and 1-2/(1-2.1 and 1-2.2 deleted)	1-1 and 1-2/(1-2.1 and 1-2.2 deleted)
1-5 through 1-18	1-4.7 through 1-18
1-21 and 1-22	1-21 through 1-22.1/(1-22.2 blank)
1-25 through 1-28	1-25 through 1-28.2
1-31 and 1-32	1-31 and 1-32
2-3 and 2-4	2-3 and 2-4
2-7 through 2-8.6	2-7 through 2-8.6
2-13 and 2-14	2-13 and 2-14
2-23 through 2-26	2-22.5 through 2-26
2-29 and 2-30	2-29 and 2-30
2-33 and 2-34	2-33 and 2-34
2-39 through 2-42	2-38.1 through 2-42
2-49 through 2-54	2-49 through 2-54
2-57 and 2-58	2-57 and 2-58
2-60.1 through 2-60.4/(2-61 blank)	2-60.1 through 2-60.4/(2-61 blank)
2-67 and 2-68	2-67 and 2-68
2-71 and 2-72	2-71 and 2-72
2-76.1 through 2-78	2-76.1 through 2-78.1/(2-78.2 blank)
2-81 and 2-82	2-81 and 2-82
2-89 and 2-90	2-89 and 2-90
2-94.1 through 2-94.6	2-94.1 through 2-94.6
2-94.11 through 2-94.14	2-94.9 through 2-94.14
2-95 and 2-96	2-95 and 2-96
2-99/(2-100 blank)	2-98.7 through 2-99/(2-100 blank)
2-101 through 2-102.2	2-101 through 2-102.2
2-122.1 and 2-122.2	2-122.1 and 2-122.2
2-135/(2-136 blank)	2-135 and 2-136
3-6.1 and 3-6.2/(3-7 blank)	3-6.1 and 3-6.2/(3-7 blank)
3-13 and 3-14	3-13 and 3-14
A-1 and A-2	A-1 and A-2
C-1 through C-4	C-1 through C-4
E-5 and E-6	E-4.1 through E-6
G-3 and G-4	G-3 and G-4
G-9 and G-10	G-9 and G-10
None	G-14.1/(G-14.2 blank)
G-15 and G-16	G-15 and G-16
Index 1 through Index 8	Index 1 through Index 8
Cover	Cover

CHANGE

NO. 5

HEADQUARTERS,
DEPARTMENTS OF THE ARMY, AIR FORCE,
AND MARINE CORPS
WASHINGTON, D.C., *30 NOVEMBER 2007*

OPERATOR'S MANUAL
FOR

TRUCK, UTILITY: S250 SHELTER CARRIER,
4X4, M1113
(2320-01-412-0143) (EIC: B6B);

TRUCK, UTILITY: UP-ARMORED
CARRIER, 4X4, M1114
(2320-01-413-3739) (EIC: B6C);

TRUCK, UTILITY: EXPANDED
CAPACITY, ARMAMENT CARRIER, M1151
(2320-01-518-7330) (EIC: BA5);

TRUCK, UTILITY: EXPANDED
CAPACITY, ARMAMENT CARRIER,
IAP/ARMOR READY, M1151A1
(2320-01-540-2038);

TRUCK, UTILITY: EXPANDED
CAPACITY, ENHANCED, M1152
(2320-01-518-7332) (EIC: BA6);

TRUCK, UTILITY: EXPANDED
CAPACITY, ENHANCED,
IAP/ARMOR READY, M1152A1
(2320-01-540-2007);

TRUCK, UTILITY: COMMAND AND CONTROL/GENERAL
PURPOSE VEHICLE, M1165
(2320-01-540-1993);

TRUCK, UTILITY: COMMAND AND CONTROL/GENERAL
PURPOSE VEHICLE, IAP/ARMOR READY, M1165A1
(2320-01-540-2017).

TM 9-2320-387-10, 17 October 1997, is changed as follows:
1. Remove old pages and insert new pages as indicated below.
2. New or changed material is indicated by a vertical bar in the margin of the page.
3. File this change sheet in front of the publication for reference purposes.

Approved for public release; distribution is unlimited.

Remove pages	**Insert pages**
a and b	a and b
A and B	A and B
i through vi	i through vi
1-1 through 1-14	1-1 through 1-14
1-31 through 1-33/(1-34 blank)	1-31 through 1-33/(1-34 blank)
2-1 through 2-4	2-1 through 2-4
2-7 through 2-8.2	2-7 through 2-8.6
2-13 through 2-18.1/(2-18.2 blank)	2-13 through 2-18.1/(2-18.2 blank)
2-22.1 through 2-24	2-22.1 through 2-24
2-27 and 2-28	2-27 and 2-28
2-35 and 2-36	2-34.1 through 2-36
2-51 through 2-54	2-51 through 2-54
2-61 and 2-62	2-60.3 through 2-62
2-65 and 2-66	2-65 and 2-66
2-73 and 2-74	2-73 through 2-74.1/(2-74.2 blank)
2-76.1 through 2-76.4	2-76.1 through 2-76.4/(2-77 blank)
2-80.1 through 2-86	2-80.1 through 2-86
2-89 and 2-90	2-89 and 2-90
2-94.1 through 2-96	2-94.1 through 2-96
2-99 and 2-100	2-98.1 through 2-99/(2-100 blank)
2-101 through 2-102.1/(2-102.2 blank)	2-101 through 2-102.3/(2-102.4 blank)
2-122.1 through 2-124	2-122.1 through 2-124
2-125 and 2-126	2-125 and 2-126
(2-130.1 blank)/2-130.2	(2-130.1 blank)/2-130.2
3-6.1 and 3-6.2	3-6.1 and 3-6.2
3-13 through 3-16	3-13 through 3-16
3-21 and 3-22	3-21 and 3-22
A-1 and A-2	A-1 and A-2
B-1 and B-2	B-1 and B-2
C-1 through C-4	C-1 through C-4
D-1 through D-4	D-1 through D-4
E-3 through E-6	E-3 through E-6
F-1 and F-2	F-1 and F-2
G-1 through G-4	G-1 through G-4
G-7 and G-8	G-7 and G-8
G-15 and G-16	G-15 and G-16
Index 1 through Index 7/(Index 8 blank)	Index 1 through Index 8

ARMY TM 9-2320-387-10
AIR FORCE TO 36A12-1A-3061-1
MARINE CORPS TM 11033-OR

CHANGE

NO. 4

HEADQUARTERS,
DEPARTMENTS OF THE ARMY, AIR FORCE,
AND MARINE CORPS
WASHINGTON, D.C., *28 FEBRUARY 2007*

OPERATOR'S MANUAL
FOR

*TRUCK, UTILITY: S250 SHELTER CARRIER,
4X4, M1113
(2320-01-412-0143) (EIC: B6B);*

*TRUCK, UTILITY: UP-ARMORED
CARRIER, 4X4, M1114
(2320-01-413-3739) (EIC: B6C);*

*TRUCK, UTILITY: EXPANDED
CAPACITY, ARMAMENT CARRIER, M1151
(2320-01-518-7330) (EIC: BA5);*

*TRUCK, UTILITY: EXPANDED
CAPACITY, ARMAMENT CARRIER,
IAP/ARMOR READY, M1151A1
(2320-01-540-2038);*

*TRUCK, UTILITY: EXPANDED
CAPACITY, ENHANCED, M1152
(2320-01-518-7332) (EIC: BA6);*

*TRUCK, UTILITY: EXPANDED
CAPACITY, ENHANCED,
IAP/ARMOR READY, M1152A1
(2320-01-540-2007);*

*TRUCK, UTILITY: COMMAND AND CONTROL/GENERAL
PURPOSE VEHICLE, M1165
(2320-01-540-1993);*

*TRUCK, UTILITY: COMMAND AND CONTROL/GENERAL
PURPOSE VEHICLE, IAP/ARMOR READY, M1165A1
(2320-01-540-2017).*

TM 9-2320-387-10, 17 October 1997, is changed as follows:
  1. Four new models have been added to the front cover. The new cover, located at the end of the change package, replaces the existing cover.
  2. Remove old pages and insert new pages as indicated below.
  3. New or changed material is indicated by a vertical bar in the margin of the page.

Approved for public release; distribution is unlimited.

Remove pages	Insert pages
A and B	A and B
i and ii	i and ii
1-1 through 1-4	1-1 through 1-4
Index 1 through Index 6	Index 1 through Index 6
Cover	Cover

4. File this change sheet in front of the publication for reference purposes.

CHANGE

NO. 3

HEADQUARTERS,
DEPARTMENTS OF THE ARMY, AIR FORCE,
AND MARINE CORPS
WASHINGTON, D.C., *30 APRIL 2006*

OPERATOR'S MANUAL
FOR

TRUCK, UTILITY: S250 SHELTER CARRIER, 4X4, M1113 (2320-01-412-0143) (EIC: B6B);

TRUCK, UTILITY: UP-ARMORED CARRIER, 4X4, M1114 (2320-01-413-3739) (EIC: B6C);

TRUCK, UTILITY: EXPANDED CAPACITY, ARMAMENT CARRIER, M1151 (2320-01-518-7330);

TRUCK, UTILITY: EXPANDED CAPACITY, ENHANCED, M1152 (2320-01-518-7332).

TM 9-2320-387-10, 17 October 1997, is changed as follows:

1. Two new models have been added to the front cover. The new cover, located at the end of the change package, replaces the existing cover.
2. Remove old pages and insert new pages as indicated below.
3. New or changed material is indicated by a vertical bar in the margin of the page.

Remove pages	*Insert pages*
warning a and warning b	warning a and warning b
A and B	A and B
i through vi	i through vi
1-1 through 1-14	1-1 through 1-14
1-31 and 1-32	1-31 through 1-33/(1-34 blank)
2-1 through 2-10	2-1 through 2-10
2-13 through 2-18	2-13 through 2-18.1/(2-18.2 blank)
2-23 and 2-24	2-22.1 through 2-24
2-33 and 2-34	2-33 and 2-34
2-37 through 2-42	2-37 through 2-42
2-51 through 2-56	2-51 through 2-56

Approved for public release; distribution is unlimited.

Remove pages	**Insert pages**
2-59 through 2-62	2-59 through 2-62
2-65 and 2-66	2-65 and 2-66
2-75 and 2-76	2-75 through 2-76.4
2-81 through 2-90	2-80.1 through 2-90
2-95 through 2-108	2-94.1 through 2-108
2-111 through 2-120	2-111 through 2-120
2-123 through 2-126	2-122.1 through 2-126
2-129 through 2-132	2-129 through 2-132
3-3 and 3-4	3-3 and 3-4
3-9 and 3-10	3-9 through 3-10.1/(3-10.2 blank)
3-25 through 3-29/(3-30 blank)	3-25 through 3-29/(3-30 blank)
(3-31 and 3-32 deleted)	(3-31 and 3-32 deleted)
B-1 through B-6	B-1 through B-6
B-9/(B-10 blank)	B-9/(B-10 blank)
C-1 through C-4	C-1 through C-4
D-1 and D-2	D-1 and D-2
E-1 through E-6	E-1 through E-6
F-1 and F-2	F-1 and F-2
F-11/(F-12 blank)	F-11 throughF-15/(F-16 blank)
G-1 through G-10	G-1 through G-10
G-15 and G-16	G-15 and G-16
Index 1 through Index 6	Index 1 through Index 7/ (Index 8 blank)
Cover	Cover

4. File this change sheet in front of the publication for reference purposes.

CHANGE

NO. 2

HEADQUARTERS,
DEPARTMENTS OF THE ARMY AND THE AIR FORCE
WASHINGTON, D.C., *30 JULY 2004*

OPERATOR'S MANUAL
FOR

TRUCK, UTILITY: S250 SHELTER CARRIER, 4X4, M1113 (2320-01-412-0143) (EIC: B6B);

TRUCK, UTILITY: UP-ARMORED CARRIER, 4X4, M1114 (2320-01-413-3739) (EIC: B6C).

TM 9-2320-387-10, 17 October 1997, is changed as follows:

1. Remove old pages and insert new pages as indicated below.
2. New or changed material is indicated by a vertical bar in the margin of the page.

Remove pages	*Insert pages*
A/(B blank)	A and B
1-3 and 1-4	1-3 and 1-4
1-11 through 1-14	1-11 through 1-14
1-31/(1-32 blank)	1-31 and 1-32
2-3 through 2-6	2-3 through 2-6
2-8.1/(2-8.2 blank)	2-8.1/(2-8.2 blank)
2-21 through 2-24	2-21 through 2-24
2-29 through 2-36	2-29 through 2-36
2-41 and 2-42	2-41 and 2-42
2-49 and 2-50	2-49 and 2-50
2-53 and 2-54	2-53 and 2-54
2-59 through 2-64	2-59 through 2-64
2-77 and 2-78	2-76.1 through 2-78
2-93 and 2-94	2-92.1 through 2-94
2-97 through 2-100	2-97 through 2-100
NONE	2-102.1/(2-102.2 blank)
2-121 through 2-124	2-121 through 2-124
3-5 through 3-8	3-5 through 3-8
3-21 through 3-24	3-21 through 3-24
A-1 and A-2	A-1 and A-2
B-3 through B-9/(B-10 blank)	B-3 through B-9/(B-10 blank)
C-1 and C-2	C-1 and C-2
F-1 through F-8	F-1 through F-8
G-3 and G-4	G-3 and G-4
G-7 and G-8	G-7 and G-8
G-13 through G-16	G-13 through G-16
Index 1 through Index 6	Index 1 through Index 6

3. File this change sheet in front of the publication for reference purposes.

Approved for public release; distribution is unlimited.

ARMY TM 9-2320-387-10
AIR FORCE TO 36A12-1A-3061-1

CHANGE   HEADQUARTERS,
                DEPARTMENTS OF THE ARMY AND THE AIR FORCE
NO. 1   WASHINGTON, D.C., *15 OCTOBER 2001*

OPERATOR'S MANUAL
FOR

TRUCK, UTILITY: S250 SHELTER CARRIER, 4X4, M1113 (2320-01-412-0143) (EIC: B6B);

TRUCK, UTILITY: UP-ARMORED CARRIER, 4X4, M1114 (2320-01-413-3739) (EIC: B6C).

TM 9-2320-387-10, 17 October 1997, is changed as follows:

1. Remove old pages and insert new pages as indicated below.
2. New or changed material is indicated by a vertical bar in the margin of the page.

Remove page	*Insert page*
warning a through warning c/(warning d blank)	warning a through warning d
None	A/(B blank) (after warning d)
i and ii	i and ii
1-1 through 1-4	1-1 through 1-4
1-7 through 1-14	1-7 through 1-14
2-1 through 2-6	2-1 through 2-6
None	2-8.1/(2-8.2 blank)
2-9 through 2-14	2-9 through 2-14
2-23 and 2-24	2-23 and 2-24
2-27 through 2-48	2-27 through 2-48
None	2-48.1 and 2-48.2
2-49 through 2-70	2-49 through 2-70
2-77 and 2-78	2-77 and 2-78
2-83 through 2-90	2-83 through 2-90
2-95 and 2-96	2-95 and 2-96
2-101 through 2-104	2-101 through 2-104
2-109 through 2-114	2-109 through 2-114

Remove page	Insert page
2-117 and 2-118	2-117 and 2-118
2-123 and 2-124	2-123 and 2-124
2-129 and 2-130	2-129 and 2-130
2-133 and 2-134	2-133 and 2-134
3-7 through 3-10	3-7 through 3-10
3-13 through 3-20	3-13 through 3-20
3-23 and 3-24	3-23 and 3-24
3-29 through 3-31/	3-29/(3-30 blank)
(3-32 blank)	(3-31 and 3-32 deleted)
A-1 and A-2	A-1 and A-2
B-1 and B-2	B-1 and B-2
B-5 and B-6	B-5 and B-6
B-9/(B-10 blank)	B-9/(B-10 blank)
D-3 and D-4	D-3 and D-4
E-3 through E-6	E-3 through E-6
G-1 through G-8	G-1 through G-8
G-13 through G-16	G-13 through G-16
Index 1 through Index 6	Index 1 through Index 6
DA Form 2028-2	DA Form 2028

3. File this change sheet in front of the publication for reference purposes.

LIST OF EFFECTIVE PAGES

NOTE: The portion of the text affected by the changes is indicated by a vertical line in the outer margins of the page.

Dates of issue for original and changed pages are:

Original . . . . 0 . . . . 17 October 1997
Change . . . . 1 . . . . 15 October 2001
Change . . . . 2 . . . . 30 July 2004
Change . . . . 3 . . . . 30 April 2006
Change . . . . 4 . . . . 28 February 2007
Change . . . . 5 . . . . 30 November2007
Change . . . . 6 . . . . .30 June 2009

TOTAL NUMBER OF PAGES IN THIS PUBLICATION IS 442
CONSISTING OF THE FOLLOWING:

Page No. . . . . . .*Change No.	Page No. . . . . . .*Change No.	Page No. . . . . . .*Change No.
Warning a 0	1-16.1 - 1-16.2 Added 6	2-18.1 5
Warning b 5	1-17 6	2-18.2 Blank Added 3
Warning c 1	1-18 - 1-21 0	2-19 - 2-21 0
Warning d 6	1-22 6	2-22 2
A - B 6	1-22.1 Added 6	2-22.1 - 2.22.4 6
C Added 6	1-22.2 Blank Added 6	2-22.5 - 2-22.8 Added 6
D Blank Added 6	1-23 - 1-24 0	2-23 6
i .6	1-25 - 1-28 6	2-24 5
ii 3	1-28.1 - 1-28.2 Added 6	2-25 6
iii 5	1-29 - 1-30 0	2-26 0
iv 6	1-31 6	2-27 1
v 5	1-32 - 1-33 5	2-28 5
vi 3	1-34 Blank Added 3	2-29 6
1-1 6	2-1 3	2-30 - 2-32 2
1-2 5	2-2 5	2-32.1 - 2-32.2 Added 2
1-2.1 - 1-2.2 Deleted 5	2-3 6	2-33 3
1-3 - 1-4 5	2-4 1	2-34 6
1-4.1 - 1-4.2 5	2-5 2	2-34.1 - 2-34.2 Added 5
1-4.3 - 1-4.6 Added 5	2-6 3	2-35 - 2-36 5
1-4.7 6	2-7 0	2-37 3
1-4.8 Added 6	2-8 5	2-38 0
1-5 Blank 6	2-8.1 3	2-38.1 - 2-38.2 Added 6
1-6 6	2-8.2 6	2-39 - 2-41 6
1-6.1 6	2-8.3 Blank Added 5	2-42 3
1-6.2 Blank Added 3	2-8.4 6	2-43 - 2-48 1
1-7 - 1-10 6	2-8.5 Added 5	2-48.1 - 2-48.2 Added 1
1-10.1 6	2-8.6 6	2-49 1
1-10.2 3	2-9 1	2-50 2
1-11 - 1-12 6	2-10 3	2-51 1
1-12.1 Added 6	2-11 - 2-12 1	2-52 6
1-12.2 Blank Added 6	2-13 6	2-52.1 6
1-13 6	2-14 3	2-52.2 Blank Added 3
1-14 1	2-15 0	2-53 - 2-54 6
1-15 - 1-30 0	2-16 - 2-18 5	2-55 1
1-16 6		

*Zero in this column indicates original page.

LIST OF EFFECTIVE PAGES

*Zero in this column indicates original page.

LIST OF EFFECTIVE PAGES

*Zero in this column indicates original page.

TECHNICAL MANUAL
NO. 9-2320-387-10
NO. 11033-OR
TECHNICALORDER
NO. 36A12-1A-3061-1

HEADQUARTERS,
DEPARTMENTS OFTHE ARMY,
THE AIR FORCE, AND MARINE CORPS
WASHINGTON, D.OCTOBER 1997

OPERATOR'S MANUAL
FOR

TRUCK, UTILITY: S250 SHELTER CARRIER, 4X4, M1113
(2320-01-412-0143) (EIC: B6B);

TRUCK, UTILITY: UP-ARMORED   CARRIER, 4X4, M1114
(2320-01-413-3739) (EIC: B6C);

TRUCK, UTILITY: EXPANDED CAPACITY, ARMAMENT   CARRIER, M1151
(2320-01-518-7330) (EIC: BA5);

TRUCK, UTILITY: EXPANDED CAPACITY, ARMAMENT CARRIER,
IAP/ARMOR READY, M1151A1
(2320-01-540-2038) (EIC: BEG);

TRUCK, UTILITY: EXPANDED CAPACITY, ENHANCEDM1152
(2320-01-518-7332) (EIC: BA6);

TRUCK, UTILITY: EXPANDED CAPACITY, ENHANCED, IAP/ARMOR READY, M1152A1
(2320-01-540-2007) (EIC: BEH);

TRUCK, UTILITY: COMMAND AND CONTROL/GENERAL PURPOSE VEHICLE, M1165
(2320-01-540-1993) (EIC: BEK);

TRUCK, UTILITY: COMMAND AND CONTROL/GENERAL PURPOSE VEHICLE, IAP/ARMOR
READY, M1165A1 (2320-01-540-2017) (EIC: BEJ);

TRUCK, UTILITY: EXPANDED CAPACITY, TOW ITAS CARRIER, M1167
(2320-01-544-9638).

Approved for public release; distribution is unlimited.

REPORTING ERRORS AND RECOMMENDING IMPROVEMENTS

(Army) You can help improve this publication. If you find any errors, or if you would like to recommend any improvements to the procedures in this publication, please let us know. The preferred method is to submit your DAForm 2028 (Recommended Changes to Publications and Blank Forms) through the Internet, on the Army Electronic Product Support (AEPS) website. The Internet address is https://aeps.ria.army.mil. The DAForm 2028 is located under the Public Applications section in the AEPS Public Home Page. Fill out the form and click on SUBMIT. Using this form on the AEPS will enable us to respond quicker to your comments and better manage the DA Form 2028 program. You may also mail, e-mail, or fax your comments or DA Form 2028 directly to the U.S. Army TACOM Life Cycle Management Command. The postal mail address is U.S. Army TACOM Life Cycle Management Command, ATTN: AMSTA-LC-LMPP / TECH PUBS, 1 Rock Island Arsenal, Rock Island, IL 61299-7630. The e-mail address is tacomlcmc.daform2028@us.army .mil. The fax number is DSN 793-0726 or Commercial (309) 782-0726. (Marine Corps) Submit notice of discrepancies or suggest changes on a NAVMC 10772. Users without CAC/PKI certificates may submit the NAVMC via the Internet using website http://www.ala.usmc.mil/navmc/part1.htm, You can then fill in and submit the automated NAVMC 10772 Form. NAVMC forms may also be submitted by electronic mail to bmatcommarlogbases@logcom.usmc.mil mailto:mbmatcommarlogbases@logcom.usmc.mil, or by mailing a paper copy of the NAVMC 10772 in an envelope addressed to Commander, Marine Corps Systems Command, Attn: Assistant Commander Acquisition and Logistics (LOG/TP), 814 Radford Blvd., Room 316E, Albany, Georgia 31704-0343. Problems or questions regarding the NA VMC 10772 program should be reported by calling DSN 567-5017 or DSN 567-6439 . In addition to electronic submittal of the NAVMC form via the above web links, forward an information copy (cc:) or mail a paper copy to the Logistics Management Specialist at the following address: Logistics Management Specialist, Code PMM151, 814 Radford Blvd., STE 310W, Albany, Georgia 31704-0343.

TABLE OF CONTENTS

HOW TO USE THIS MANUAL

ABOUT YOUR MANUAL

Spend some time looking through this manual. You'll find that it takes a positive approach and clearly states only what you can do. Before attempting any questionable operation which is not specifically authorized in this manual, clearance must be obtained from your supervisor.

Features added to improve the convenience of this manual and increase your efficiency are:

a. Accessing Information – Troubleshooting guides for specific systems lead directly to step-by-step directions for problem solving. Maintenance tasks use preventive maintenance checks and services for each vehicle.

b. Illustrations – A variety of methods are used to make locating and identifying components easier. Locator illustrations with keyed text, exploded views, and cut-away diagrams make the information in this manual easier to understand and follow.

c. Keying Text With Illustrations – Instructions/text are located together with figures that illustrate the specific task you are working on. Generally, the task steps and figures are located side by side.

USING YOUR MANUAL: EXAMPLE 1

TASK: You are starting your ECV vehicle engine and are in need of instructions to complete this procedure.

OPERATING INSTRUCTION STEPS:

1. Look for OPERATING INSTRUCTIONS in the table of contents (page ii) of this manual.

2. Look through the list of section titles until you find section III, Operation Under Usual Conditions.

3. Turn to page 2-60.3 as indicated.

TABLE OF CONTENTS

HOW TO USE THIS MANUAL (Cont'd)

4. On page 2-60.3 look through the Operation Under Usual Conditions reference index for starting the engine.

5. Starting on page 2-62 and continuing through page 2-64, you will find directions for starting the engine listed in progressive order.

6. Before performing the operating instruction steps, take time to examine and familiarize yourself with the complete operation procedure and required PMCS.

7. Procedures include everything you must do to accomplish a basic operations task.

8. Numbered callouts, found with the art and text and arranged in a clockwise pattern, will make it easier for you to identify and locate instruments and controls.

9. Pay particular attention to all notes, cautions, and warnings. They are designed to assist you with your task, prevent damage to the vehicle and its components, and protect you from injury.

TM 9-2320-387-10

Section III. OPERATION UNDER USUAL CONDITIONS

2-9. GENERAL

This section provides instructions for vehicle operations under moderate temperature, humidity, and terrain conditions. For vehicle operation under unusual conditions, refer to section IV of this chapter.

2-10. OPERATION UNDER USUAL CONDITIONS REFERENCE INDEX

TM 9-2320-387-10

2-11. BREAK-IN SERVICE

CAUTION

Do not tow trailer during the first 500 mi (805 km) of operation. Damage to equipment may occur.

Break-in precautions are no longer required during the first 500 mi (805 km) of operation, with the exception of trailer towing.

2-12. STARTING THE ENGINE

WARNING

- The automatic transmission has a PARK position. Never use the shift lever in place of the parking brake. Set the parking brake. Make sure the transmission shift lever is in the P (park) position and transfer case shift lever is NOT in the N (neutral) position. Damage to equipment and injury to personnel may occur if these instructions are not followed.
- Chock blocks will be used when parking a vehicle with inoperative parking brakes, operating in extreme cold conditions, parking on inclines, or whenever and wherever maintenance is being performed. Failure to do so may result in injury to personnel or damage to equipment.
- Hearing protection is required for driver and passengers when engine is running. Noise levels produced by these vehicles exceed 85 dBA, which may cause injury to personnel.

a. Ensure transmission shift lever (3) is in P (park) position and transfer case shift lever (4) is NOT in N (neutral) position.

NOTE

To apply parking brake, grasp handle firmly and pull upward until handle is locked in a straight-up position.

a.1. Ensure parking brake (5) is applied.

b. Adjust driver's seat (para. 2-16).

USING YOUR MANUAL: EXAMPLE 2

TASK: Your ECV engine cranks but does not start.

TROUBLESHOOTING STEPS:

1. Look for MAINTENANCE INSTRUCTIONS in the of table of contents (page ii) of this manual.

2. Look through the list of section titles until you find section II, Troubleshooting.

3. Turn to page 3-2 as indicated.

4. Starting on page 3-3 and continuing on page 3-4, you will find the troubleshooting table 3-1 instructions that will identify and correct simple engine malfunctions. Look at item 3. ENGINE CRANKS BUT DOES NOT START.

TM 9-2320-387-10

Section II. TROUBLESHOOTING

3-5. GENERAL

Troubleshooting, table 3-1, contains instructions that will help the operator identify and correct simple vehicle malfunctions. The table also helps the operator identify major mechanical difficulties that must be referred to unit maintenance. The listing of possible malfunctions come under major vehicle headings. They are:

- Engine
- Heating system
- Transmission
- Transfer case
- Brakes
- Wheels and tires
- Steering
- Winch
- Up-armored carrier

3-6. TROUBLESHOOTING PROCEDURES

a. Table 3-1 lists the common malfunctions which you may find during the operation or maintenance of the vehicles or their components. You should perform the tests/inspections and corrective actions in the order listed.

b. This manual cannot list all malfunctions that may occur, nor all tests or inspections and corrective actions. If a malfunction is not listed or is not corrected by listed actions, notify your supervisor.

NOTE

- Hydrostatic lock is caused by the entry of substantial amounts of water into the engine through the air intake system and subsequent contamination of the fuel system. Hydrostatic lock most frequently occurs during or just after fording. Water is forced into the air intake system, is drawn into the engine, and effectively locks up the engine.
- Notify unit maintenance if you suspect hydrostatic lock and they will test the engine.

3-2

TM 9-2320-387-10

Table 3-1. Troubleshooting.

| MALFUNCTION |
| TEST OR INSPECTION |
| CORRECTIVE ACTION |

ENGINE

1. ENGINE FAILS TO CRANK

Step 1. Check to see if transmission shift lever is in P (park).
   If not, place lever in P (park).
Step 2. Check battery fluid level and check battery cable connections for looseness, damage, or corrosion.
   If any of these conditions exist, notify unit maintenance.
Step 3. Attempt to slave-start vehicle (para. 2-23).
Step 4. Other causes.
   Notify unit maintenance.

2. ENGINE CRANKS SLOWLY

Step 1. Check battery fluid level and check battery cable connections for looseness, damage, or corrosion.
   If any of these conditions exist, notify unit maintenance.
Step 2. Attempt to slave-start vehicle (para. 2-23).
Step 3. Other causes.
   Notify unit maintenance.

3. ENGINE CRANKS BUT DOES NOT START

Step 1. Check to see if fuel gauge indicates E (empty).
   Fill fuel tank, and start engine.
Step 2. Purge fuel system of air (para. 3-11).
Step 3. Check to see if wait-to-start lamp assembly fails to light or does not go out.
   Notify unit maintenance if wait-to-start lamp assembly fails to light or does not go out.
Step 4. Other causes.
   Notify unit maintenance.

4. VEHICLE NOT CHARGING ACCORDING TO VOLTMETER

Step 1. Check battery cable connections for looseness, damage, or corrosion.
   Notify unit maintenance of any damage to battery cables.
Step 2. Check for broken or missing drivebelt.
   Notify unit maintenance if drivebelt is broken or missing.
Step 3. Other causes.
   Notify unit maintenance.

3-3

HOW TO USE THIS MANUAL (Cont'd)

MAINTENANCE PROCEDURES:

5. Procedures include everything you must do to accomplish a basic maintenance task.

6. Before beginning a maintenance task, familiarize yourself with the entire maintenance procedure.

7. Pay particular attention to all notes, cautions, and warnings. They are designed to assist you with your task, prevent damage to the vehicle and its components, and protect you from injury or death.

8. An exploded diagram of the component, removed from the vehicle, shows part locations, attachments, and assembly relationships.

9. Numbered callouts, found with the art and text and arranged in a clockwise pattern, will make it easier for you to identify parts and locations.

10. Examine this manual and you will discover it is easier to use when you understand its design. We hope it will encourage you to use it often.

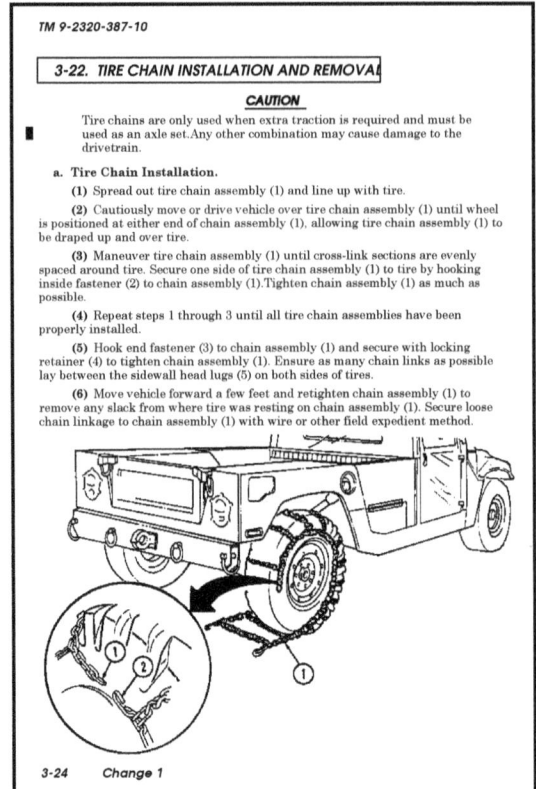

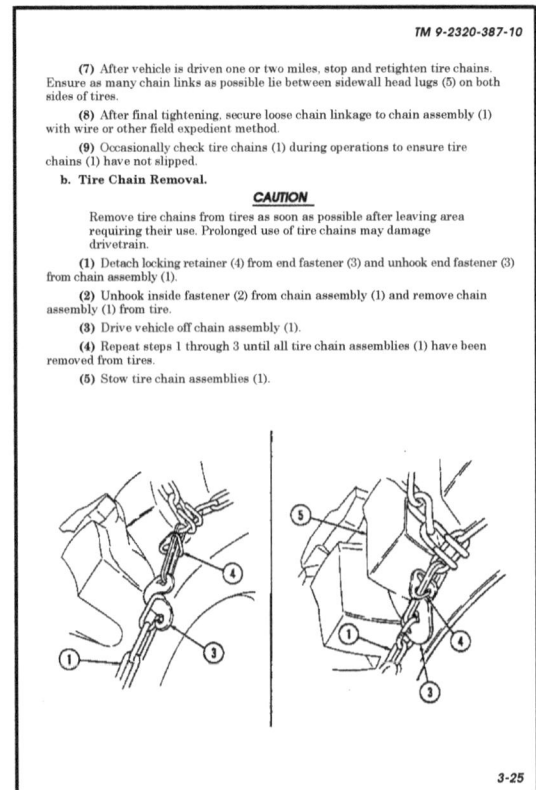

TM 9-2320-387-10

3-22. TIRE CHAIN INSTALLATION AND REMOVAL

CAUTION

Tire chains are only used when extra traction is required and must be used as an axle set. Any other combination may cause damage to the drivetrain.

a. Tire Chain Installation.

(1) Spread out tire chain assembly (1) and line up with tire.

(2) Cautiously move or drive vehicle over tire chain assembly (1) until wheel is positioned at either end of chain assembly (1), allowing tire chain assembly (1) to be draped up and over tire.

(3) Maneuver tire chain assembly (1) until cross-link sections are evenly spaced around tire. Secure one side of tire chain assembly (1) to tire by hooking inside fastener (2) to chain assembly (1). Tighten chain assembly (1) as much as possible.

(4) Repeat steps 1 through 3 until all tire chain assemblies have been properly installed.

(5) Hook end fastener (3) to chain assembly (1) and secure with locking retainer (4) to tighten chain assembly (1). Ensure as many chain links as possible lay between the sidewall head lugs (5) on both sides of tires.

(6) Move vehicle forward a few feet and retighten chain assembly (1) to remove any slack from where tire was resting on chain assembly (1). Secure loose chain linkage to chain assembly (1) with wire or other field expedient method.

3-24   Change 1

TM 9-2320-387-10

(7) After vehicle is driven one or two miles, stop and retighten tire chains. Ensure as many chain links as possible lie between sidewall head lugs (5) on both sides of tires.

(8) After final tightening, secure loose chain linkage to chain assembly (1) with wire or other field expedient method.

(9) Occasionally check tire chains (1) during operations to ensure tire chains (1) have not slipped.

b. Tire Chain Removal.

CAUTION

Remove tire chains from tires as soon as possible after leaving area requiring their use. Prolonged use of tire chains may damage drivetrain.

(1) Detach locking retainer (4) from end fastener (3) and unhook end fastener (3) from chain assembly (1).

(2) Unhook inside fastener (2) from chain assembly (1) and remove chain assembly (1) from tire.

(3) Drive vehicle off chain assembly (1).

(4) Repeat steps 1 through 3 until all tire chain assemblies (1) have been removed from tires.

(5) Stow tire chain assemblies (1).

3-25

CHAPTER 1
INTRODUCTION

Section I. GENERAL INFORMATION

1-1. SCOPE

a. This manual contains instructions for operating and servicing the following HMMWVs (High Mobility Multipurpose Wheeled Vehicles):

- M1113 S250 Shelter Carriers
- M1114 Up-Armored Carriers
- M1151 Armament Carriers
- M1151A1 Armament Carriers, IAP/Armor Ready
- M1152 Expanded Capacity Utility Trucks
- M1152A1 Expanded Capacity Utility Trucks, IAP/Armor Ready
- M1165 Command and Control/General Purpose Vehicles
- M1165A1 Command and Control/General Purpose Vehicles, IAP/Armor Ready
- M1167 Expanded Capacity TOW ITAS Carriers

b. The material presented here provides operators with information and procedures needed to provide the safest and most efficient operation of these vehicles.This information includes:

(1) Operator forms and records.

(2) Descriptions of each vehicle and its operation.

(3) The purpose of each vehicle.

(4) Vehicle limitations.

(5) The function of all controls and indicators.

(6) Operating instructions for each vehicle.

(7) Cautions and warnings to operators regarding safety to personnel and equipment.

(8) How and when to use special purpose kits.

(9) Operator maintenance checks and service procedures.

(10) Troubleshooting procedures to be followed by operators if the vehicle malfunctions.

1-2. MAINTENANCE FORMS AND RECORDS

Department of the Army forms and procedures used for equipment maintenance will be those prescribed by DA Pam 750-8,The Army Maintenance Management System (TAMMS) Users Manual (Marine Corps) refer to 4700-13/1 series.

1-3. HAND RECEIPT MANUAL

This operator's manual has a companion document with a TM number followed
by -HR (which stands for Hand Receipt).TM 9-2320-387-10-HR consists of
preprinted hand receipts (DA Form 2062) that list end item related equipment
(i.e., COEI, BII, and AAL) you must account for.As an aid to property
accountability, additional -HR manuals may be requisitioned from the following
source in accordance with procedures in chapter 12,AR 25-30:

> Commander
> U.S.Army Publications Distribution Center
> 2800 Eastern Blvd.
> Baltimore, MD 21220-2896

1-4. CORROSION PREVENTION AND CONTROL (CPC)

Corrosion Prevention and Control (CPC) of Army materiel is a continuing concern. It
is important that any corrosion problems with this item be reported so that the
problem can be corrected and improvements made to prevent the problem in future
items.While corrosion is typically associated with rusting of metals, it can also
include deterioration of other materials, such as rubber and plastic. Unusual
cracking, softening, swelling, or breaking of these materials may be a corrosion
problem. If a corrosion problem is identified, it can be reported using SF 368 Product
Quality Deficiency Report (PQDR). Use of key words such as corrosion, rust,
deterioration, or cracking will ensure that the information is identified as a CPC
problem.The form should be submitted to the address specified in DA Pam 750-8.

1-5. REPORTING EQUIPMENT IMPROVEMENT RECOMMENDATIONS (EIRs)

If your vehicle needs improvement,let us know.Send us an EIR.You,the user,are the
only one who can tell us what you don't like about your equipment.Let us know why you
don't like the design or performance.The preferred method for submitting QDRs is
through the Army Electronic Product Support (AEPS) website under the Electronic
Deficiency Reporting System (EDRS).The web address is:https://aeps.ria.army.mil.This is
a secured site requiring a password which can be applied for on the front page of the
website.If the above method is not available to you,put it on an SF 368,Product Quality
Deficiency Report (PQDR),and mail it to us at:U.S.Army Tank-automotive and
Armaments Command,ATTN:AMSTA-TR-E/PQDR MS 267,6501 E.11 Mile Road,
Warren,MI 48397-5000.We'll send you a reply.

1-6. EQUIPMENT IMPROVEMENT REPORT AND MAINTENANCE DIGEST (EIR MD)

The quarterly Equipment Improvement Report and Maintenance Digest, TB 43-0001-62 series, contains valuable field information on equipment covered in this manual. Information in TB 43-0001-62 series is compiled from some of the Equipment Improvement Reports you prepared on vehicles covered in this manual. Many of these articles result from comments, suggestions, and improvement recommendations that you submitted to the EIR program.The TB 43-0001-62 series contains information on equipment improvements, minor alterations, proposed Modification Work Orders (MWOs), warranties (if applicable), actions taken on some of your DA Form 2028s (Recommended Changes to Publications), and advance information on proposed changes that may affect this manual.The information will help you in doing your job better and will help in keeping you advised of the latest changes to this manual.Also refer to DA Pam 25-30, Consolidated Index of Army Publications and Blank Forms, and appendix A, References, of this manual.

1-7. BREAK-IN SERVICE

CAUTION

Do not tow trailer during the first 500 mi (805 km) of operation. Damage to equipment may occur.

Break-in precautions are no longer required during the first 500 mi (805 km) of operation, with the exception of trailer towing.

Section II. EQUIPMENT DESCRIPTION

1-8. EQUIPMENT CHARACTERISTICS, CAPABILITIES, AND FEATURES

The 1-1/4 ton, 4x4, ECV series of vehicles are tactical vehicles designed for use over all types of roads, as well as cross-country terrain, in all weather conditions.The vehicles have four driving wheels, powered by a V-8 liquid-cooled turbocharged diesel engine. Four-wheel hydraulic service brakes and a mechanical parking brake are standard.All vehicles are equipped with a pintle hook for towing.Tiedowns and lifting eyes are provided for air, rail, or sea shipment.

1-9. METRIC SYSTEM

The equipment/system described herein contains metric components and requires metric, common, and special tools; therefore, both metric and standard units will be used throughout this publication. In addition, a metric conversion table is located on the inside back cover of this publication.

1-10. S250 SHELTER CARRIER (M1113)

a. Purpose of Vehicle. The M1113 shelter carrier provides the capability to secure and transport the S250 electrical equipment shelter.The optional winch permits recovery operations of similar vehicles. Refer to table 1-10.1 for payload.

b. Performance. Fully-loaded M1113 shelter carriers will climb road grades as steep as 40% (22°) and traverse a side slope of up to 30% (17°).The vehicle fords hard bottom water crossings up to 30 inches (76 centimeters) without a deep water fording kit and 60 inches (152 centimeters) with the kit. Refer to table 1-10 for cruising range.

c. Special Limitations. None.

d. Special Instructions in the Manual.

   (1) Refer to para. 2-2, Controls, Indicators, and Equipment.

   (2) Refer to chapter 2, section V, S250 Shelter Carrier Operation.

M1113

1-11. UP-ARMORED CARRIER (M1114)

a. Purpose of Vehicle. The M1114 up-armored vehicle carrier provides added ballistic protection for armament components, crew, and ammunition.The optional rear winch permits recovery operations of similar vehicles. Refer to table 1-10.1 for payload.

b. Performance. The M1114 up-armored vehicle carrier will climb road grades as steep as 40% (22°) and traverse a side slope of up to 30% (17°).The vehicle fords hard bottom water crossings up to 30 inches (76 centimeters). Refer to table 1-10 for cruising range.

c. Special Limitations. None.

d. Special Instructions in the Manual.

    (1) Refer to para. 2-2, Controls, Indicators, and Equipment.

    (2) Refer to para. 2-33, Up-Armored Weapon Station Operation.

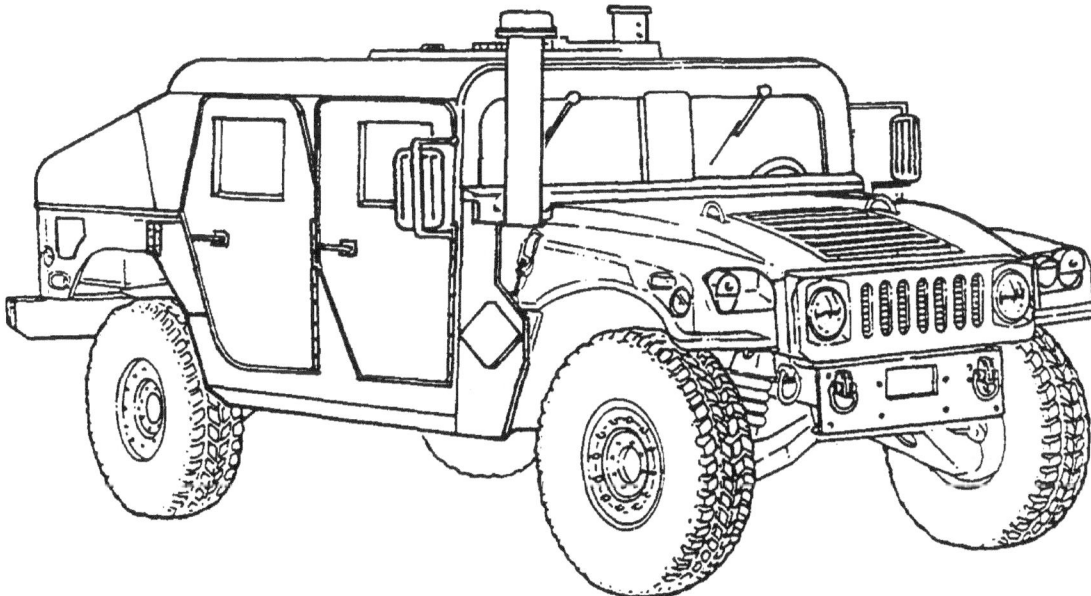

M1114

1-11.1. EXPANDED CAPACITY, ARMAMENT CARRIER (M1151)

a. **Purpose of Vehicle.** The M1151 expanded capacity armament carrier provides mounting and firing of the MK19 automatic grenade launcher, M2, caliber .50 machine gun; M60, 7.62 mm machine gun; M240B, 7.62 mm machine gun; and M249, 5.56 mm Squad Assault Weapon (SAW); ring-mounted with a 360° arc of fire, with armor protection for crew, weapon components, and ammunition. For higher payload capacity, the M1151 is equipped with a reinforced frame, crossmemebers, lifting shackles, heavy-duty variable rate rear springs, shock absorbers, reinforced control arms, heavy-duty tires and rims, and a transfer case and differential with a modified gear ratio.The optional front winch permits recovery operations of similar vehicles. Refer to table 1-10.1 for payload.

b. **Performance.** Fully-loaded M1151 armament carriers will climb road grades as steep as 40% (22°) and traverse a side slope of up to 30% (13.5°).The vehicles ford hard bottom water crossings up to 30 inches (76 centimeters) without a deep water fording kit and 60 inches (152 centimeters) with the kit. Refer to table 1-10 for cruising range.

c. **Special Limitations.** None.

d. **Special Instructions in the Manual.**

(1) Refer to paragraph 2-2, Controls, Indicators, and Equipment.

(2) Refer to chapter 2, section III, Operation Under Usual Conditions.

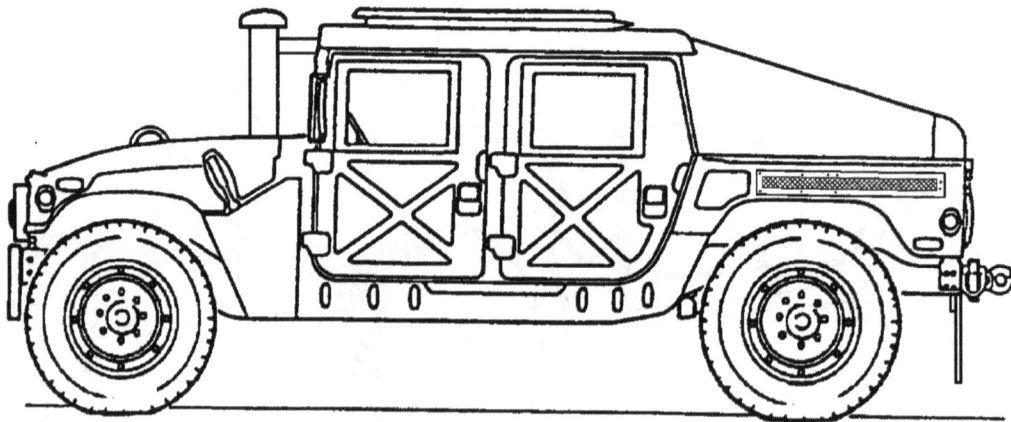

M1151

1-11.2. EXPANDED CAPACITY, ARMAMENT CARRIER IAP/ARMOR READY (M1151A1)

a. **Purpose of Vehicle.** The M1151 expanded capacity armament carrier provides mounting and firing of the MK19 automatic grenade launcher, M2, caliber .50 machine gun; M60, 7.62 mm machine gun; M240B, 7.62 mm machine gun; and M249, 5.56 mm Squad Assault Weapon (SAW); ring-mounted with a 360° arc of fire, with armor protection for crew, weapon components, and ammunition. For higher payload capacity, the M1151 is equipped with a reinforced frame, crossmemebers, lifting shackles, heavy-duty variable rate rear springs, shock absorbers, reinforced control arms, heavy-duty tires and rims, and a transfer case and differential with a modified gear ratio.The optional front winch permits recovery operations of similar vehicles. Refer to 1-10.1 for payload.

b. **Performance.** Fully-loaded M1151A1 armament carrier will climb road grades as steep as 40% (22°) and traverse a side slope of up to 30% (13.5°) The vehicle fords hard bottom water crossings up to 30 inches (76 centimeters) without a deep water fording kit and 60 inches (152 centimeters) with the kit. Refer to table 1-10 for cruising range.

c. **Special Limitations.** None.

d. **Special Instructions in the Manual.**

(1) Refer to paragraph 2-2, Controls, Indicators, and Equipment.

(2) Refer to chapter 2, section III, Operation Under Usual Conditions.

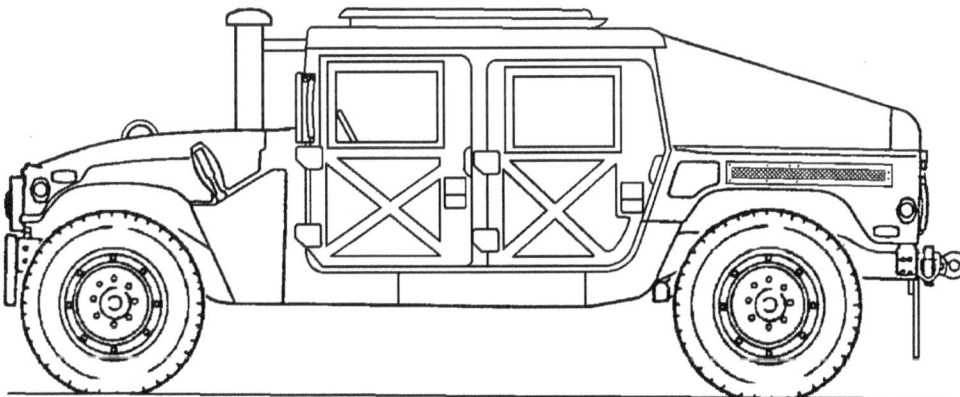

M1151A1

1-11.3. EXPANDED CAPACITY, ENHANCED (M1152)

a. Purpose of the Vehicle. The M1152 expanded capacity, enhanced truck is used to transport personnel.The M1152 provides the capability to secure and transport the S250 electrical equipment shelter. For higher payload capacity, the M1152 is equipped with a reinforced frame, crossmembers, lifting shackles, heavy-duty variable rate rear springs, shock absorbers, reinforced control arms, heavy-duty tires and rims, and a transfer case and differential with a modified gear ratio.The optional front winch permits recovery operations of similar vehicles. Refer to table 1-10.1 for payload.

b. Performance. Fully-loaded M1152 utility trucks will climb road grades as steep as 40% (22°) and traverse a side slope of up to 30% (13.5°).The vehicle fords hard bottom water crossings up to 30 inches (76 centimeters) without a deep water fording kit and 60 inches (152 centimeters) with the kit. Refer to table 1-10 for cruising range.

c. Special Limitations. None.

d. Special Instructions in the Manual.

    (1) Refer to paragraph 2-2, Controls, Indicators, and Equipment.

    (2) Refer to paragraph 2-53.1,Troop Seat Kit Operation.

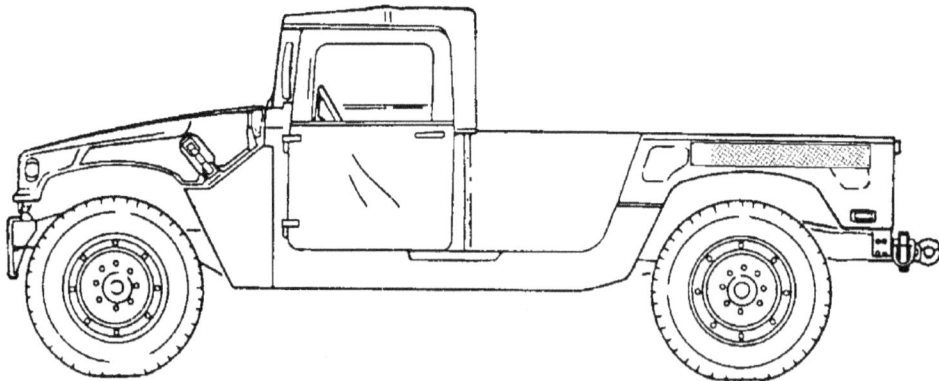

M1152

1-11.4. EXPANDED CAPACITY, ENHANCED, IAP/ARMOR READY (M1152A1)

a. Purpose of the Vehicle. The M1152A1 expanded capacity, enhanced, IAP/armor ready is used to transport personnel.The M1152A1 come equipped with Integrated Armor protection (IAP) which provides added ballistic protection for armament components, crew, and ammunition.The M1152A1 vehicles are capable of transporting a two-man or four-man crew and eight passengers.The M1152A1 provides the capability to secure and transport the S250 electrical equipment shelter. For higher payload capacity, the M1152A1 is equipped with a reinforced frame, crossmembers, lifting shackles, heavy-duty variable rate rear springs, shock absorbers, reinforced control arms, heavy-duty tires and rims, and a transfer case and differential with a modified gear ratio.The optional front winch permits recovery operations of similar vehicles. Refer to table 1-10.1 for payload.

b. Performance. Fully loaded M1152A1 expanded capacity, enhanced, IAP/armor ready will climb road grades as steep as 40% (22°) and traverse a side slope of up to 30% (13.5°) The vehicle fords hard bottom water crossings up to 30 inches (76 centimeters) without a deep water fording kit and 60 inches (152 centimeters) with the kit. Refer to table 1-10 for cruising range.

c. Special Limitations. None.

d. Special Instructions in the Manual.

(1) Refer to paragraph 2-2, Controls, Indicators, and Equipment.

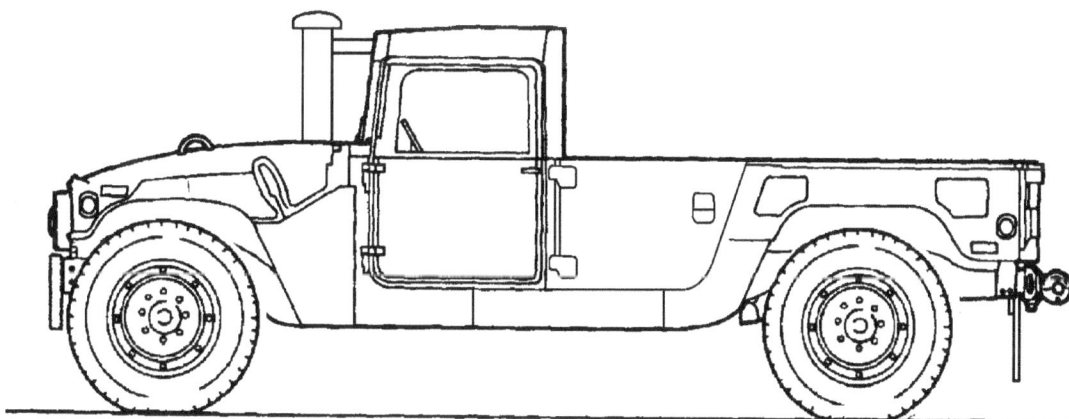

M1152A1

1-11.5. COMMAND AND CONTROL/GENERAL PURPOSE VEHICLE (M1165)

a. Purpose of the Vehicle. The M1165 command and control/general purpose vehicle is used to transport personnel. The M1165 vehicles are capable of transporting a two-man or four-man crew and eight passengers. For higher payload capacity, the M1165 is equipped with a reinforced frame, crossmembers, lifting shackles, heavy-duty variable rate rear springs, shock absorbers, reinforced control arms, heavy-duty tires and rims, and a transfer case and differential with a modified gear ratio. The optional front winch permits recovery operations of similar vehicles. Refer to table 1-10.1 for payload.

b. Performance. Fully loaded M1165 expanded capacity, enhanced, will climb road grades as steep as 40% (22°) and traverse a side slope of up to 30% (13.5°) The vehicle fords hard bottom water crossings up to 30 inches (76 centimeters) without a deep water fording kit and 60 inches (152 centimeters) with the kit. Refer to table 1-10 for cruising range.

c. Special Limitations. None.

d. Special Instructions in the Manual.

(1) Refer to paragraph 2-2, Controls, Indicators, and Equipment.

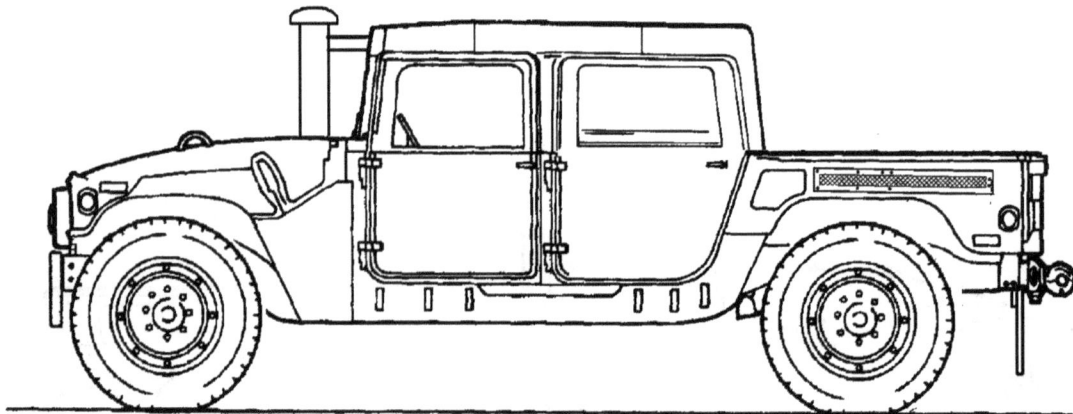

M1165

1-11.6. COMMAND AND CONTROL/GENERAL PURPOSE VEHICLE, IAP/ARMOR READY (M1165A1)

a. Purpose of the Vehicle. The M1165A1 command and control/general purpose vehicle is used to transport personnel.The M1165A1 comes equipped with Integrated Armor protection (IAP) which provides added ballistic protection for armament components, crew, and ammunition.The M1165A1 vehicles are capable of transporting a four-man crew. For higher payload capacity, the M1165A1 is equipped with a reinforced frame, crossmembers, lifting shackles, heavy-duty variable rate rear springs, shock absorbers, reinforced control arms, heavy-duty tires and rims, and a transfer case and differential with a modified gear ratio. The optional front winch permits recovery operations of similar vehicles. Refer to table 1-10.1 for payload.

b. Performance. Fully loaded M1165A1 expanded capacity, enhanced, will climb road grades as steep as 40% (22°) and traverse a side slope of up to 30% (13.5°) The vehicle fords hard bottom water crossings up to 30 inches (76 centimeters) without a deep water fording kit and 60 inches (152 centimeters) with the kit. Refer to table 1-10 for cruising range.

c. Special Limitations. None.

d. Special Instructions in the Manual.

(1) Refer to paragraph 2-2, Controls, Indicators, and Equipment.

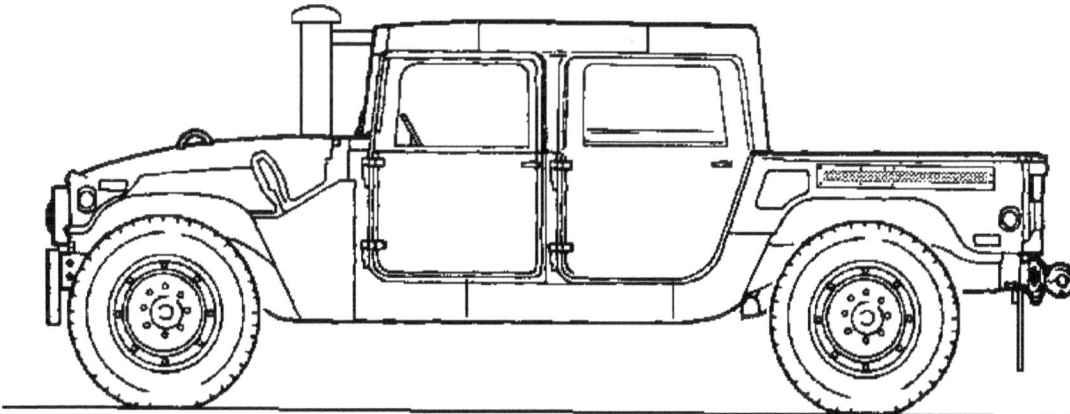

M1165A1

1-11.7. EXPANDED CAPACITY VEHICLE, TOW IMPROVED TARGET ACQUISITION SYSTEM (ITAS): M1167

a. Purpose of the Vehicle. The M1167 expanded capacity vehicle is equipped with an improved target acquisition system (ITAS) used to mount and operate the missile launcher system with armor ballistic protection for crew, missile stowage, secondary weapons mount for close range, and ammunition These tactical vehicles are designed for use over all types of roads, as well as cross-country terrain, in all weather conditions.All vehicles are able to stow a minimum of six missiles mounted in the cargo area and are equipped with a pintle hook for towing, tiedowns, and lifting eyes for air, rail, or sea shipment.The optional front winch permits recovery operations of similar vehicles. Refer to table 1-10.1 for payload.

b. Performance. Fully loaded M1167 expanded capacity TOW ITAS carrier will climb road grades as steep as 40% (22°) and traverse a side slope of up to 30% (13.5°) The vehicle fords hard bottom water crossings up to 30 inches (76 centimeters) without a deep water fording kit and 60 inches (152 centimeters) with the kit. Refer to table 1-10 for cruising range.

c. Special Limitations. None.

d. Special Instructions in the Manual.

(1) Refer to paragraph 2-2, Controls, Indicators, and Equipment.

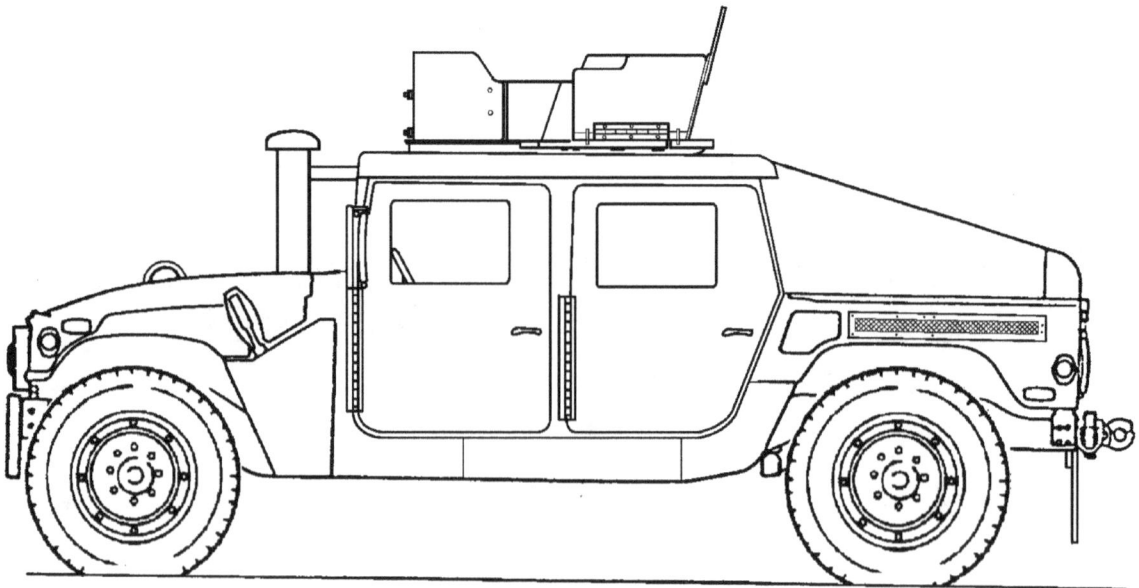

M1167

1-12. TABULATED DATA

This paragraph organizes vehicle specifications, special equipment, and model differences in table form for easy reference by operators.

Table 1-1. Differences Between Models.

EQUIPMENT/ FUNCTION	VEHICLE								
	M1113	M1114	M1151	M1151A1	M1152	M1152A1	M1165	M1165A1	M1167
Armament Mounting		X		X		X		X	X
S250 Shelter Configuration	X				X	X			
Vehicle Winch (if equipped)	X	X	X	X	X	X	X	X	X
Basic Armor			X						
Up-Armor		X							
Integrated Armor Protection (IAP)				X		X		X	X
TOW ITAS Launcher Mounting									X

Table 1-2. Capacities.

Description	Capacity	
	Standard	Metric
Cooling system	27.25 qt	25.8 L
Engine (crankcase only)	7 qt	6.6 L
Engine (crankcase with new filter)	8 qt	7.6 L
Differential (each)	2 qt	1.9 L
Transmission (drain and refill)	7.7 qt	7.3 L
Transfer case	3.35 qt	3.17 L
Fuel tank	25 gal.	94.6 L
Brake master cylinder (Serial numbers 299999 and below)	1.12 pt	0.53 L
Brake master cylinder (Serial numbers 300000 and above)	2.36 pt	1.12 L
Total brake system (Serial numbers 299999 and below)	1.63 pt	0.77 L
Total brake system (Serial numbers 300000 and above)	3.47 pt	1.64 L
Windshield washer	1 qt	0.95 L
Geared hub	1 pt	0.47 L
Geared fan drive	1.2 pt	0.47 L

1-12. TABULATED DATA (Cont'd)

Table 1-2.1. Steering System Capacities.

System	Description	Capacity	
		Standard	Metric
Serial Numbers 196901 and Above (M1113) (M1151/M1151A1/ M1152/M1152A1/ M1165/M1165A1/ M1167 without winch only)	Without front hydraulic winch (with rear differential cooler)	3.3 qt	3.12 L
	With front hydraulic winch (with rear differential cooler)	3.7 qt	3.50 L
Serial Numbers 196901 and Above (M1114)	Without rear hydraulic winch (with rear differential cooler)	3.3 qt	3.12 L
	With rear hydraulic winch (with rear differential cooler)	4.0 qt	3.78 L
Serial Numbers 196900 and Below (M1113)	Without front hydraulic winch (without rear differential cooler)	1.8 qt	1.70 L
	Without front hydraulic winch (with rear differential cooler)	2.7 qt	2.55 L
	With front hydraulic winch (without rear differential cooler)	2.2 qt	2.10 L
	With front hydraulic winch (with rear differential cooler)	3.1 qt	2.93 L
Serial Numbers 196900 and Below (M1114)	Without rear hydraulic winch (without rear differential cooler)	1.8 qt	1.70 L
	Without rear hydraulic winch (with rear differential cooler)	2.7 qt	2.55 L
	With rear hydraulic winch (without rear differential cooler)	2.5 qt	2.37 L
	With rear hydraulic winch (with rear differential cooler)	3.4 qt	3.22 L

Table 1-3. General Service Data.

Description	Expected Temperatures		
	Above +15° F (above -9°C)	+40° to -15°F (+4° to -26°C)	+40° to -65°F (+4° to -54°C)
Cooling system	1/2 ethylene glycol, 1/2 water	1/2 ethylene glycol, 1/2 water	3/5 ethylene glycol, 2/5 water
Engine	OE/HDO 30	OE/HDO 10	OEA -30**
Fuel tank	DF1, DF2	DF1	DF1, DFA*
	All Temp. JP-8		
Differentials	GO 80/90	GO 80/90	GO 75
Geared hubs	GO 80/90	GO 80/90	GO 75
Geared Fan Drive	GO 80/90 GO 75	N/A	All Temperatures +40° to -65°F (+4° to -54°C)
Transmission	Dexron® VI	Dexron® VI	Dexron® VI
Transfer case	Dexron® VI	Dexron® VI	Dexron® VI
Steering system (Serial Numbers) 196900 and Below)	Dexron® VI	Dexron® VI	Dexron® VI
Steering system (Serial Numbers 196901 and Above)	Dexron® VI	Dexron® VI	Dexron® VI
Brake system (Serial Numbers 299999 and Below)	Brake Fluid Silicone (BFS)	Brake Fluid Silicone (BFS)	Brake Fluid Silicone (BFS)
Brake system (Serial Numbers 300000 and Above)	Brake Fluid Silicone (BFS)	Master Cylinder: 2.36 pt (1.12 L) Complete Systems 3.09 pt (1.46 L)	All Temperatures
Windshield washer reservoir	1/3 cleaning compound 2/3 water	1/2 cleaning compound 1/2 water	2/3 cleaning compound 1/3 water

*   Use below -20°F (-29°C)
** All temperatures +120° to -55°F

1-12. TABULATED DATA (Cont'd)

Table 1-4. Engine Data.

Type	6.5 liter diesel, turbocharged, liquid-cooled
Cylinders	8(V)
Brake horsepower	190 horsepower @ 3,400 rpm
Idle speed (engine rpm)	700 ± 25 rpm
Operating speed (engine rpm)	1,500–2,300 rpm
Oil pressure at idle	1 0psi (69 kPa)
Normal operating oil pressure	30-50 psi (207-345 kPa)

Table 1-5. Cooling System Data.

Surge tank cap pressure	1 5psi (103 kPa)
Thermostat:	
Starts to open	190°F (88°C)
Fully open	212°F (100°C)
Radiator	Downflow type
Fan (serial numbers 299999 and below)	1 0blade, 19 in. (48 cm)
Fan (serial numbers 300000 and above)	9blade, 23 in. (58 cm)
Normal operating coolant temperature	185°–250°F (85°-120°C)

Table 1-6. Transmission Data.

General Information	
Model	Turbo Hydra-Matic 4L80-E
Type	F o uspeed automatic
Oil type	Dexron® VI

Transmission Range Selection	
Recommended Shift Lever Positi	**Operating Condition**
P (park)	Vehicle stopped with parking brake applied.
R (reverse)	Clear of traffic and obstructions, using ground guide.
N (neutral)	Vehicle stopped with parking brake applied.
Ⓓ (overdrive)	Normal driving and fording.
D (drive)	Hilly terrain and towing a trailer.
2 (second)	Hill climbing and engine braking to slow vehicle when descending steep hills.
1 (first)	Maximum engine braking when descending very steep hills, climbing steep hills, or driving through deep mud, sand, or snow.

1-12. TABULATED DATA (Cont'd)

Table 1-7. Transfer Case Data (Serial Numbers 299999 and Below). ∎

General Information
Model. . . . . New Venture Gear
Model. . . . . Magna Powertrain (Serial Numbers 300000 and Above)
Type . . . . Two-speed, locking, chain-driven
Oil type. . . . . Dexron® VI

CAUTION
Damage to drivetrain will occur if transfer case ranges are not selected properly. Refer to paragraph 2-13 for specific instructions on placing vehicle in motion, and to paragraph 2-40 for operating on unusual terrain.

Transfer Case Range Selection

Recommended Shift Lever Position	Operating Condition
H (high range)	This drive range shall be selected whenever possible. High range should be used when operating on all primary, secondary, and off-road surfaces, where little or no wheel slippage exists.This range is also to be used when encountering sharp, continuous turns on high traction surfaces.
H/L (high/lock range)	This drive range shall be selected for off-highway hilly terrain or when continuous wheel slippage is evident; i.e., when operating in mud, snow, loose sand, or on ice, and increased control or additional traction is required.
L (low range)	This drive range shall be selected only when high ranges do not provide sufficient power to negotiate steep hills or downgrades.This range shall also be used when the vehicle is mired and cannot be extracted using the high/ lock range.
N (neutral)	Vehicle is disabled and must be towed.

1-12. TABULATED DATA (Cont'd)

Table 1-8. Maximum Vehicle Operating Speeds.

Transmission Range Selection	Transfer Case Range Selection		
	L (Low Range)	H (High Range)	H/L (High/Lock Range)
R (reverse)	9 mph (14 kph)	23 mph (37 kph)	9 mph (14 kph)
Ⓓ (overdrive)	21 mph (34 kph)	55 mph (88 kph)	55 mph (88 kph)
D (drive)	21 mph (34 kph)	55 mph (88 kph)	55 mph (88 kph)
2 (second)	15 mph (24 kph)	40 mph (64 kph)	40 mph (64 kph)
1 (first)	9 mph (14 kph)	23 mph (37 kph)	23 mph (37 kph)

Table 1-9. Vehicle Dimensions.

Vehicle	Length Overall		Height Overall		HeightMinimum Reducible	
	Inches	Centimeters	Inches	Centimeters	Inches	Centimeters
M1113	196.5	499	76	193	56	142
M1114	196.5	499	76	193	76	193
M1151	194	493	79	201	N/A	N/A
M1151A1	194	493	80	203	N/A	N/A
M1152	194	493	75	191	N/A	N/A
M1152A1	194	493	76	193	N/A	N/A
M1165	194	493	75	191	N/A	N/A
M1165A1	194	493	76.25	193	N/A	N/A
M1167	194	493	79	201	N/A	N/A

	Width Overall		Ground Clearance Under Axle	
	Inches	Centimeters	Inches	Centimeters
M1113	86	218	17.5	56
M1114	90.5	230	15.5	39
M1151	86	218	16.5	42
M1151A1	87	221	18.1	46
M1152	86	218	16.75	43
M1152A1	87	221	18	46
M1165	86	218	16.5	42
M1165A1	87	221	18.2	46
M1167	86	218	18.1	46

1-12. TABULATED DATA (Cont'd)

Table 1-10.Vehicle Cruising Range.

NOTE
When vehicle is driven on hard surface and hilly terrain at a speed of 30-40 mph (48-64 kph), the cruising range shown for each vehicle can be expected.

Vehicle	Gross Vehicle Weight (GVW)	Cruising Range
M1113	11,500 lb (5,221 kg)	250 mi (402 km)
M1114	12,100 lb (5,493 kg)	250 mi (402 km)
M1151	11,500 lb (5,221 kg)	250 mi (402 km)
M1151A1	12,100 lb (5,493 kg)	250 mi (402 km)
M1152	11,500 lb (5,221 kg)	250 mi (402 km)
M1152A1	12,100 lb (5,493 kg)	250 mi (402 km)
M1165	11,500 lb (5,221 kg)	250 mi (402 km)
M1165A1	12,100 lb (5,493 kg)	250 mi (402 km)
M1167	13,100 lb (5,942 kg)	250 mi (402 km)

Table 1-10.1. Vehicle Payload (Including Crew)

Vehicle	Payload	
	Standard	Metric
M1113	5,100 lb	2,313 kg
M1114	2,300 lb	1,043 kg
M1151	4,000 lb	1,814 kg
M1151A1	3,950 lb	1,792 kg
M1151A1 w/Perimeter B Kit	1,800 lb	816 kg
M1152	5,100 lb	2,313 kg
M1152A1	5,000 lb	2,268 kg
M1152A1 w/Perimeter B Kit	3,340 lb	1,515 kg
M1165	4,950 lb	2,245 kg
M1165A1	4,870 lb	2,209 kg
M1165A1 w/Perimeter B Kit	2,230 lb	1,012 kg
M1167	2,300 lb	1,043 kg

1-12. TABULATED DATA (Cont'd)

Table 1-11. 9,000 lb Winch Data.

Description	Capacities	
	Standard	Metric
Max. Load (Fifth Layer)	6,200 lb	2,815 kg
Max. Load (Fourth Layer)	7,000 lb	3,178 kg
Max. Load (Third Layer)	8,000 lb	3,632 kg
Max. Load (Second Layer)	8,500 lb	3,859 kg
Max. Load (First Layer)	9,000 lb	4,086 kg

Table 1-11.1. 10,500 lb Winch Data.

Description	Capacities	
	Standard	Metric
Minimum Load Max. Load (Fifth Layer)	4,700 lb 5,200 lb	2,134 kg 2,361 kg
Minimum Load Max. Load (Fourth Layer)	5,600 lb 6,000 lb	2,542 kg 2,724 kg
Minimum Load Max. Load (Third Layer)	6,800 lb 8,500 lb	3,087 kg 3,859 kg
Minimum Load Max. Load (Second Layer)	8,700 lb 9,600 lb	3,950 kg 4,358 kg
Minimum Load Max. Load (First Layer)	9,620 lb 10,500 lb	4,367 kg 4,767 kg

1-12. TABULATED DATA (Cont'd)

Table 1-12. Curb Weight.

Vehicle	Curb Weight
M1113	6,400 lb (2,906 kg)
M1114	9,800 lb (4,449 kg)
M1151	7,500 lb (3,402 kg)
M1151A1	8,150 lb (3,697 kg)
M1152	6,400 lb (2,903 kg)
M1152A1	7,100 lb (3,221 kg)
M1165	6,500 lb (2,971 kg)
M1165A1	7,230 lb (3,279 kg)
M1167	10,800 lb (4,899 kg)

Table 1-13. Vehicle Bridge Classification.

Model	Empty	Loaded	
		Cross-Country	Highway
M1113	2	4	4
M1114	3	4	4
ALL OTHER VEHICLES	2	5	5

1-12. TABULATED DATA (Cont'd)

Table 1-14. Axle Weight.

Vehicle	Less Payload and Crew		
	Front Axle Lbs (kgs)	Rear Axle Lbs (kgs)	Total Lbs (kgs)
M1113	3,450 (1,566)	2,950 (1,339)	6,400 (2,906)
M1114	4,700 (2,134)	5,100 (2,315)	9,800 (4,449)
M1151	3,730 (1,692)	3,770 (1,710)	7,500 (3,402)
M1151A1	4,110 (1,864)	4,040 (1,833)	8,150 (3,697)
M1151A1 W/B1	4,980	5,320	10,300 (4,672)
M1152	3,450 (1,565)	2,950 (1,338)	6,400 (2,903)
M1152A1	3,850 (1,746)	3,250 (1,474)	7,100 (3,221)
M1152A1 W/B2	4,790	3,970	8,760 (3,973)
M1165	3,450 (1,565)	1,100 (1,406)	6,550 (2,971)
M1165A1	3,870 (1,755)	3,360 (1,524)	7,230 (3,279)
M1165A1 W/B3	4,900	4,970	9,870 (4,477)
M1167	5,360	5,890	11,250 (5,103)

1-12. TABULATED DATA (Cont'd)

Table 1-15. Tire Pressure (Radial Tire) (Load Range D Tire).

Vehicle	Front		Rear	
	Standard psi	Metric kPa	Standard psi	Metric kPa
Unloaded* M1113, M1151, M1152, M1165	20	138	20	138
M1114, M1151A1, M1151A1 W/B1, M1152A1, M1152A1 W/B2, M1165A1, M1165A1 W/B3, M1167	35	241	45	310
At GVW M1113, M1151, M1152, M1165	30	207	40	138
M1114, M1151A1, M1151A1 W/B1, M1152A1, M1152A1 W/B2, M1165A1, M1165A1 W/B3, M1167	40	276	50	345
Mud, Sand, and Snow (30 mph [48 km] max. speed)	20	138	30	207

*Driver plus one passenger

Table 1-15.1. Tire Pressure (Radial Tire) (Load Range E Tire).

Vehicle	Front		Rear	
	Standard psi	Metric kPa	Standard psi	Metric kPa
Unloaded* M1113, M1151, M1152, M1165	18	124	25	172
M1114, M1151A1, M1151A1 W/B1, M1152A1, M1152A1 W/B2, M1165A1, M1165A1 W/B3, M1167	35	241	45	310
At GVW M1113, M1151, M1152, M1165	30	207	45	310
M1114, M1151A1, M1151A1 W/B1, M1152A1, M1152A1 W/B2, M1165A1, M1165A1 W/B3, M1167	40	138	50	345
Mud, Sand, and Snow (30 mph [48 km] max. speed)	20	138	30	207

*Driver plus one passenger

Section III. PRINCIPLES OF OPERATION

1-13. GENERAL

This section explains how components of the ECV series vehicles work together. The systems (functional groups) covered are listed in the Principles of Operation Reference Index, paragraph 1-14.

1-14. PRINCIPLES OF OPERATION REFERENCE INDEX

1-15. DRIVETRAIN OPERATION

The drivetrain converts horsepower into mechanical force to move the vehicle. Major components of the drivetrain are:

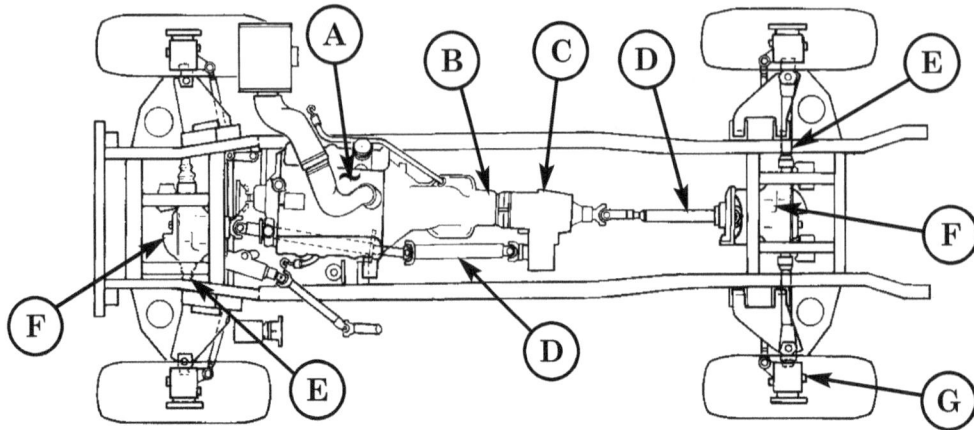

(A) **ENGINE** – The 6.5 liter V-8 turbocharged engine develops approximately 190 horsepower at 3,400 rpm. Engines equipped for deep water fording have a specially-sealed dipstick, dipstick tube, and vented CDR valve.

(B) **TRANSMISSION** – Adapts engine power to meet different driving conditions.The automatic transmission has four forward speeds, a reverse, a neutral and a park.A neutral safety switch prevents the vehicle from being started with the transmission in any selector lever position except park and neutral.

(C) **TRANSFER CASE** – Directs engine-to-transmission power to front and rear differentials simultaneously.This condition means the vehicle is always in four-wheel drive.The transfer case allows for selection of three drive ranges and a neutral position. (See table 1-7).

(D) **PROPELLER SHAFTS** – Link transfer case to differentials. Universal joints, located at either end of the front and rear propeller shafts, permit inline driving power between the transfer case and differentials even though they are mounted at different angles.

(E) **HALFSHAFTS** – Transmit power from differentials to geared hubs.

(F) **DIFFERENTIALS** – Transmit driving power, via halfshafts and geared hubs, to left and right wheels.The differential ensures power is applied to the wheel having traction.This feature is called torque biasing.

(G) **GEARED HUBS** – Serve as the front wheel steering spindle and act as the final drive components to front and rear wheels.

1-16. FUEL SYSTEM OPERATION

The diesel fuel system stores, cleans, and supplies fuel for the engine. Major components of the fuel system are:

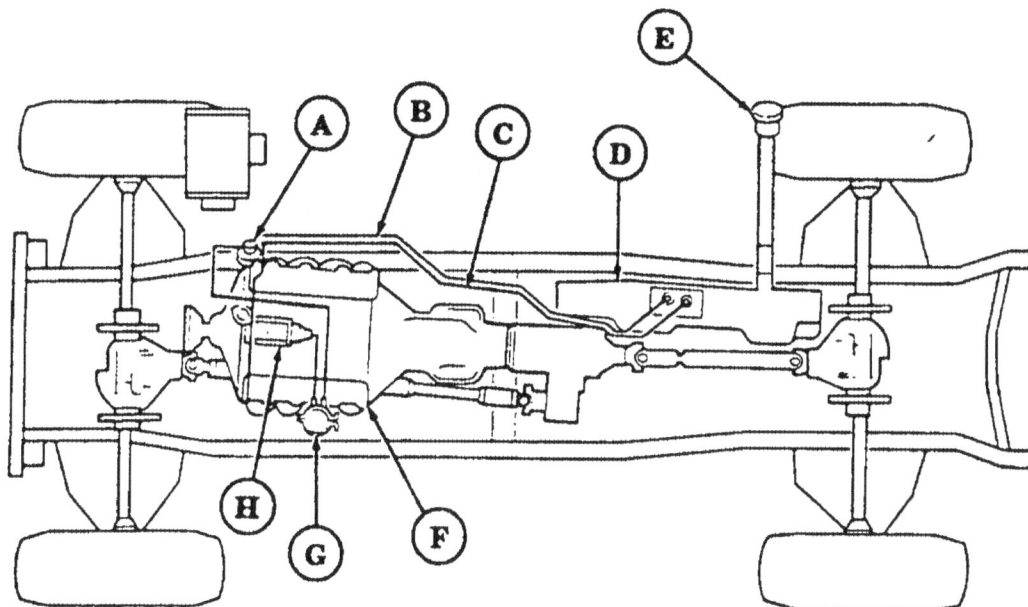

(A) **FUEL PUMP** – Draws fuel from fuel tank through the supply line and pumps it to the fuel filter.

(B) **FUEL RETURN LINE** – Directs unused fuel from the injection pump back to the fuel tank.

(C) **FUEL SUPPLY LINE** – Directs fuel from fuel tank to the system.

(D) **FUEL TANK** – Stores 25 gallons (95 liters) of diesel fuel.

(E) **FUEL FILLER CAP** – Located at right rear side of vehicle, the cap is removed to permit fuel tank servicing.

(F) **FUEL INJECTORS** – Receive metered fuel from the injection pump and sprays fuel into the combustion chamber.

(G) **FUEL FILTER/WATER SEPARATOR** – Filters water and sediment from fuel before fuel enters the injection pump.

(H) **INJECTION PUMP** – Directs metered and pressurized fuel to the eight injector nozzles. It is mounted on top of the engine under the intake manifold.

1-17. COOLING SYSTEM OPERATION (SERIAL NUMBERS 299999 AND BELOW)

The cooling system removes excess heat from the engine, engine oil, transfer case oil, and transmission oil. Major components of the cooling system are:

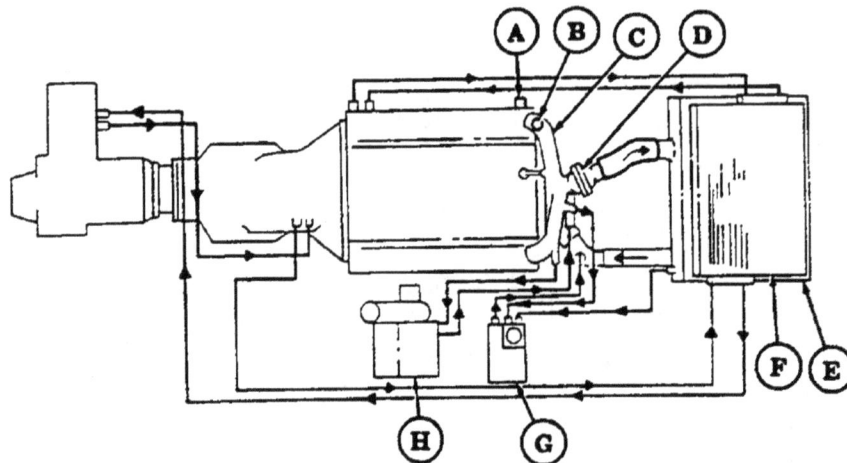

(A) **ENGINE TEMPERATURE SENDING UNIT** – Sends signal indicating coolant temperature to gauge on instrument cluster.

(B) **ENGINE TEMPERATURE SWITCH** – Sends signal to activate control valve system to operate fan when engine temperature exceeds 220°F (104°C) and deactivates control valve system when engine temperature drops below 190°F (88°C).

(C) **WATER CROSSOVER** – Collects coolant from cylinder heads and channels it to the thermostat housing where it is redirected through the cooling system.

(D) **THERMOSTAT** – Shuts off coolant return flow to radiator until temperature reaches 190°F (88°C). Coolant is then directed to the radiator through the radiator inlet hose.

(E) **RADIATOR** – Directs coolant through a series of fins and baffles so outside air can dissipate excess engine heat before the coolant is recirculated through the engine.

(F) **OIL COOLER** – Directs engine oil (lower half of cooler) and transmission oil (upper half of cooler) through a series of fins or baffles so outside air can remove heat from oil.

(G) **SURGE TANK** – Filling and expansion point for cooling system.

(H) **PERSONNEL HEATER** – Provides heat for personnel and interior of vehicle.

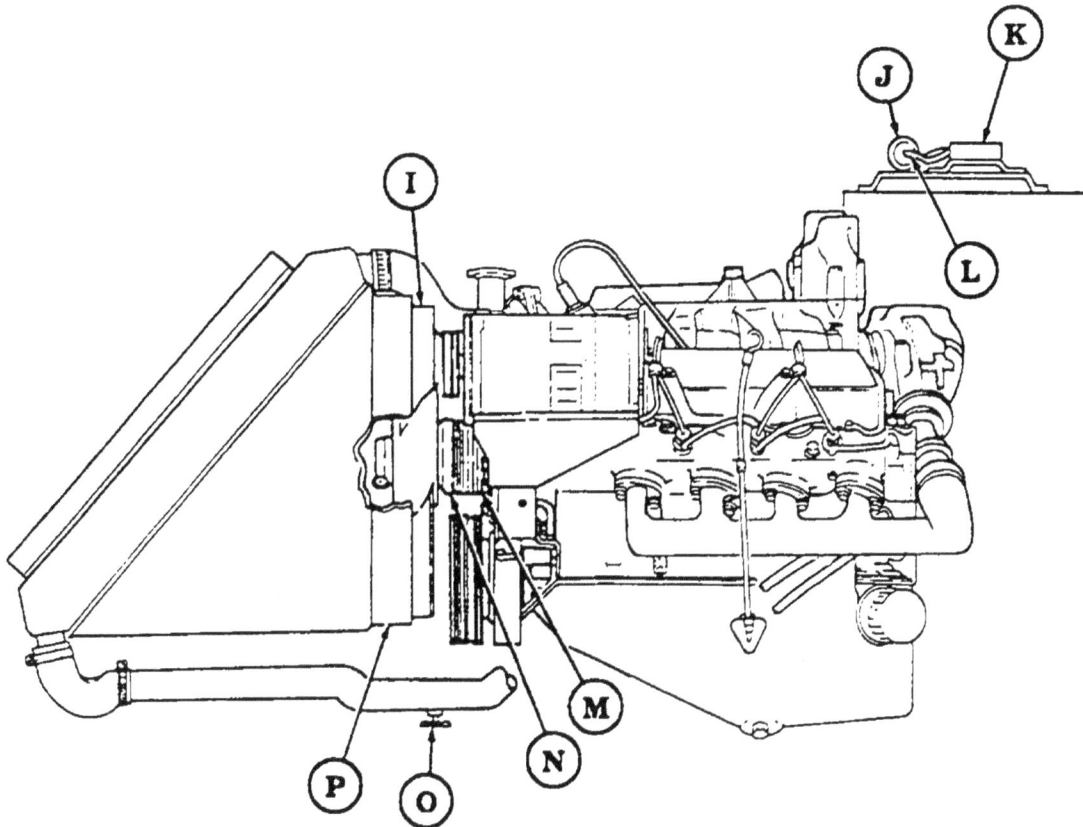

(I) **FAN** – Pulls outside air through radiator to remove heat from coolant.

(J) **HYDRAULIC CONTROL VALVE** – Directs hydraulic fluid to provide required pressure to actuate fan clutch as required by engine temperature. Hydraulic pressure supplied by power steering pump.

(K) **TIME DELAY MODULE** – Sends delayed signal to fan clutch solenoid for delay of fan actuation to provide needed horsepower for engine acceleration.

(L) **FAN CLUTCH SOLENOID** – Actuates hydraulic control valve as required by coolant temperature.

(M) **WATER PUMP** – Driven by serpentine belt, provides circulation of coolant through cooling system.

(N) **FAN CLUTCH** – Hydraulically actuated by pressure from hydraulic control valve to control operation of fan. Hydraulic pressure supplied by power steering pump.

(O) **DRAINVALVE** – Draining point for radiator and cooling system.

(P) **RADIATOR SHROUD** – Permits a greater concentration of air to be pulled through the radiator.

1-17.1. COOLING SYSTEM OPERATION (SERIAL NUMBERS 300000 AND ABOVE)

The cooling system removes excess heat from the engine, engine oil, transfer case oil, and transmission oil. Major components of the cooling system are:

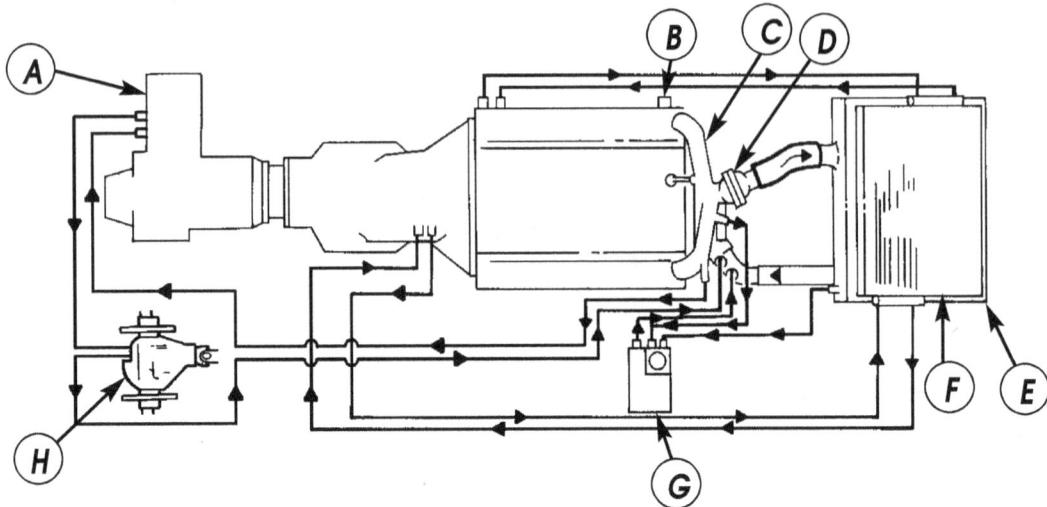

(A) **TRANSFER CASE** – Directs engine coolant through a series of fins or baffles so coolant can remove heat from transfer case oil.

(B) **ENGINE TEMPERATURE SENDING UNIT** – Sends signal indicating coolant temperature to gauge on instrument cluster.

(C) **WATER CROSSOVER** – Collects coolant from cylinder heads and channels it to the thermostat housing where it is redirected through the cooling system.

(D) **THERMOSTAT** – Shuts off coolant return flow to radiator until temperature reaches 190°F (88°C). Coolant is then directed to the radiator through the radiator inlet hose.

(E) **RADIATOR** – Directs coolant through a series of fins and baffles so outside air can dissipate excess engine heat before the coolant is recirculated through the engine.

(F) **OIL COOLER** – Directs engine oil (lower half of cooler) and transmission oil (upper half of cooler) through a series of fins or baffles so outside air can remove heat from oil.

(G) **SURGE TANK** – Filling and expansion point for cooling system.

(H) **DIFFERENTIAL COOLER** – Directs engine coolant through a series of fins or baffles so coolant can remove heat from differential oil.

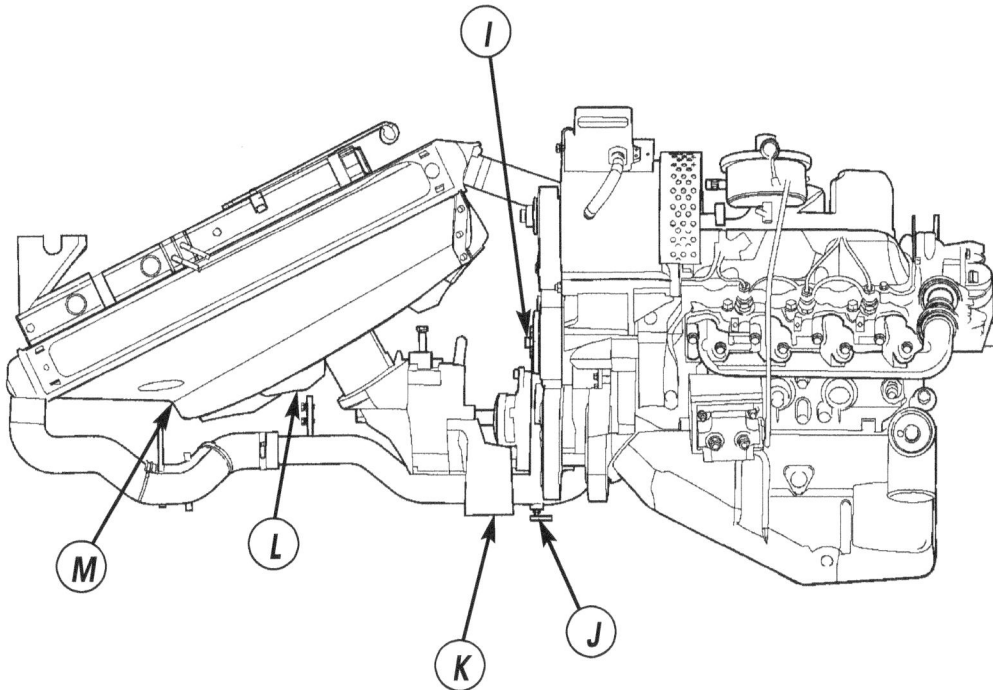

(I) **WATER PUMP** – Driven by serpentine belt, provides circulation of coolant through cooling system.

(J) **DRAINVALVE** – Draining point for radiator and cooling system.

(K) **GEARED FAN DRIVE** – Transmits engine power and torque to drive the radiator cooling fan and clutch.

(L) **FAN** – Pulls outside air through radiator to remove heat from coolant.

(M) **TWO-PIECE RADIATOR SHROUD** – Permits a greater concentration of air to be pulled through the radiator.

1-18. STARTING SYSTEM OPERATION

The starting system consists of the following components and circuits:

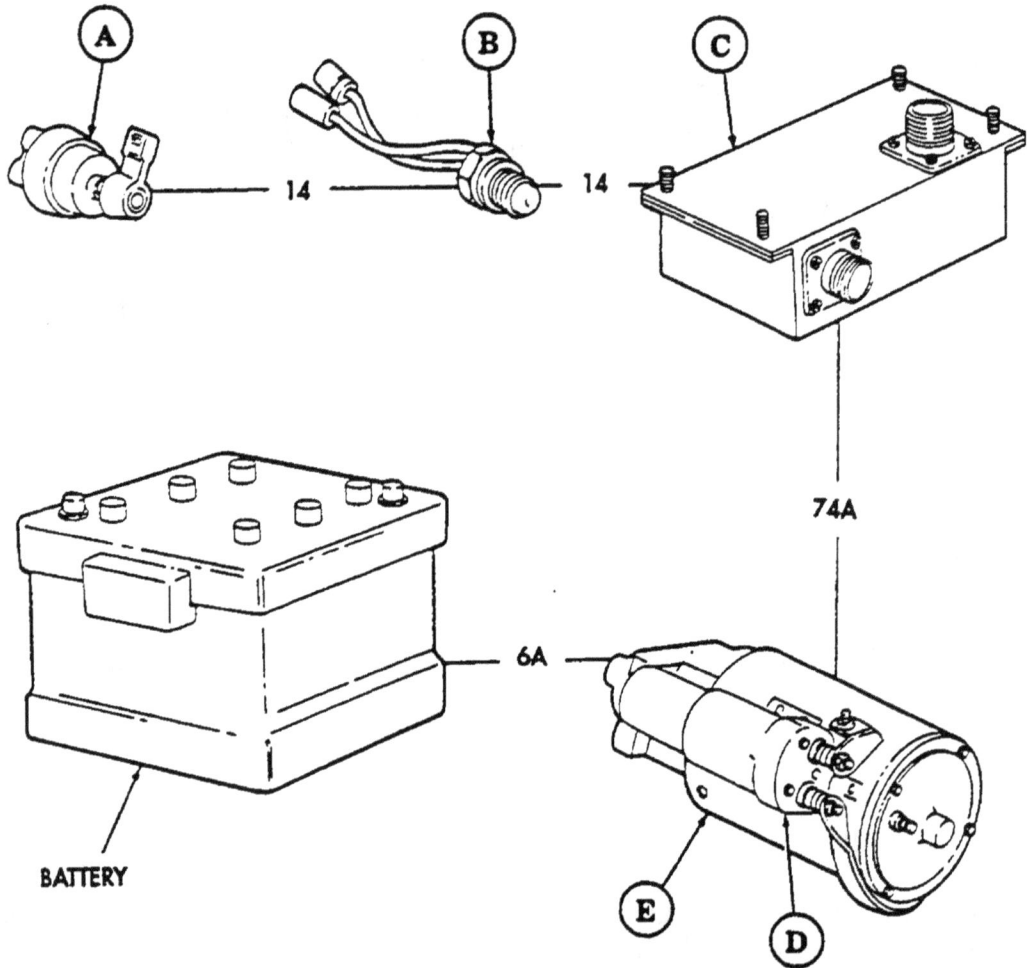

BATTERY

(A) **ROTARY SWITCH** – When in START position, provides battery power to the starter solenoid and to the neutral start switch through circuit 14.

(B) **NEUTRAL STARTER SWITCH** – When transmission shift lever is in P (park) or N (neutral) position, this switch closes, allowing battery power to reach the starter solenoid.

(C) **PROTECTIVE CONTROL BOX** – Acts as a link between body harness and engine harness.

(D) **STARTER SOLENOID** – A magnetic relay that transmits 24-volt battery power to the starter motor.

(E) **STARTER MOTOR** – Cranks the engine for starting, and is supplied 24-volt battery power through circuit 6A.

1-19. GENERATING SYSTEM OPERATION (200-AMPERE ALTERNATOR)

The 200-ampere generating system maintains battery charge and provides electrical power to operate vehicle circuits. Major components of the generating system are:

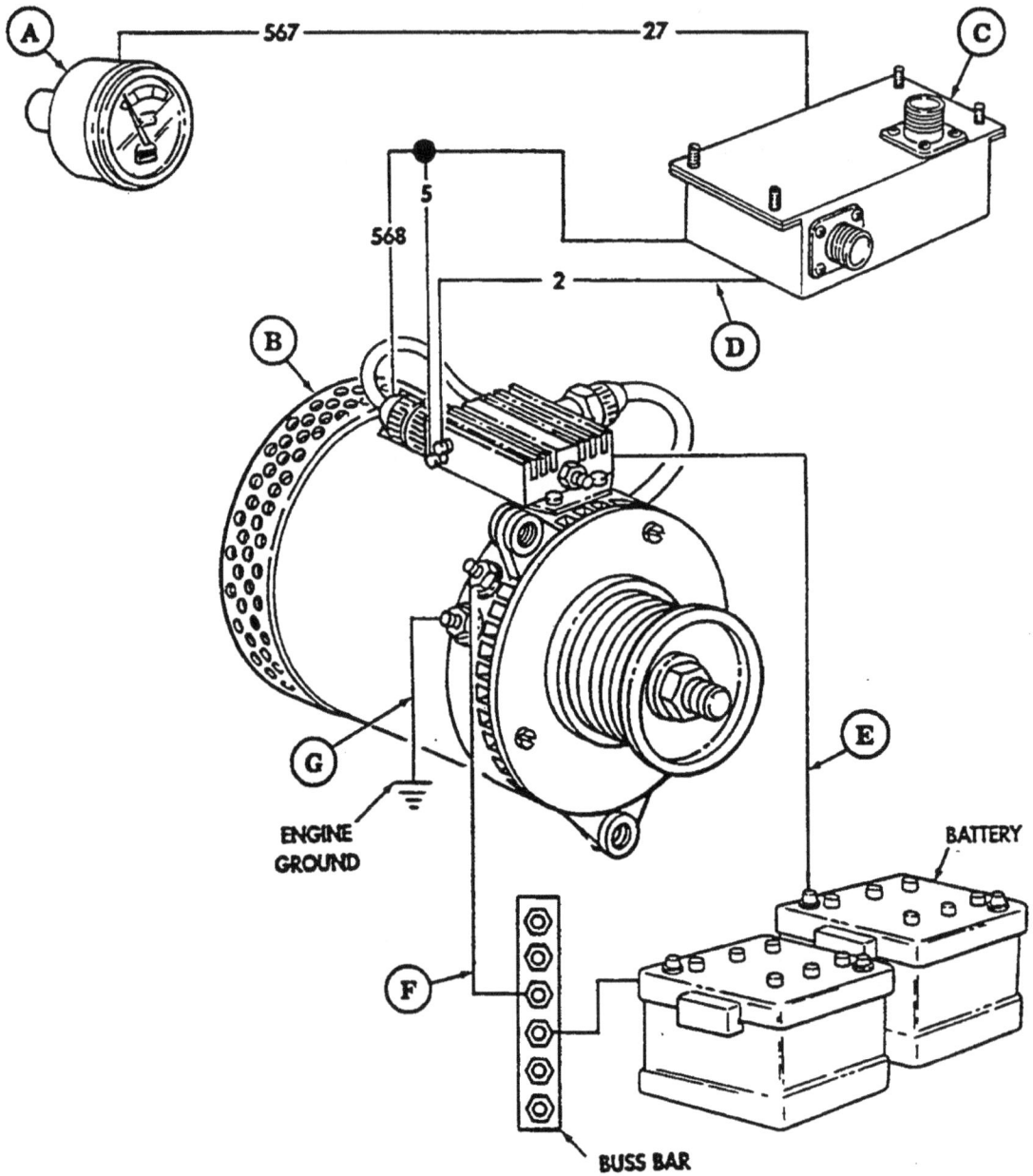

(A) **BATTERY GAUGE** – Indicates electrical system voltage. It is connected to the electrical system through circuit 567.

(B) **ALTERNATOR (200-AMPERE)** – Is a dual volt, rated at 14/28 volts, 200 amperes, with external regulator.The alternator assists and recharges vehicle batteries during operation.

(C) **PROTECTIVE CONTROL BOX** – Protects vehicle electrical system in the event battery polarity is reversed. Provides load dump, glow plug operation, and interfacing of engine and body harnesses.

(D) **CIRCUIT 2** – Sends AC signal, indicating alternator shaft rpm, to frequency switch in protective control box to prevent operation of starter solenoid when engine is running.

(E) **POSITIVE CABLE 6** – Provides 28-volt alternator output to maintain charge across two batteries.

(F) **POSITIVE CABLE 68A** – Provides 14-volt alternator output to maintain charge across the lower battery.

(G) **GROUND STRAP** – Connects negative stud on alternator with engine ground strap to provide a ground circuit to alternator.

1-20. BATTERY SYSTEM OPERATION

The battery system consists of the following circuits and components:

(A) **CIRCUIT 6A** – Connects batteries (through buss bar) to starter and protective control box through circuit 74A.

(B) **BATTERIES** – Two 6TN batteries are connected in a series to provide 24 volts DC for the electrical starting system and to assist alternator during operation.

(C) **BATTERY POSITIVE CABLE** – Connects battery positive to buss bar.

(D) **24 VOLT MAIN POSITIVE BUSS BAR** – Provides a primary junction for 24 volt circuits.

(E) **12 AND 24 VOLT TERMINAL BOARDS** – (Optional on M114) Provides circuit protected 12 or 24 volt auxiliary power for accessories.

(F) **SLAVE RECEPTACLE** – Links an external power source directly to slaved vehicle's batteries to assist in cranking engine when vehicle's batteries are not sufficiently charged.

(G) **12 VOLT AUXILIARY POWER OUTLET** – (Optional) Provides circuit protected 12 volt power to compatible accessories.

(H) **SHUNT** – Used when measuring current draw from batteries utilizing STE/ICE-R.

(I) **PROTECTIVE CONTROL BOX** – Protects vehicle electrical system in the event battery polarity is reversed. Provides frequency lockout, load dump, glow plug operation, and interfacing of engine and body harness.

(J) **NEGATIVE BUSS BAR** – Provides adequate grounding points when using external accessories.

(K) **ROTARY SWITCH** – When in START position, actuates starter solenoid through circuits 11A and 74A.When in RUN position, closes circuit 29A to activate instrument cluster gauges through circuit 27.

(L) **CIRCUIT 7A** – Connects starter negative circuit terminal to engine ground while 7E connects shunt to engine ground.

(M) **STARTER SOLENOID** – Actuates starter motor gear to crank vehicle engine.

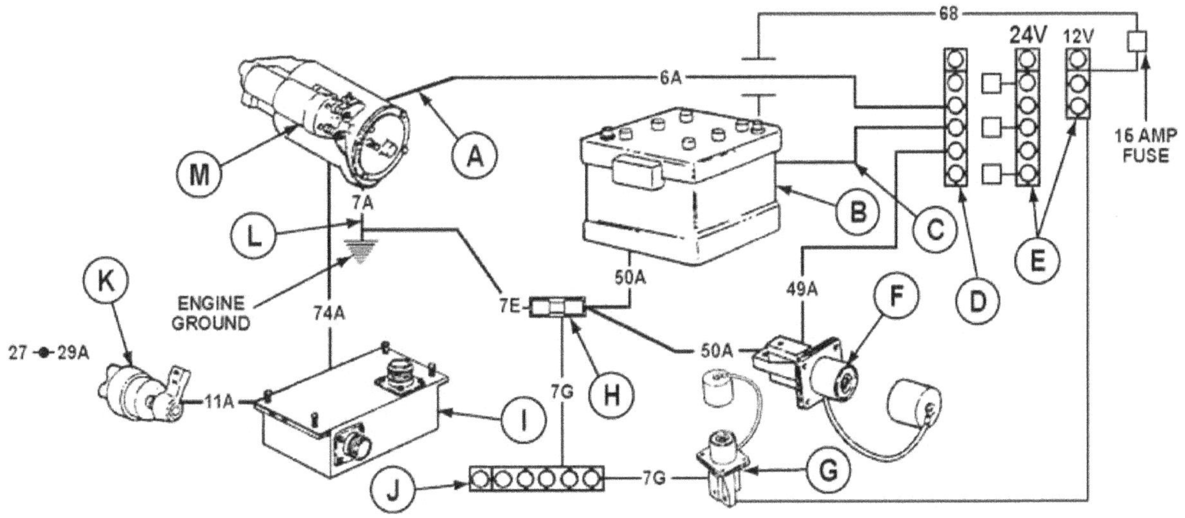
68
24V
12V
6A
16 AMP
FUSE
A
M
B
C
7A
L
E
D
K
ENGINE
GROUND
74A
7E
50A
49A
F
27 -●- 29A
50A
11A
7G
H
I
G
J
7G

1-21. WINDSHIELD WIPER/WASHER SYSTEM OPERATION

The following miscellaneous components and circuits are not covered in any of the other electrical systems:

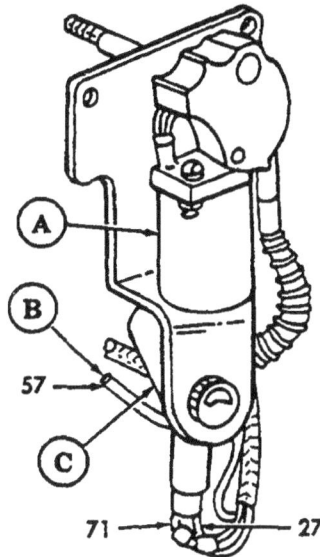

(A) **WINDSHIELD WIPER MOTOR** – When knob is turned to LOW or HIGH position, circuit 27 carries battery power to wiper motor to activate windshield wipers.

(B) **CIRCUIT 57** – Provides a ground circuit for wiper motor.

(C) **WINDSHIELD WASHER MOTOR** – When knob is pushed, the washer motor is activated through circuit 71 to spray water onto windshield.

1-22. SERVICE/PARKING BRAKE SYSTEM OPERATION

The parking brake system is a mechanically-actuated system that keeps the vehicle still once it stops. It also assists in emergency stopping if there is a complete service brake system failure. Major components of the parking brake system are:

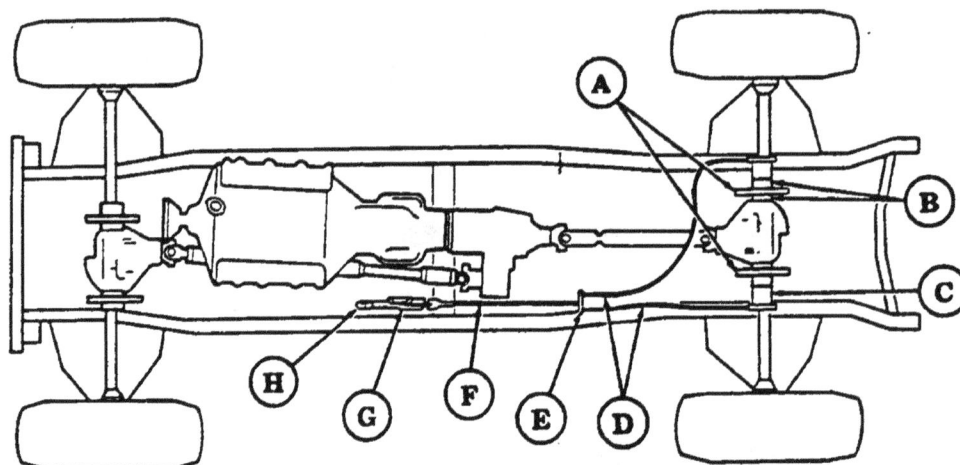

(A) PARKING BRAKE ROTORS – Attached to output flanges on rear differential, rotors prevent output flanges from turning when parking brake is applied.

(B) BRAKE PADS – Apply friction to rotors when hand lever is applied.

(C) PARKING BRAKE CALIPERS – Force brake pads against rotors when hand lever is applied.

(D) PARKING BRAKE CABLES – Connect parking brake hand lever to equalizer bar.

(E) EQUALIZER BAR – Evenly distributes braking pressure to the rear brake rotors.

(F) PARKING BRAKE ROD – Connects parking brake hand lever to equalizer bar.

(G) PARKING BRAKE HAND LEVER – Permits operator to engage the parking brake.

(H) PARKING BRAKE HAND LEVER ADJUSTING CAP – Permits operator to make minor tension adjustment of parking brake.

1-23. SERVICE BRAKE SYSTEM OPERATION

The service brake system is an inboard-mounted, four-wheel, disc brake, hydraulically-assisted system. Major components of the braking system are:

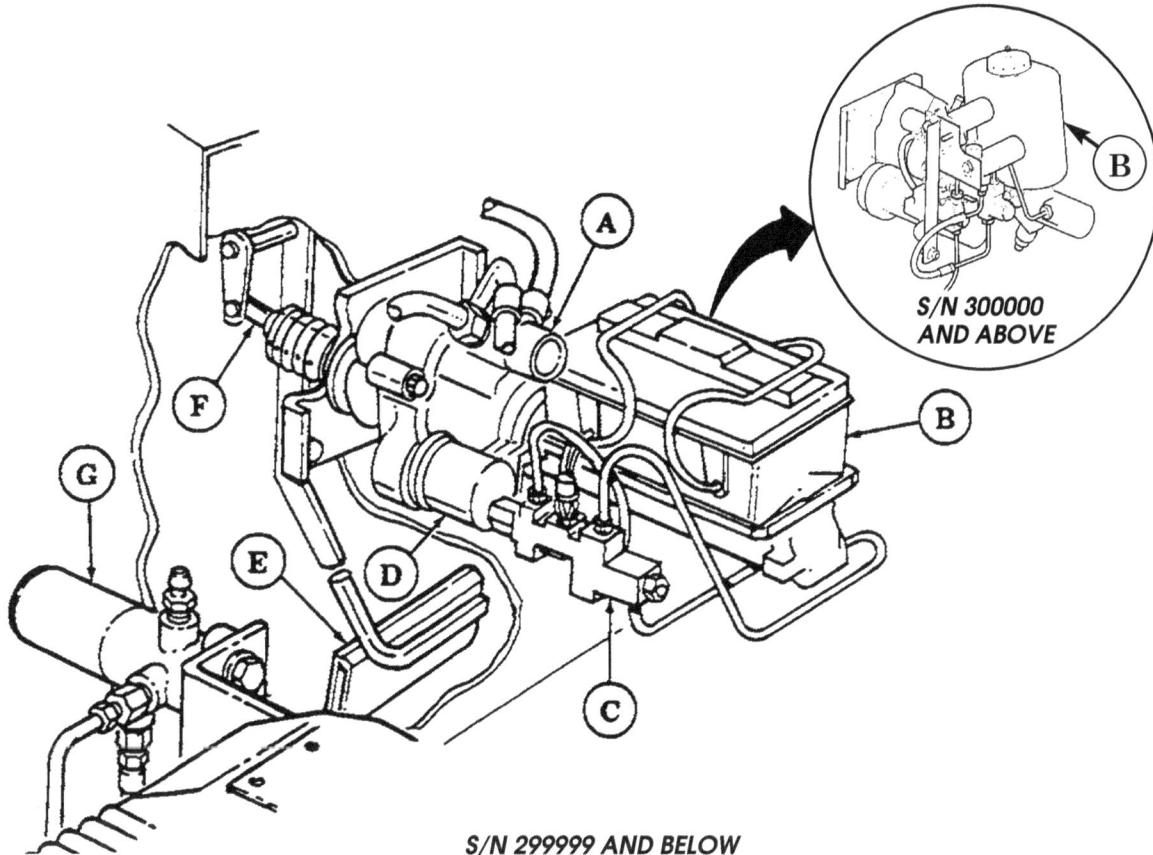

S/N 300000 AND ABOVE

S/N 299999 AND BELOW

(A) **HYDRO-BOOSTER** – Converts hydraulic power from the steering pump to mechanical power to the master cylinder, providing power assist during braking.

(B) **MASTER CYLINDER/RESERVOIR** – Stores brake fluid and converts mechanical pedal pressure to hydraulic pressure.

(C) **PROPORTIONING VALVE** – Provides balanced front-to-rear braking and activates brake warning lamp in case of brake system malfunction.

(D) **ACCUMULATOR** – Stores hydraulic pressure for additional power-assisted braking in case of loss of pressure in steering system.

(E) **BRAKE PEDAL** – Provides operator control for stopping vehicle.

(F) **BRAKE LINKAGE** – Directs brake pedal pressure to hydro-booster.

(G) **BRAKE PRESSURE LIMITER** – Limits front brake line pressure to prevent brake lockup.

1-23. SERVICE BRAKE SYSTEM OPERATION (Cont'd)

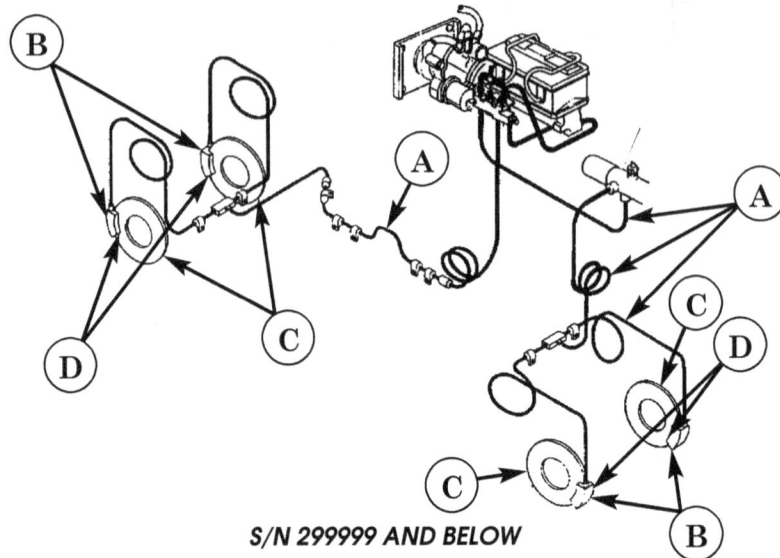

S/N 299999 AND BELOW

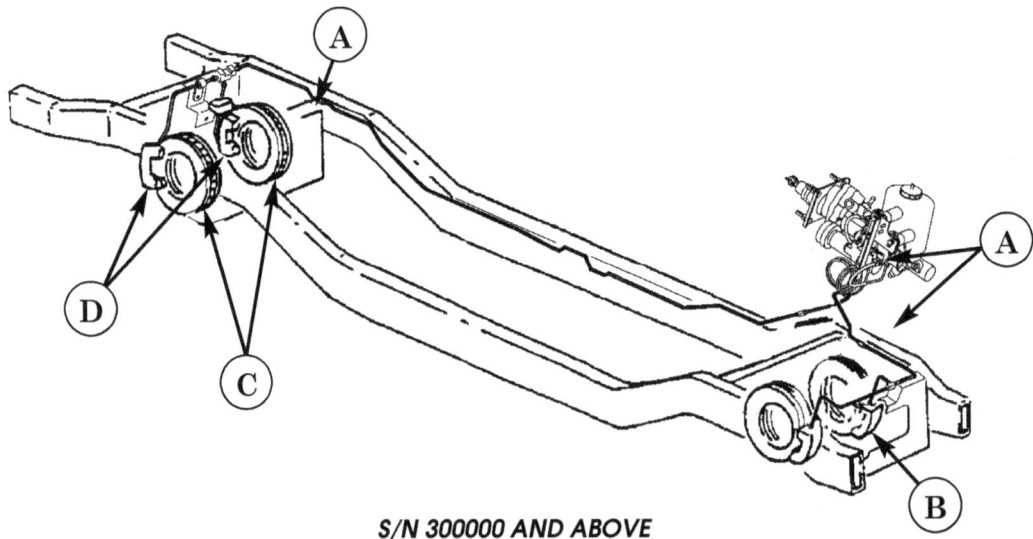

S/N 300000 AND ABOVE

(A) **HYDRAULIC BRAKE LINES** – Direct brake fluid under pressure to all four brake calipers from master cylinder.

(B) **BRAKE CALIPER** – Converts hydraulic pressure to mechanical force to compress brake pads against brake rotors.

(C) **BRAKE ROTOR** – Attached to output flange on front and rear differentials, rotor prevents output flange from turning when brakes are applied.

(D) **BRAKE PADS** – Apply friction to brake rotor when brake pedal is depressed.

1-24. STEERING CONTROL SYSTEM OPERATION (SERIAL NUMBERS 299999 AND BELOW)

Major components of the steering system are:

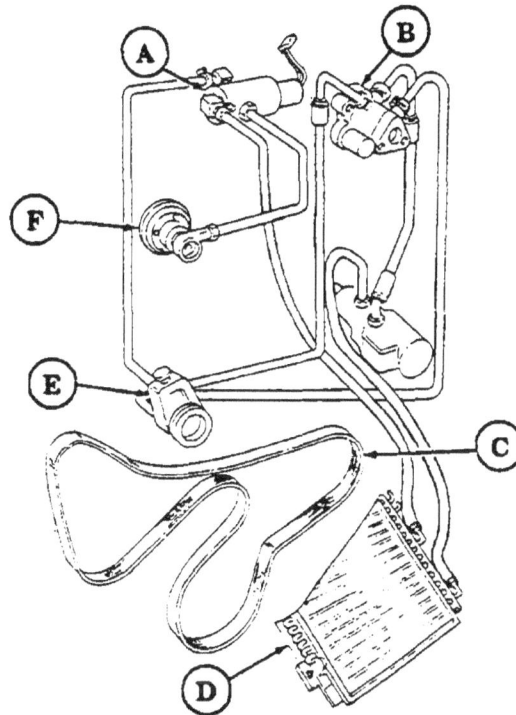

(A) **HYDRAULIC CONTROL VALVE** – Directs hydraulic fluid to provide required pressure to actuate and deactuate fan clutch as required by engine temperature. Hydraulic pressure supplied by power steering pump.

(B) **HYDRO-BOOSTER** – Converts hydraulic power from the steering pump to mechanical power to the master cylinder, providing power assist during braking.

(C) **SERPENTINE BELT** – Transmits mechanical driving power from crankshaft drive pulley to steering pump pulley which drives the steering pump.

(D) **POWER STEERING COOLER** – Directs power steering fluid through a series of fins or baffles so outside air can dissipate excess heat before the fluid is recirculated through the steering system.

(E) **OIL RESERVOIR AND STEERING PUMP** – Combined in one unit, the reservoir serves as an oil filling point and the pump supplies the oil under pressure throughout the steering system.

(F) **FAN CLUTCH** – Hydraulically-actuated by pressure from hydraulic control valve to control operation of fan. Hydraulic pressure is supplied by the power steering pump.

1-24. STEERING CONTROL SYSTEM OPERATION (SERIAL NUMBERS 299999 AND BELOW) (Cont'd)

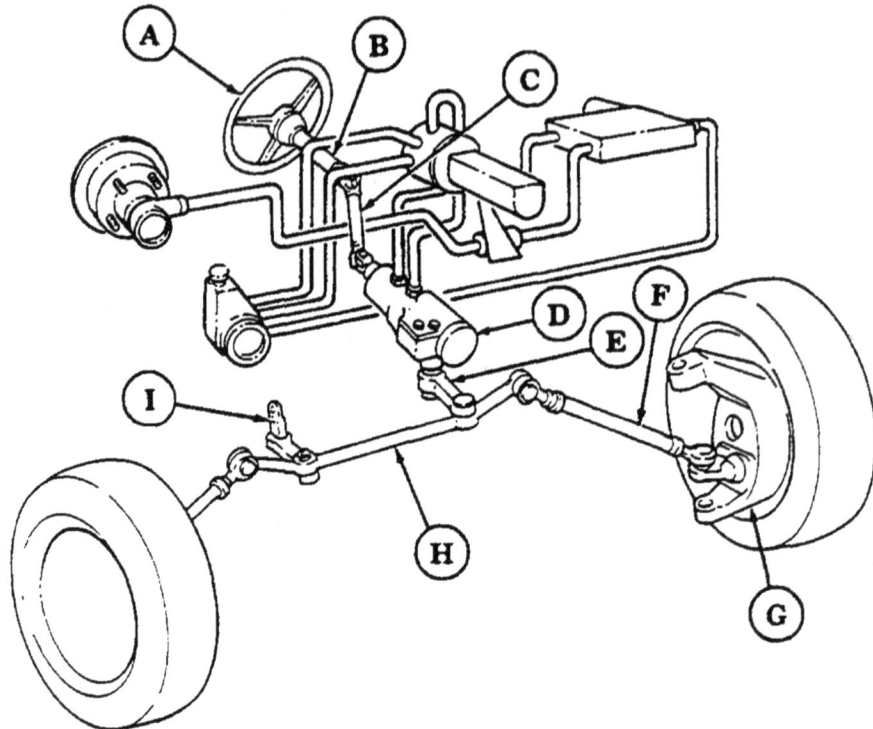

(A) **STEERING WHEEL** – Serves as manual steering control for the operator.

(B) **STEERING COLUMN** – Transmits turning effort from steering wheel to intermediate steering shaft.

(C) **INTERMEDIATE STEERING SHAFT** – Permits angle of torque from steering column to input shaft of power steering gear.

(D) **STEERING GEAR** – Converts hydraulic power from steering pump to mechanical power at pitman arm.

(E) **PITMAN ARM** – Transfers steering torque from power steering gear to center link.

(F) **TIE ROD ASSEMBLY** – Transmits movement from center link to geared hub.

(G) **GEARED HUB** – Serves as the pivot point and link for the front wheels via the tie rod assembly.

(H) **CENTER LINK** – Transmits movement from pitman arm to tie rods.

(I) **IDLER ARM** – Supports right side of center link.

1-24.1. STEERING CONTROL SYSTEM OPERATION (SERIAL NUMBERS 300000 AND ABOVE)

Major components of the steering system are:

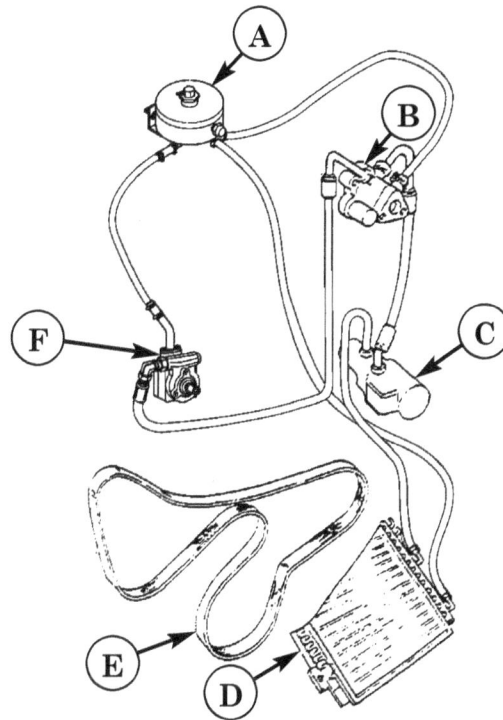

(A) **OIL RESERVOIR** – The oil reservoir serves as an oil filling point.

(B) **HYDRO-BOOSTER** – Converts hydraulic power from the steering pump to mechanical power to the master cylinder, providing power assist during braking.

(C) **STEERING GEAR** – Converts hydraulic power from steering pump to mechanical power at pitman arm.

(D) **POWER STEERING COOLER** – Directs power steering fluid through a series of fins or baffles so outside air can dissipate excess heat before the fluid is recirculated through the steering system.

(E) **SERPENTINE BELT** – Transmits mechanical driving power from crankshaft drive pulley to steering pump pulley which drives the steering pump.

(F) **STEERING PUMP** – Supplies the oil under pressure throughout the steering system.

1-24.1. STEERING CONTROL SYSTEM OPERATION (SERIAL NUMBERS 300000 AND ABOVE) (Cont'd)

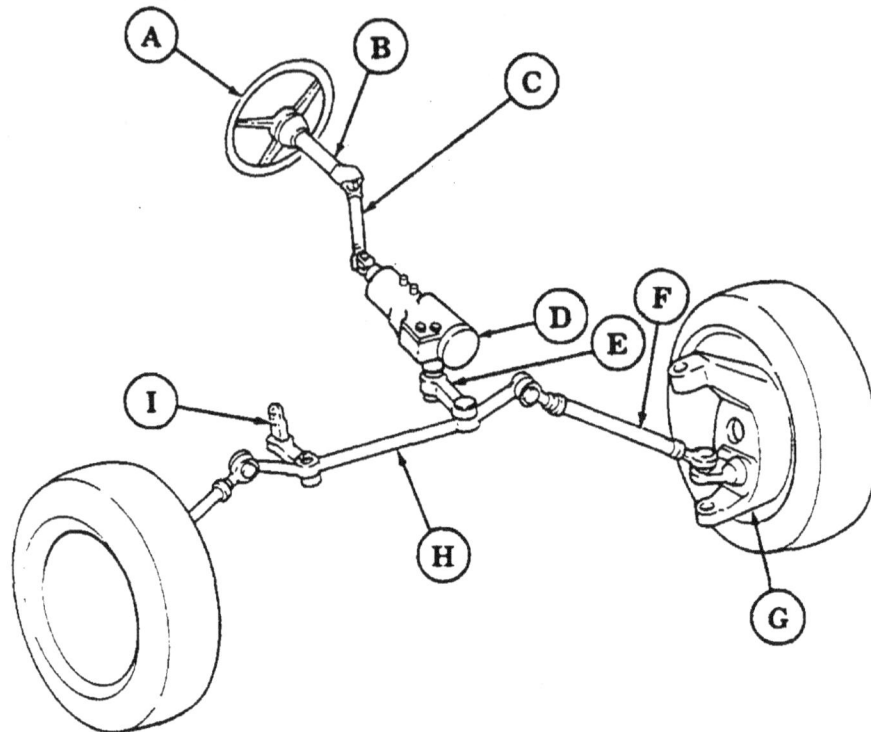

(A) **STEERING WHEEL** – Serves as manual steering control for the operator.

(B) **STEERING COLUMN** – Transmits turning effort from steering wheel to intermediate steering shaft.

(C) **INTERMEDIATE STEERING SHAFT** – Permits angle of torque from steering column to input shaft of power steering gear.

(D) **STEERING GEAR** – Converts hydraulic power from steering pump to mechanical power at pitman arm.

(E) **PITMAN ARM** – Transfers steering torque from power steering gear to center link.

(F) **TIE ROD ASSEMBLY** – Transmits movement from center link to geared hub.

(G) **GEARED HUB** – Serves as the pivot point and link for the front wheels via the tie rod assembly.

(H) **CENTER LINK** – Transmits movement from pitman arm to tie rods.

(I) **IDLER ARM** – Supports right side of center link.

1-25. SUSPENSION SYSTEM OPERATION

The suspension system is an independent coil spring-type system. Major components of the suspension system are:

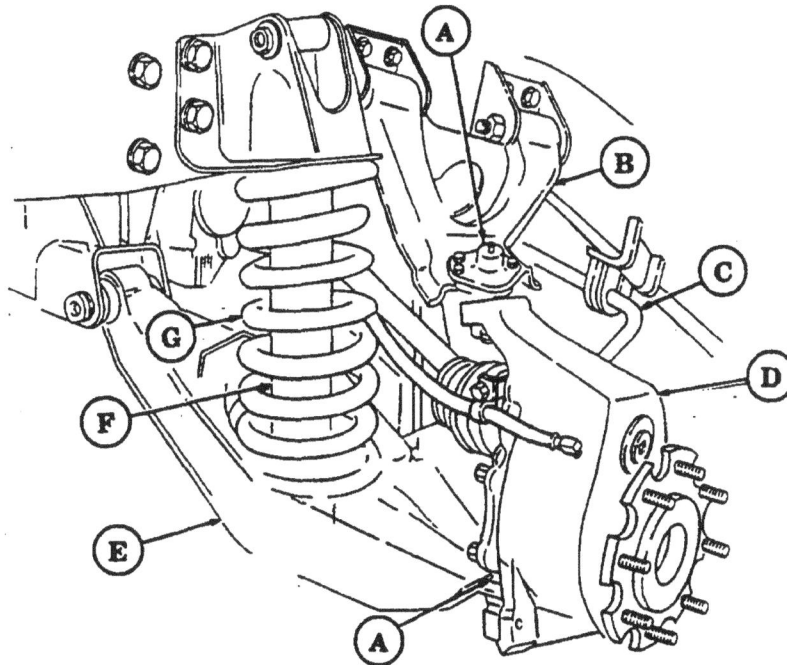

(A) **BALL JOINTS** – Connect geared hub to control arms, and allow change of angle between geared hub and control arms during suspension movement.

(B) **UPPER CONTROL ARM** – Connects geared hub to frame rail.

(C) **STABILIZER BAR (FRONT ONLY)** – Prevents vehicle sway when cornering.

(D) **GEARED HUB** – Serves as a mounting point for wheel and tire assembly and provides 1.92:1 gear reduction to increase torque to wheel and tire assembly.

(E) **LOWER CONTROL ARM** – Connects geared hub to crossmember.

(F) **SHOCK ABSORBER** – Dampens suspension movement and limits amount of suspension travel.

(G) **COIL SPRING** – Supports weight of vehicle and allows suspension travel to vary depending on terrain and vehicle loading.

1-25. SUSPENSION SYSTEM OPERATION (Cont'd)

(A) **RADIUS ROD (REAR ONLY)** – Connects geared hub to frame to maintain rear end alignment.

1-26. 200-AMPERE UMBILICAL POWER CABLE

The 200-ampere umbilical power cable covered in this manual applies to the M1113, M1152, and M1152A1 and consists of the following major component: ∎

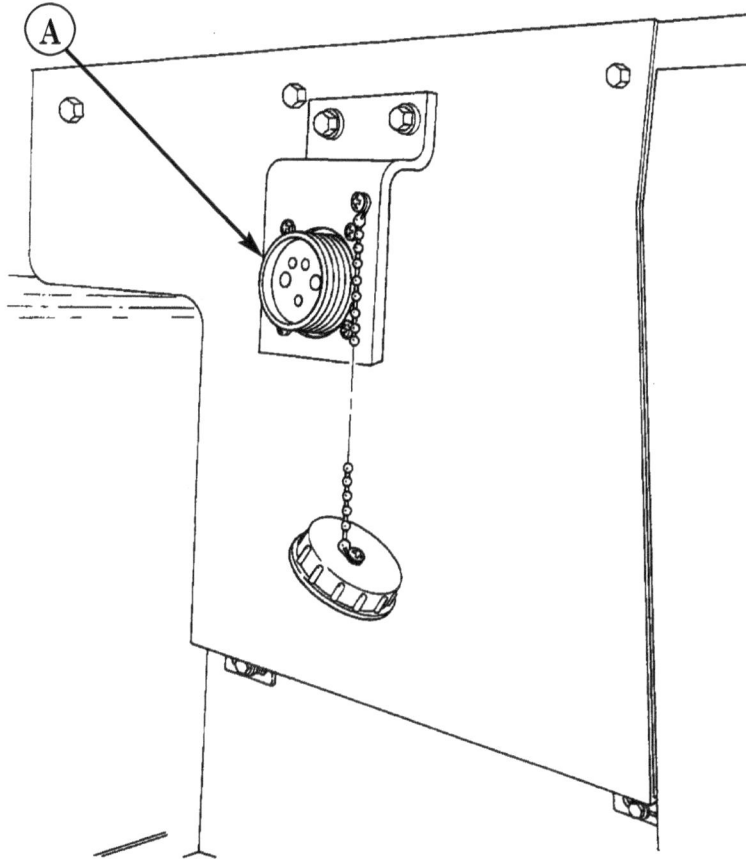

(A) **POWER CABLE** – Located behind the companion seat, provides power for shelter equipment.

1-27. AIR CONDITIONING SYSTEM OPERATION (M1114)

Major components of the air conditioning system are:

(A) **COMPRESSOR** – Inside the compressor, low-pressure gas refrigerant is compressed into a high-pressure gas that is pushed into the condenser by the compressor.

(B) **CONDENSER** – Refrigerant enters the condenser as a high-pressure gas. When condensed, it gives up its heat to the outside air and becomes a high-pressure liquid.

(C) **EVAPORATOR** – Refrigerant enters evaporator as a liquid spray. It absorbs heat from the air in the crew compartment and vaporizes into a low-pressure gas.

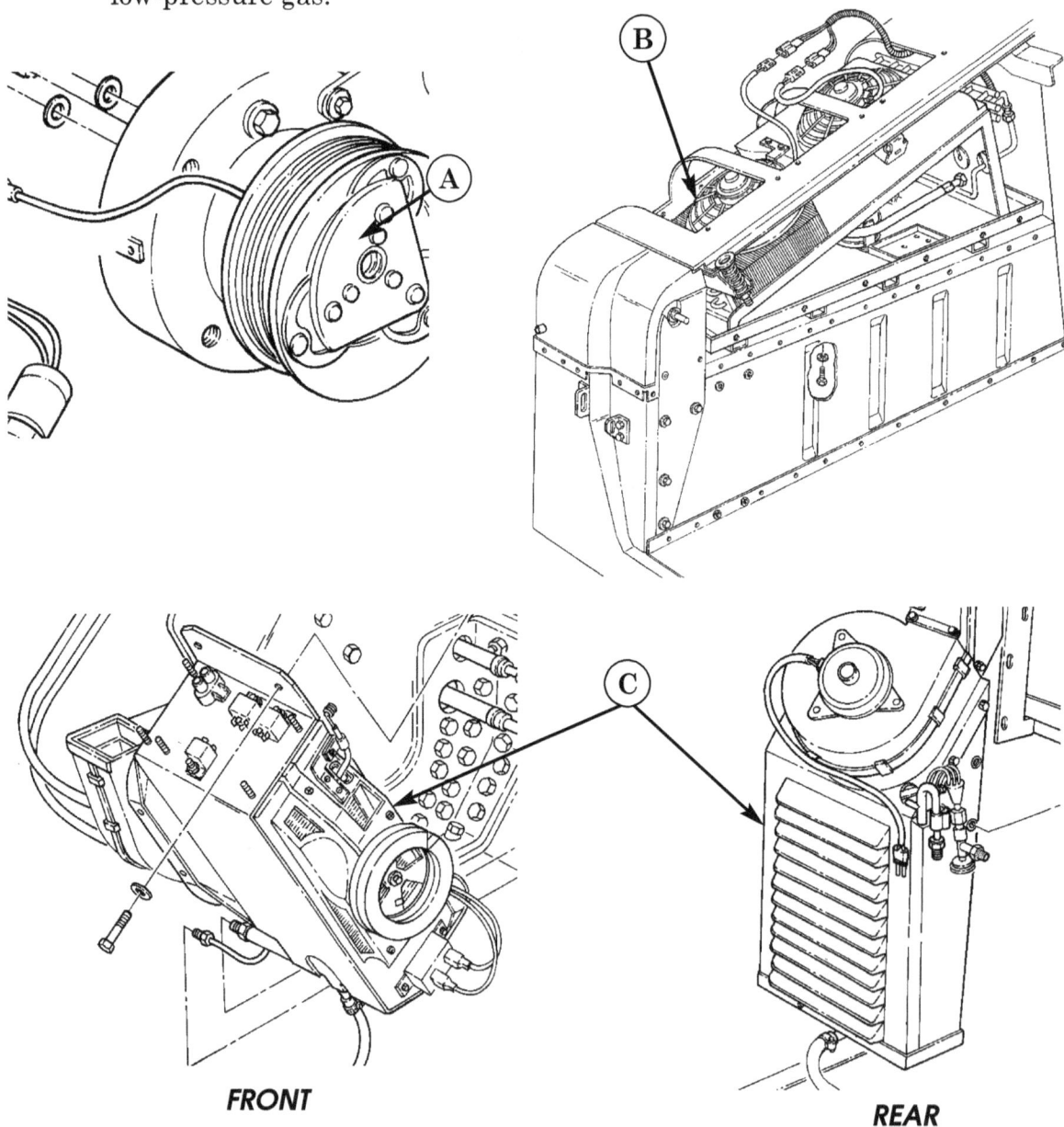

FRONT

REAR

1-28. AIR CONDITIONING SYSTEM OPERATION (TWO-MAN AND FOUR-MAN)

Major components of the air conditioning system are:

(A) **COMPRESSOR** – Inside the compressor, low-pressure gas refrigerant is compressed into a high-pressure gas that is pushed into the condenser by the compressor.

(B) **CONDENSER** – Refrigerant enters the condenser as a high-pressure gas. When condensed, it gives up its heat to the outside air and becomes a high-pressure liquid.

(C) **EVAPORATOR** – Refrigerant enters evaporator as a liquid spray. It absorbs heat from the air in the crew compartment and vaporizes into a low-pressure gas.

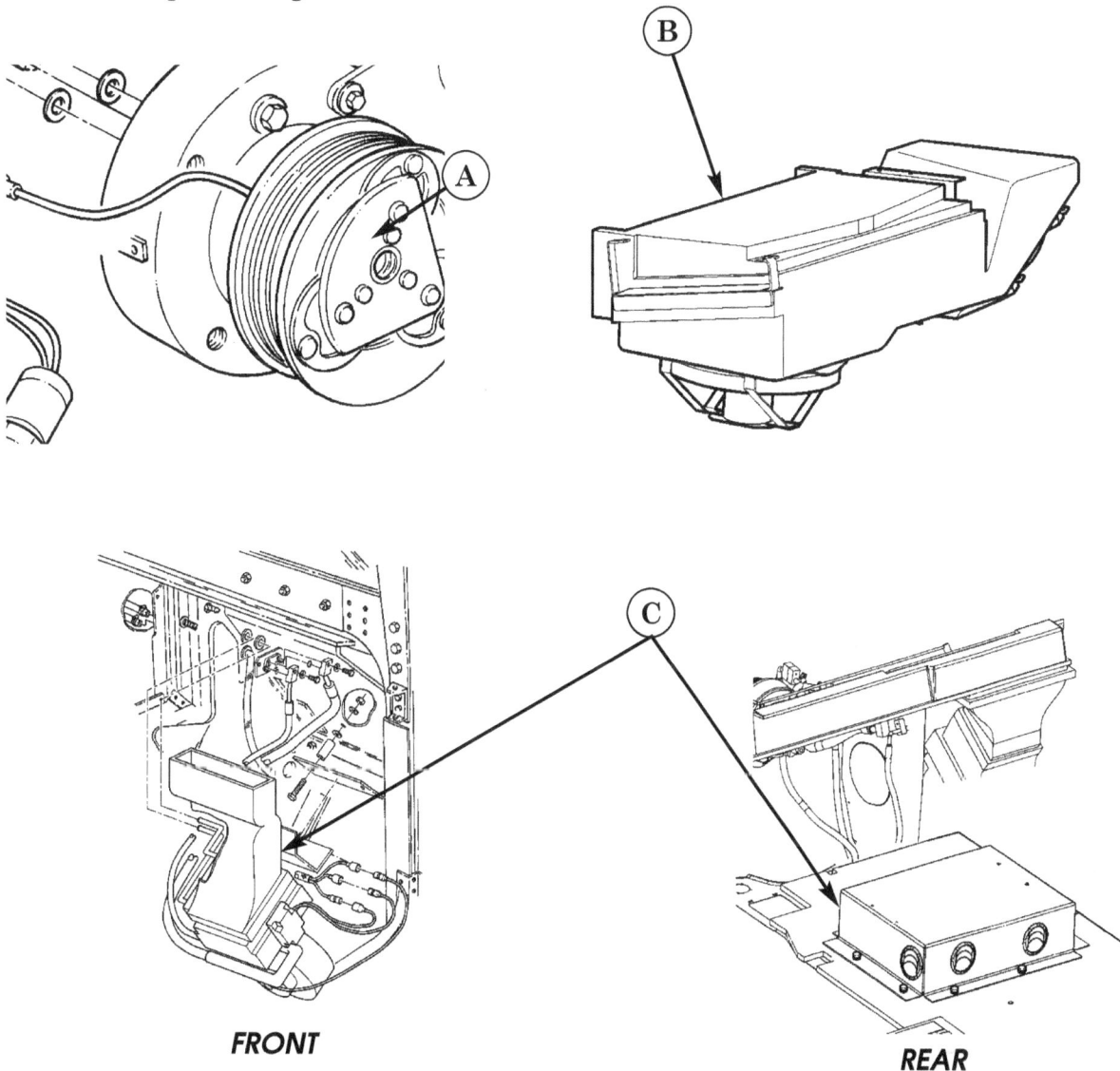

FRONT

REAR

CHAPTER 2
OPERATING INSTRUCTIONS

Section I. DESCRIPTION AND USE OF OPERATOR'S CONTROLS AND INDICATORS

2-1. KNOW YOUR CONTROLS AND INDICATORS

Before you attempt to operate your equipment, make sure you are familiar with the location and function of all controls and indicators.The location and function of controls and indicators are described in this section.

WARNING

• This vehicle has been designed to operate safely and efficiently within the limits specified in this TM. Operation beyond these limits is prohibited IAW AR 750-1 without written approval from the Commander, U.S.Army Tank-automotive and Armaments Command,ATTN:AMSTA-CM-S,Warren, MI 48397-5000.

• Do not use hand throttle as an automatic speed or cruise control. The hand throttle does not automatically disengage when brake is applied, resulting in increased stopping distances and possible hazardous and unsafe operation. Injury to personnel or damage to equipment may result.

NOTE

• Except where noted, the controls and indicators in this section are applicable to all vehicle models covered in this manual.

• In this manual, left side indicates the driver's side of the vehicle. Right side indicates the opposite side of the vehicle.

2-2. CONTROLS, INDICATORS, AND EQUIPMENT

a. Driver's Compartment.

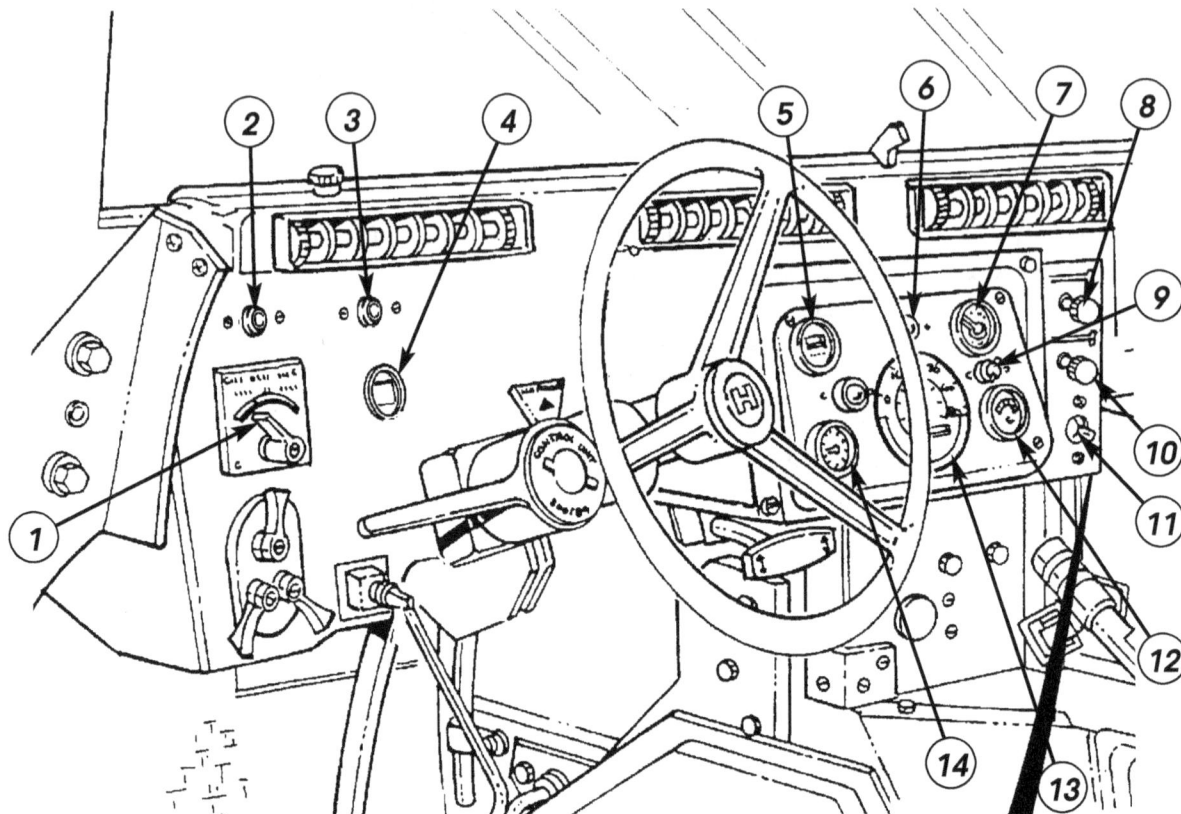

M1113, M1114

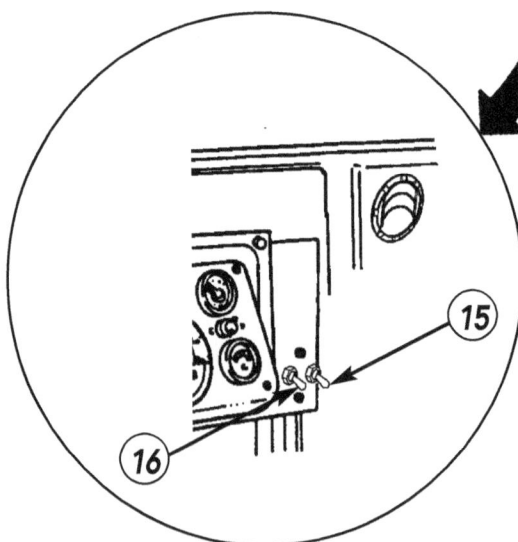

ALL OTHER VEHICLES

KEY   ITEM AND FUNCTION

1   *Rotary switch* has ENG STOP, RUN, and START positions.When positioned to RUN, glow plugs are activated to heat engine precombustion chambers. When positioned to START, starter will engage and crank engine.

2   *Wait-to-start lamp assembly* illuminates when glow plugs are activated; goes out when engine is ready to be started.

3   *Brake warning lamp assembly* illuminates if parking brake is applied or if a hydraulic imbalance (leak, etc.) exists between the front and rear brake systems.

4   *Air restriction gauge* signals the operator that a restriction exists in the air cleaner.

5   *Engine oil pressure gauge* indicates oil pressure when engine is running.

6   *High beam lamp assembly* illuminates when front headlights are switched to HIGH beam position.

7   *Coolant temperature gauge* indicates engine coolant temperature.

NOTE

Item 8 applies to M1113/M1114 vehicles only.

8   *Defroster control knob* directs heated air to windshield on the M1113. For the M1114, control knob directs outside air in for circulation.

9   *Instrument panel lights* illuminate instrument panel gauges.

NOTE

Items 10 and 11 apply to M1113/M1114 vehicles only.

10   *Heater control knob* controls the temperature of the air output.

11   *Heater fan switch* controls blower motor fan speed.The switch has HI (high), OFF, and LO (low) settings to regulate defroster and heater air flow into vehicle.

12   *Voltmeter* indicates the charging level and activity of the battery charging system.

13   *Speedometer/odometer* indicates vehicle speed and accumulated mileage.

14   *Fuel gauge* indicates amount of fuel in fuel tank.

NOTE

Items 15 and 16 apply to M1151/M1151A1/M1152/M1152A1/ M1165/M1165A1/M1167 vehicles only.

15.   *A/C /Heatswitch* controls the temperature of air output.

16.   *Fan switch* controls blower motor fan speed.

2-2. CONTROLS, INDICATORS, AND EQUIPMENT (Cont'd)

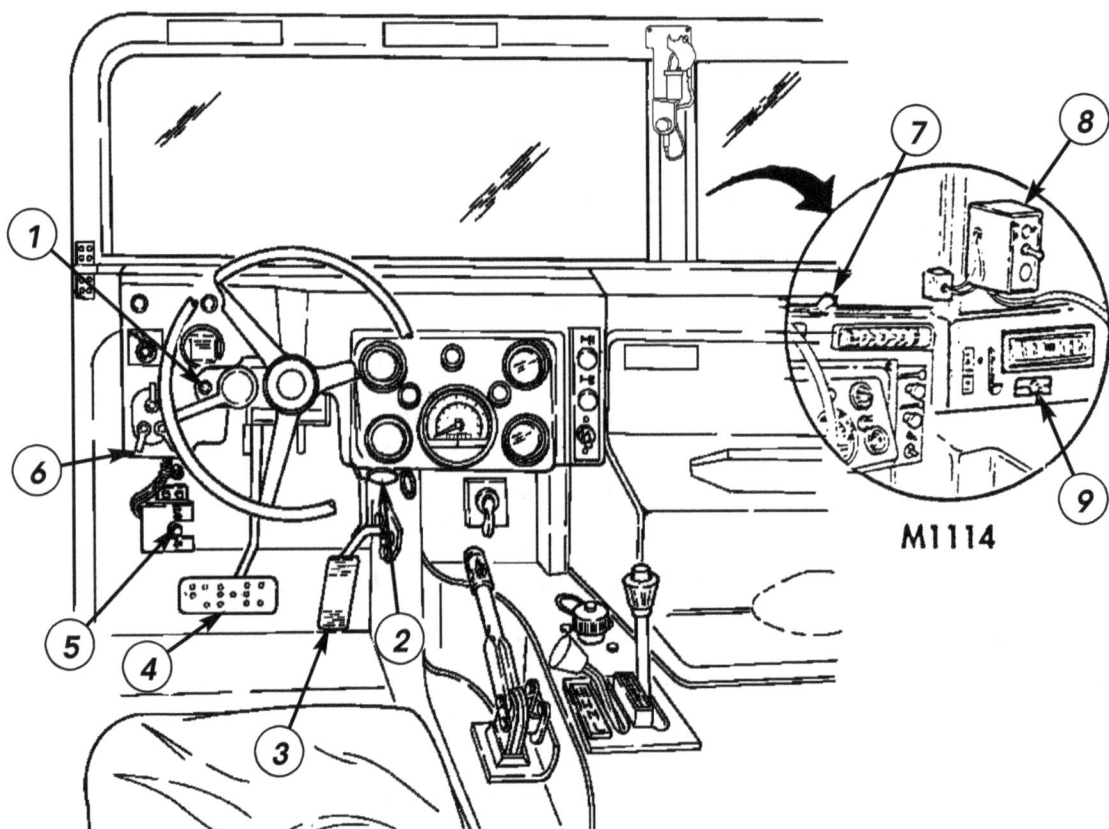

M1114

KEY   ITEM AND FUNCTION

1   *Directional signal/emergency flasher indicator light* illuminates during directional signal or emergency flasher use.

2   *Hand throttle* increases engine speed for use during cold weather starting, winch operation, deep water fording, and to obtain maximum alternator output for communications/electrical requirements; is not to be used as an automatic speed or cruise control.

3   *Accelerator pedal* is the operator's foot control for varying engine speed.

4   *Service brake pedal* is depressed to slow or stop vehicle.

5   *Dimmer switch* permits the operator to select high or low headlight beam.

6   *Light switch* controls operation of vehicle service lights and blackout lights.

NOTE
Items 7 through 9 apply to the M1114 vehicle only.

7   *Defroster register* opens and closes the defroster to direct heated air to windshield for defrosting.

8   *Windshield de-icer switch* operates the windshield de-icer by moving the switch to on or off position.

9   *Air conditioner on/off switch* allows operation of the air conditioning compressor, condenser fans, and rear blower fan (rear blower fan will only operate if fan speed is on high).

2-2. CONTROLS, INDICATORS, AND EQUIPMENT (Cont'd)

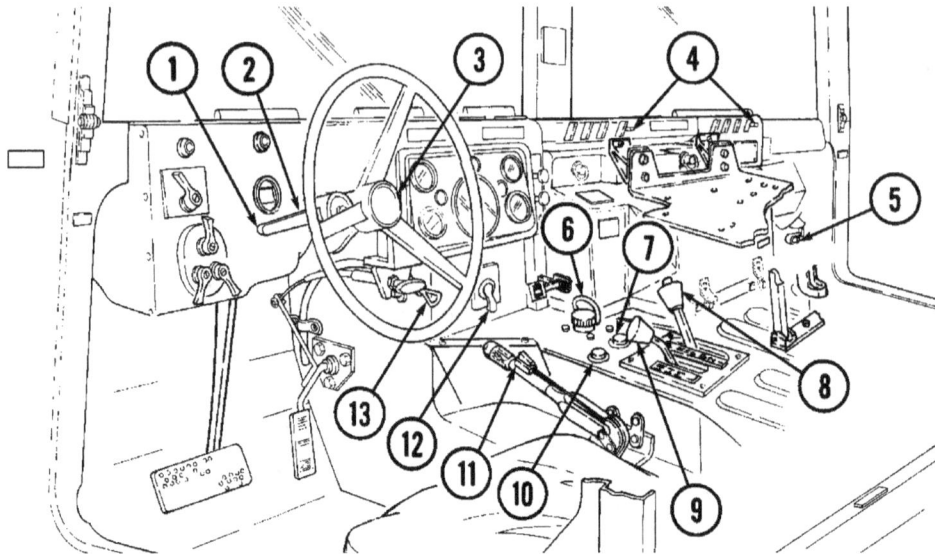

KEY ITEM AND FUNCTION

1   *Directional signal lever* activates turn signal lights.

2   *Warning hazard control* activates the warning flashers.

3   *Horn button* activates vehicle horn.

4   *Baffle operating rods* slide open to allow heated air into crew compartment.

5   *Fresh air intake lever* (M1113/M1114) is pulled back to allow fresh air into crew compartment, or pushed forward to close.

6   *Simplified Test Equipment/Internal Combustion Engine-Reprogrammable (STE/ICE-R) diagnostic connector* is attaching point for the Vehicle Test Meter (VTM) to facilitate vehicle electrical and engine systems diagnoses.

7   *Transmission indicator lamp* comes on when ignition switch is turned on, and goes off when vehicle is started and is also used to flash diagnostic trouble codes when Transmission Control Module (TCM) is placed in diagnostic mode.

8   *Transmission shift lever* is used to select vehicle driving range, P (park), R (reverse), N (neutral), Ⓓ (overdrive), D (drive), 2 (second), and 1 (first).

9   *Transfer case shift lever* is used to select vehicle driving range, H/L (high/lock range), H (high range), N (neutral), and L (low range).

10 *Transfer case indicator lamp* illuminates when transfer case has completed the low-range shift and high/lock-range shift.

11 *Parking brake lever* is used to apply parking brake. Safety release button must be depressed to release parking brake.

12 *Fording control switch* (M1113/M1151/M1152) (deep water fording kit only) allows operator to select VENT during normal operating conditions or DEEP FORD for deep water fording.

13 *Steering wheel lock cable* permits steering wheel to be locked to prevent unauthorized use of vehicle.

KEY ITEM AND FUNCTION

14 *Windshield washer/wiper control knob* operates a two-speed electric wiper motor; when depressed, operates windshield washer.

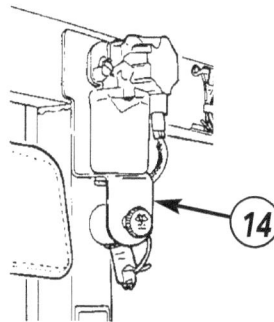

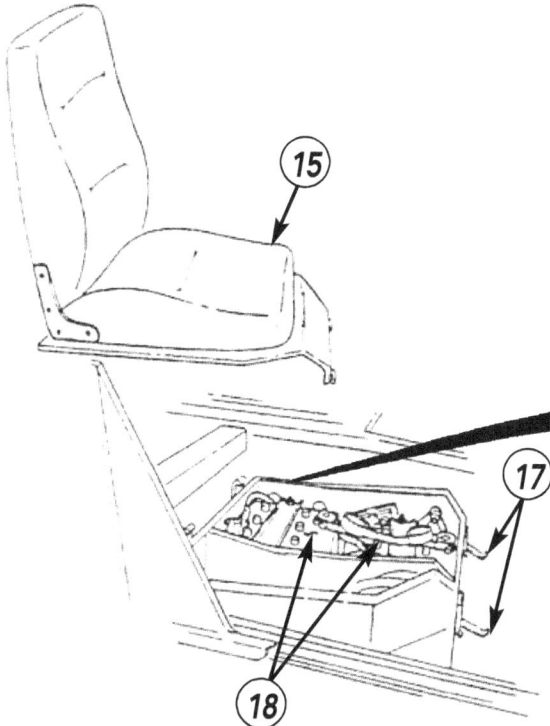

15 *Companion seat* is removed to provide access to batteries.

16 *Optional 12-volt auxiliary power outlet* provides 12-volt power to accessories.

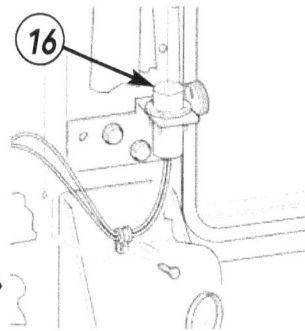

17 *Battery box latches* release to permit removal of companion seat for access to batteries.

18 *Batteries* provide 24-volt power to vehicle electrical system.

19 *Slave receptacle* is located at outside front of battery box. It is the connecting point for the slave cable for slave-starting the vehicle.

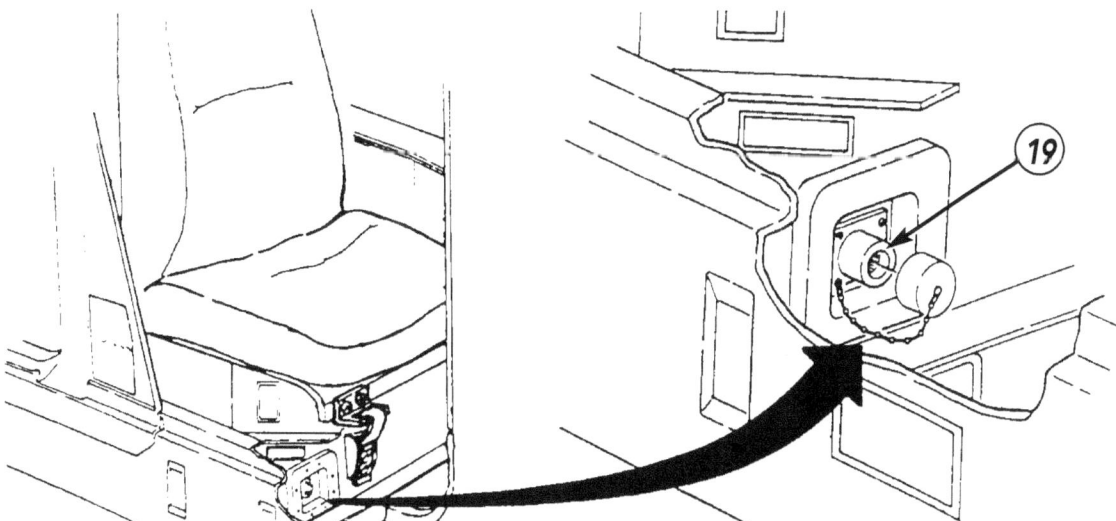

2-2. CONTROLS, INDICATORS, AND EQUIPMENT (Cont'd)

KEY   ITEM AND FUNCTION

1   *Engine access cover* is removed to provide access to rear of engine.

2   *Radio rack* serves as mounting point for AN/GRC-121.3 or AN/VRC-91 radio.

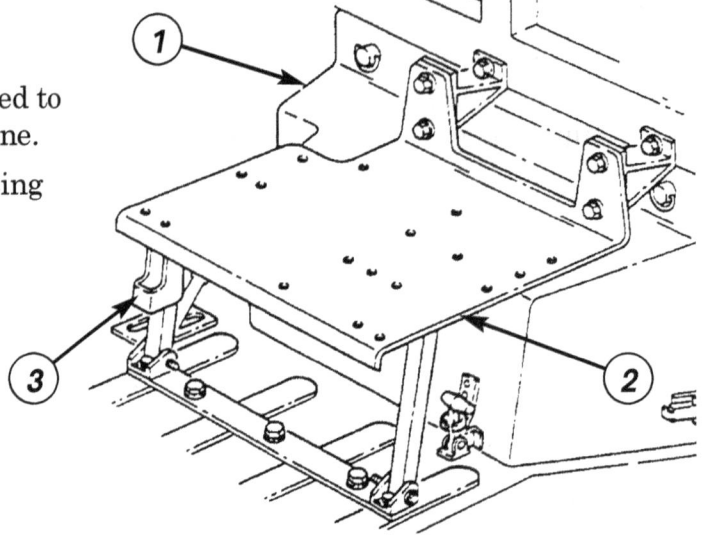

3   *Microphone bracket* serves as mounting point for microphone.

4   *Catch assembly* holds M16A2/A4 rifle safely in place for travel. Adjustable catch assembly configuration holds M16A2/A4 rifle when positioned all the way in, or M203 grenade launcher when extended out.

5   *Stock brace* holds stock end of M16A2/A4 rifle or M203 grenade launcher in position for travel.

6   *Driver's seat adjusters* permit driver's seat to be adjusted in forward, rearward, up, or lower positions.

7   *First aid kit* is used for crew emergency first aid treatment. Actual location shall be determined by mission profile.

8   *Fire extinguisher* is located on the side of the driver's seat.

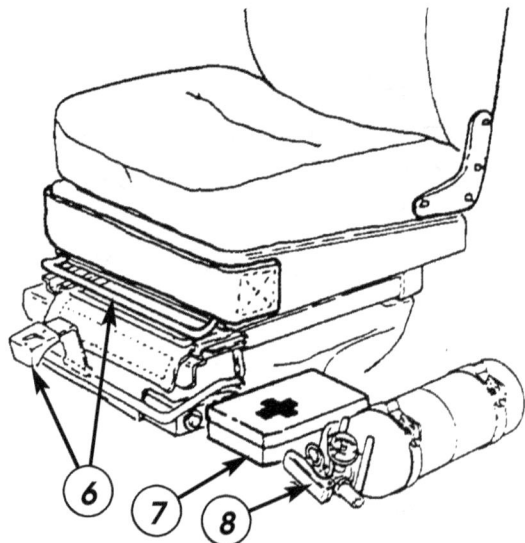

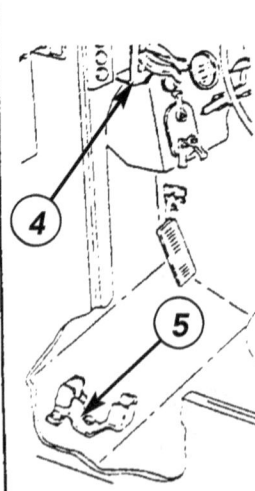

ALL OTHER VEHICLES

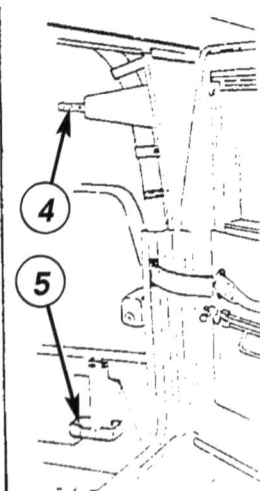

M1114

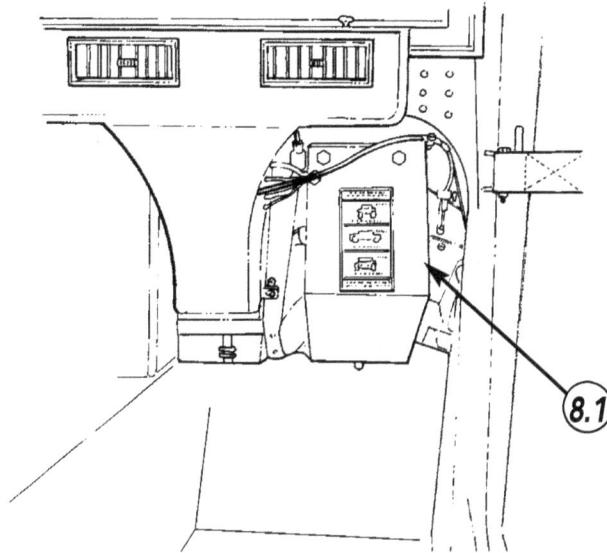

KEY   ITEM AND FUNCTION

8.1.   *Front Blower Assembly for M1114* consists of a fan blower and evaporator
that blows cool air through the A/C vents located in the front of the passenger
compartment.

8.2.   *Rear Blower Assembly for M1114* consists of a fan blower and evaporator that
blows cool air from the rear of the passenger compartment when the A/C
switch is on and blower switch is on high.

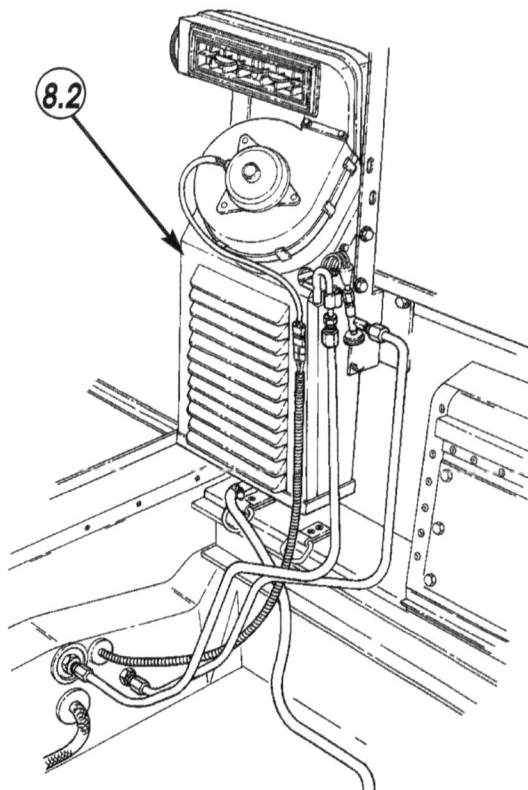

2-2. CONTROLS, INDICATORS, AND EQUIPMENT (Cont'd)

KEY   ITEM AND FUNCTION

8.3.   *Front Blower Assembly for M1151/M1151A1/M1152/M1152A1/
M1165/M1165A1/M1167* consists of a fan blower, heater and evaporator that
blows air through the vents located in the front of the passenger
compartment when the heater switch is on high or low.

8.4.   *Rear Blower Assembly for M1151/M1151A1/M1152/M1152A1/
M1165/M1165A1/M1167* consists of a fan blower and evaporator that blows
cold air through the A/C vents located on the rear blower assembly when the
heater switch is on high or low.

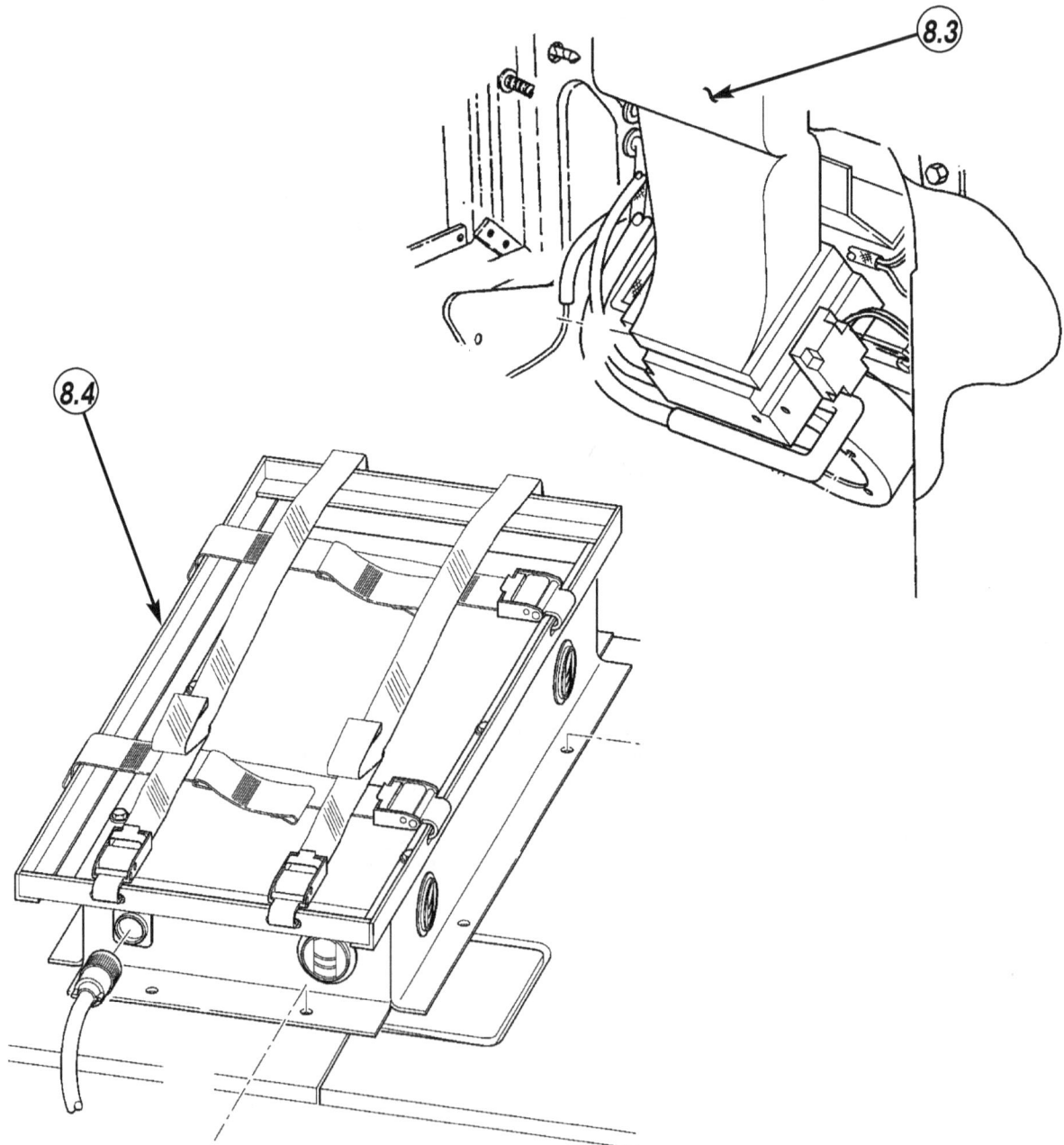

2-2. CONTROLS, INDICATORS, AND EQUIPMENT (Cont'd)

KEY   ITEM AND FUNCTION

8.5. *Automatic fire extinguishing system master control module for M1114/
■   M11151A1/M1152A1/M1165A1/M1167* is located on the driver's side of the
       radio rack; allows for visual and physical access to check system status, reset
       after discharge and power removal during maintenance operations.

8.6. *Automatic fire extinguishing system sensors for M1114/M11151A1/
■   M1152A1/M1165A1/M1167* are located in the crew area (two-man, four-man)
       and cargo area (four-man) of the vehicle.

8.7. *Automatic fire extinguishing system extinguishers for M1114/M11151A1/
■   M1152A1/M1165A1/M1167* are located in the crew area (two-man, four-man)
       and cargo area (four-man) of the vehicle.

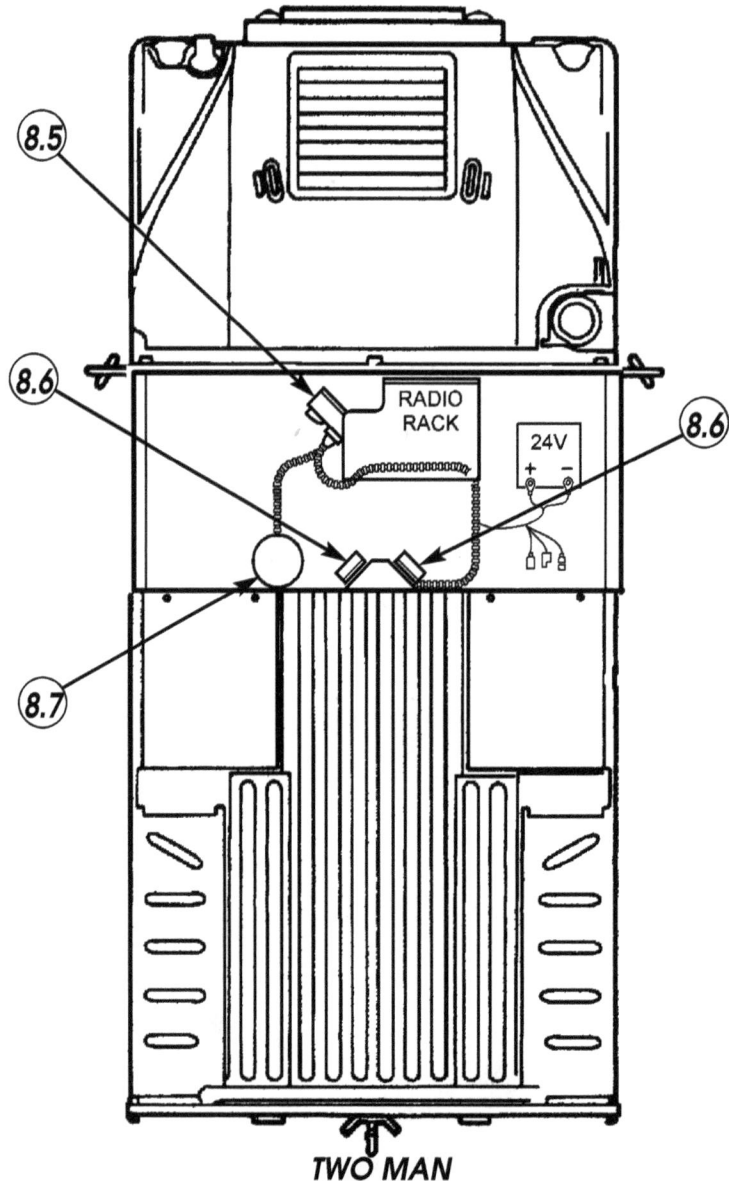

TWO MAN

2-2. CONTROLS, INDICATORS, AND EQUIPMENT (Cont'd)

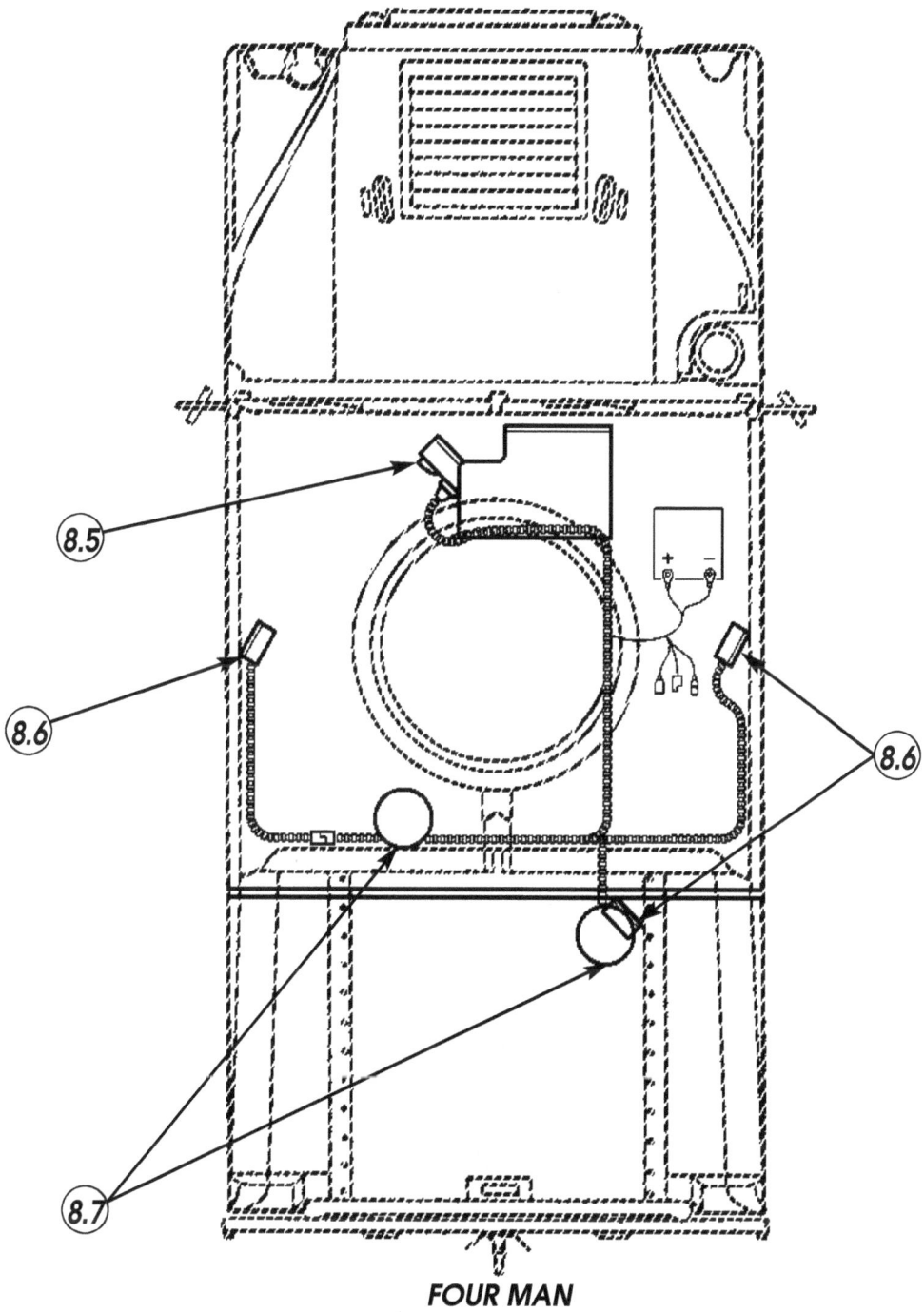

FOUR MAN

2-2. CONTROLS, INDICATORS, AND EQUIPMENT (Cont'd)

■ 8.8. The AFES system is powered only when the engine run switch is in the ON position.When all sensors and extinguishers are properly connected, the control module SYSTEM OK lamp will show continuous green.

■ 8.9. A gauge on the valve indicates extinguisher pressure. Normal extinguisher pressure is 900 psi at 70° F. Minimum operational pressures at various temperatures are indicated on the extinguisher and in table 2-2.

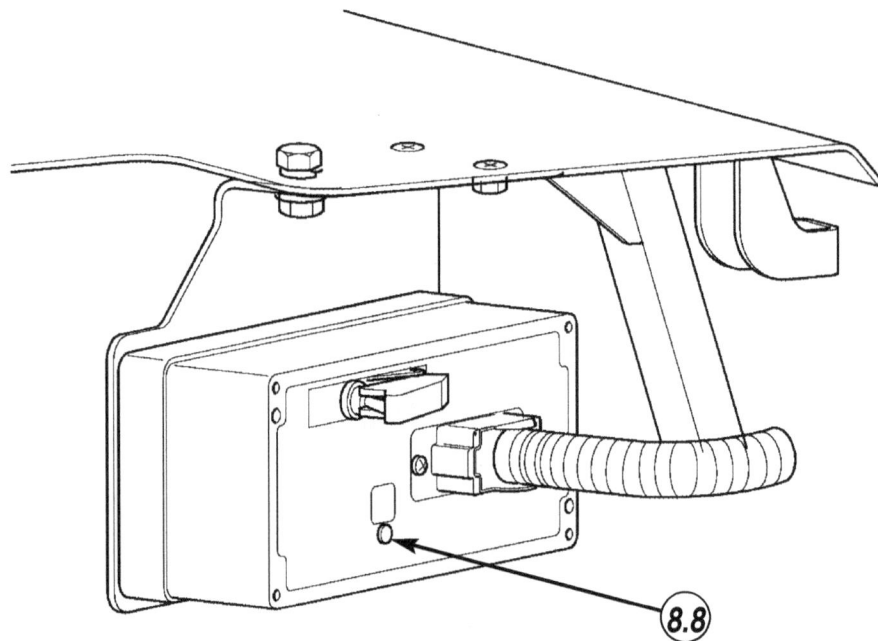

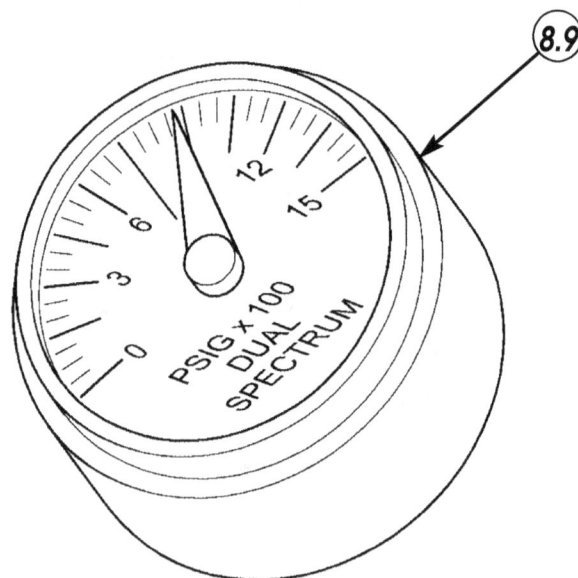

b. Engine Compartment.

KEY   ITEM AND FUNCTION

9   *Transmission oil dipstick* is located right rear of engine; removed to check transmission fluid level.

10 *Transmission oil dipstick tube* (located right rear of engine) is fill point for transmission fluid.

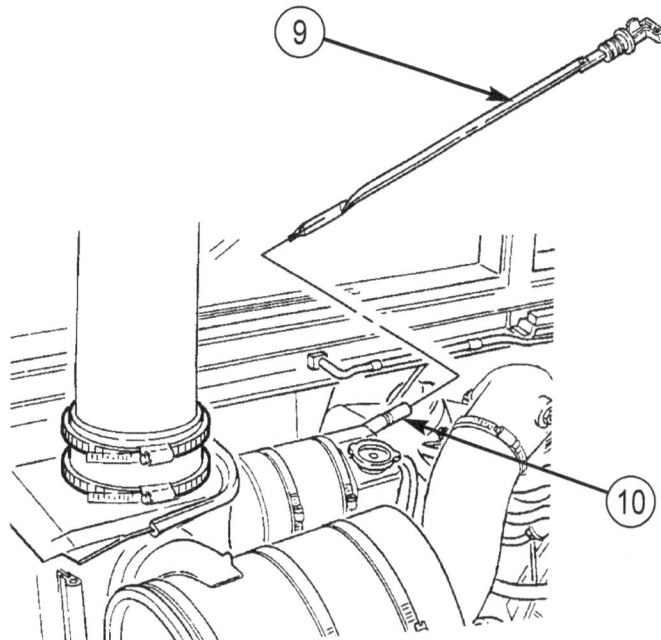

11 *Power steering fluid reservoir cap/dipstick* (located left front of engine) is removed to fill and/or check power steering fluid level.

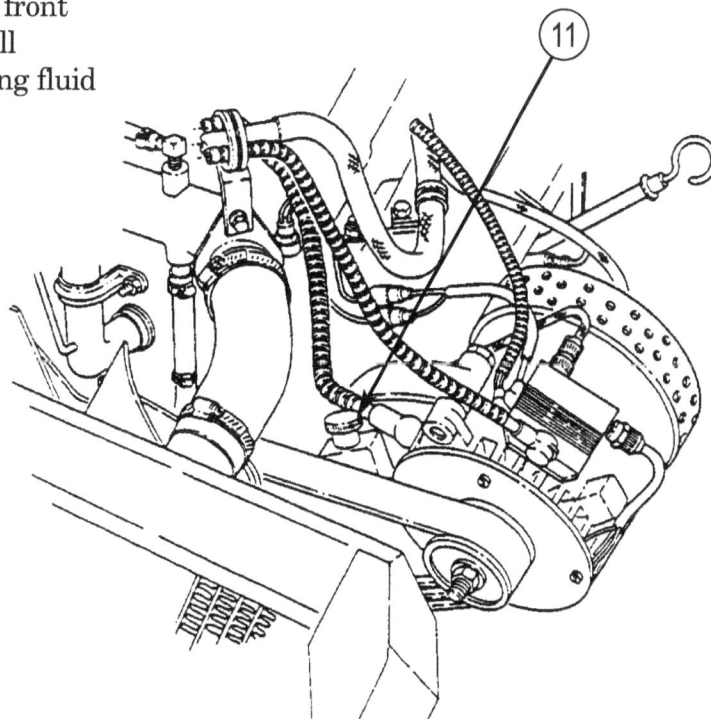

2-2. CONTROLS, INDICATORS, AND EQUIPMENT (Cont'd)

KEY   ITEM AND FUNCTION

1   *Engine oil filler cap* (located front center of engine) is removed from oil filler neck to add oil to engine.

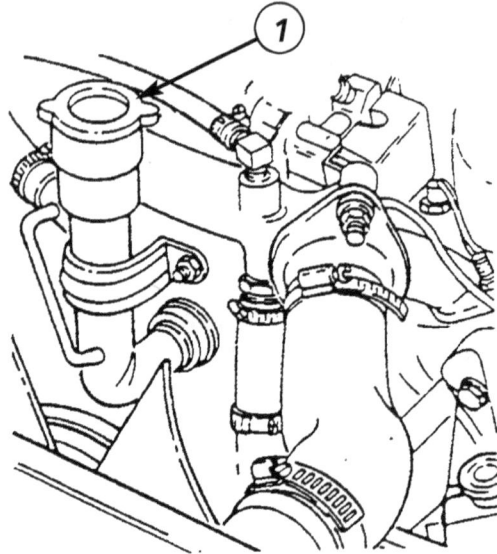

2   *Engine oil dipstick* is located behind alternator on left side of engine; is removed to check engine oil level.

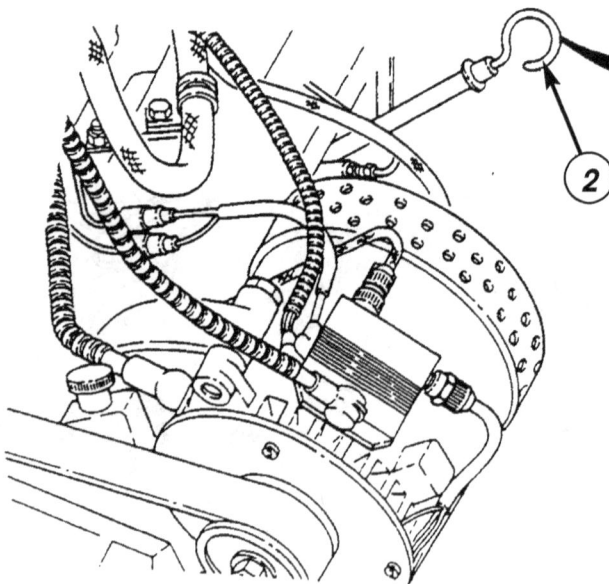

M1113, M1151, AND M1152 VEHICLES WITH DEEP WATER FORDING KIT

KEY	ITEM AND FUNCTION
3	*Radiator drainvalve* (located beneath right front of engine on lower radiator crossover pipe) is turned counterclockwise to drain coolant from radiator.

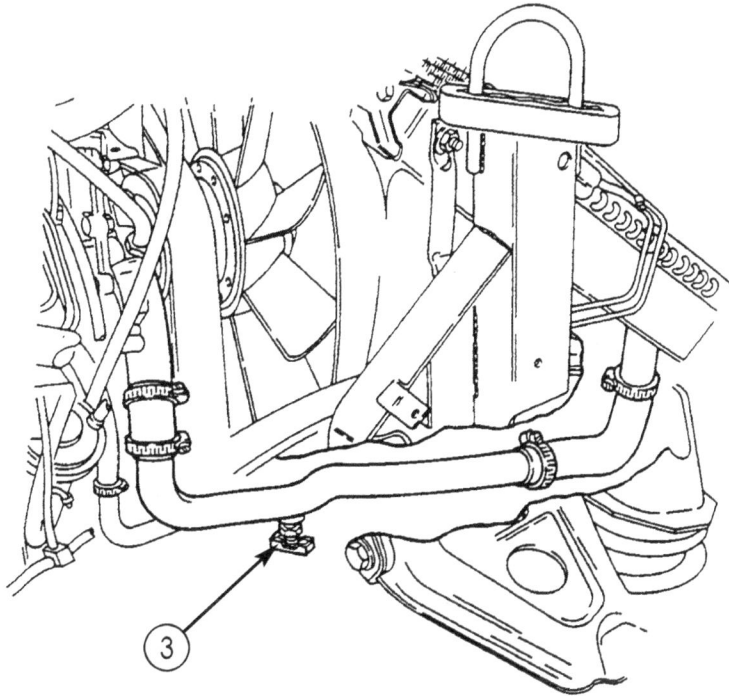

| 4 | *Coolant surge tank cap* (located right rear of engine) is removed from surge tank to add coolant to cooling system. |

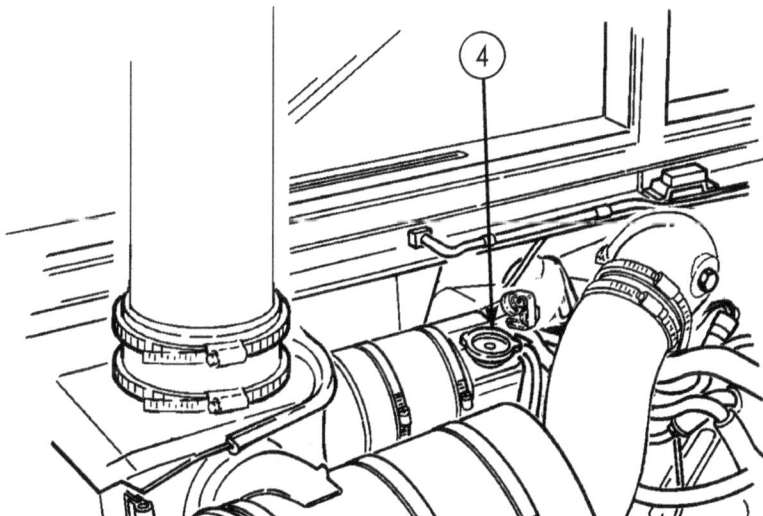

2-2. CONTROLS, INDICATORS, AND EQUIPMENT (Cont'd)

KEY   ITEM AND FUNCTION

1   *Fuel filter* (located left rear of engine on firewall) filters water from fuel system.

2   *Windshield washer reservoir cap* (located left rear of engine) unsnaps to add windshield washer fluid to reservoir.

3   *Fuel filter drainvalve* (located left rear of engine compartment on cowl, beneath and in front of windshield washer reservoir) is draining point for water collected in fuel filter.

4   *Master cylinder cover* (located left of engine) is removed to fill and/or check brake fluid level.

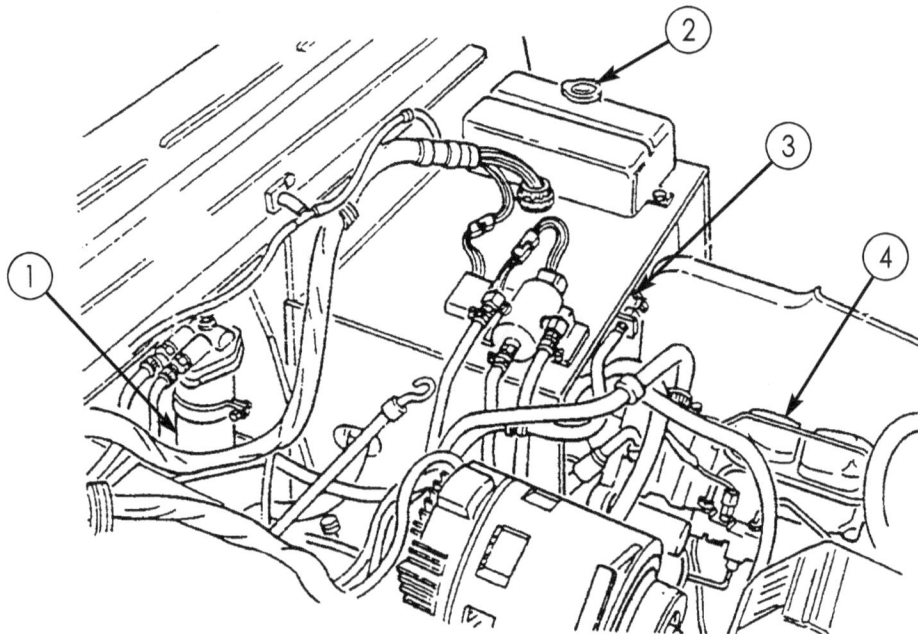

5   *Air cleaner assembly* (located right rear of engine) houses air cleaner element which filters dirt and dust from air before it enters combustion chamber.

6   *Air cleaner dump valve,* when squeezed, releases dirt, mud, or water from air cleaner body assembly.

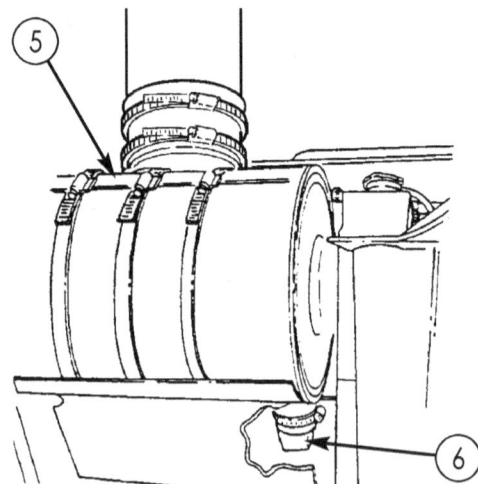

c. **Vehicle Exterior.**

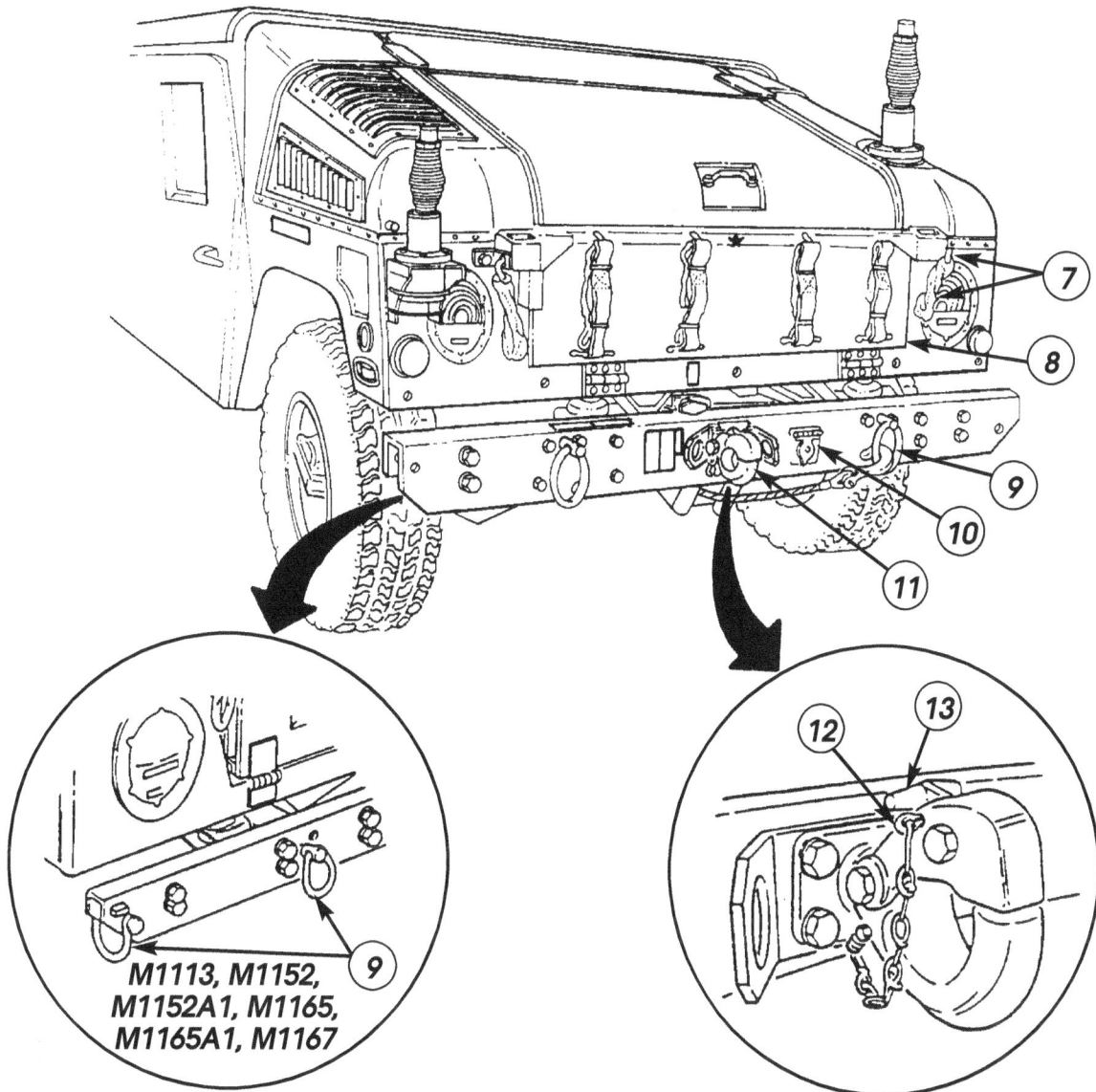

M1113, M1152,
M1152A1, M1165,
M1165A1, M1167

KEY	ITEM AND FUNCTION
7	*Tailgate chains and hooks* secure tailgate to rear of vehicle body.
8	*Tailgate* opens and closes to allow access to vehicle cargo area.
9	*Lifting shackles* (located at front and rear of vehicle) are used to lift or tie down vehicle.
10	*Trailer receptacle* provides electrical power for towed equipment.
11	*Towing pintle* (rear bumper) provides connection point for towing.
12	*Pintle pin* locks pintle latch to towing pintle.
13	*Pintle latch* pulls up to open towing pintle; pushes down to lock towing pintle.

2-2. CONTROLS, INDICATORS, AND EQUIPMENT (Cont'd)

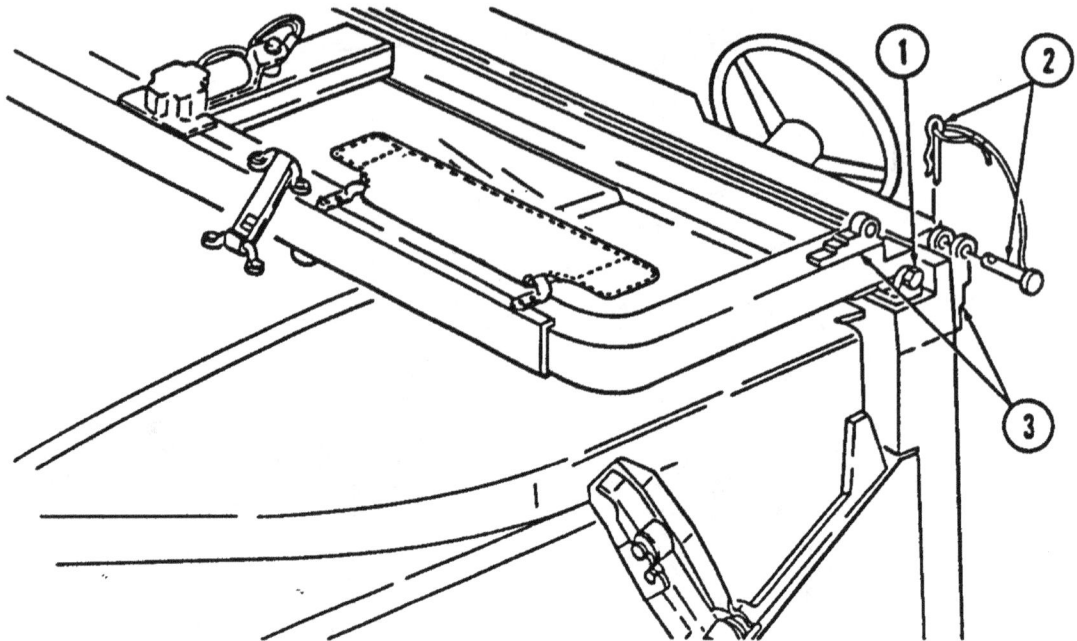

KEY   ITEM AND FUNCTION

1   *Windshield folddown hinges* (M1113, M1152) are used as a hinge point when lowering windshield. Hinge pins are removed when detaching windshield assembly.

2   *Hitch pin and hinge pin* (M1113, M1152) are removed to allow windshield to be lowered or detached.

3   *Windshield hinges* (M1113, M1152) secure windshield in the raised (up) position with hinge pins installed. Hinge pins are removed when lowering or detaching windshield.

4   *Fuel tank filler cap* (located at right rear side of vehicle) is removed to permit fuel servicing.

5   *Fuel door* (M1114) (located at right rear side of vehicle) covers fuel tank filler cap.

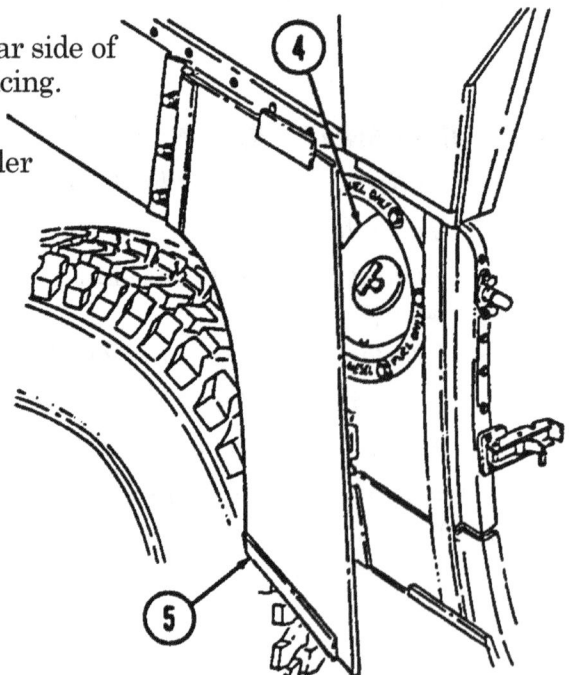

KEY   ITEM AND FUNCTION

6   *Hood latches* (one on each side of hood) unlatch to release hood.

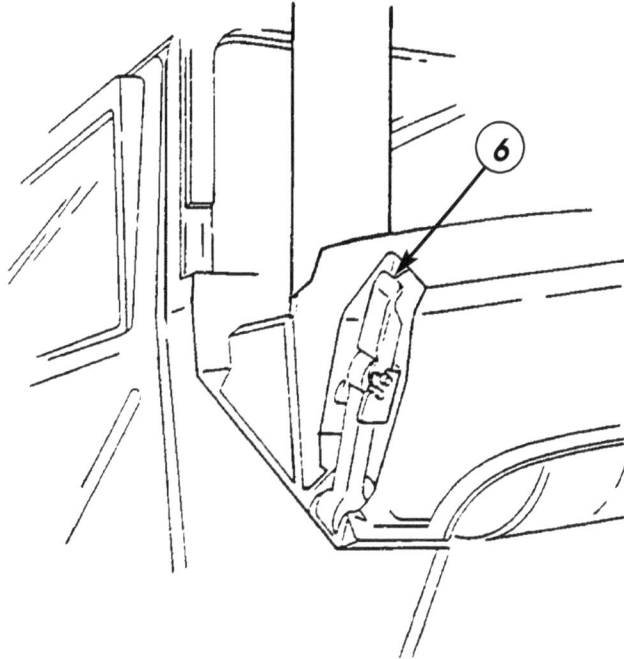

7   *Hood support rod* (on left side of vehicle) supports hood in the raised position.

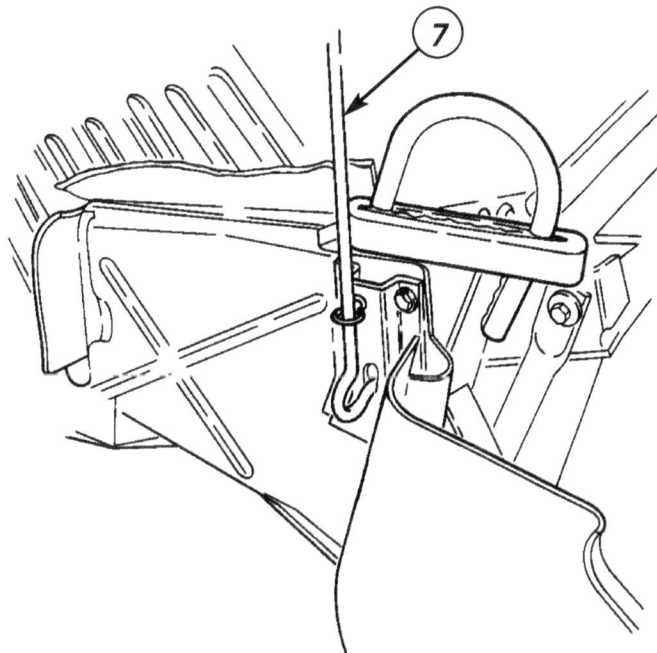

2-2. CONTROLS, INDICATORS, AND EQUIPMENT (Cont'd)

WARNING

Do not attempt to operate cargo shell door forward latch. The cargo shell door is not to be opened from inside the vehicle. Opening cargo shell door from inside the vehicle may cause damage to equipment or injury to personnel.

KEY   ITEM AND FUNCTION

NOTE

Items 1 through 7 apply to M1114 and M1151A1 vehicles w/perimeter B-Kit armor only.

1   *Cargo shell door strap* serves as a grab strap to lower cargo shell door.

2   *Cargo shell door assist cylinders* provide a lift boost for raising and holding cargo shell door open.

3   *Cargo shell door* when door rear latch is released, door opens forward and permits access to cargo area from rear of vehicle for stowing.

4   *Cargo shell door rear latch* is pulled upward to open cargo shell door from rear of vehicle.

5   *Cargo shell door retaining cables* limit cargo shell door travel when opened from either end.

5.1   *Rear hatch support rod* serves as an additional support for keeping cargo shell door open.

6   *C-pillar/partition door handle(s) and locking mechanism(s)* serve(s) to open or close, lock or unlock the cargo C-pillar/partition door.

7   *C-pillar/partition door(s)* provide(s) access from the crew compartment to cargo shell.

2-2. CONTROLS, INDICATORS, AND EQUIPMENT (Cont'd)

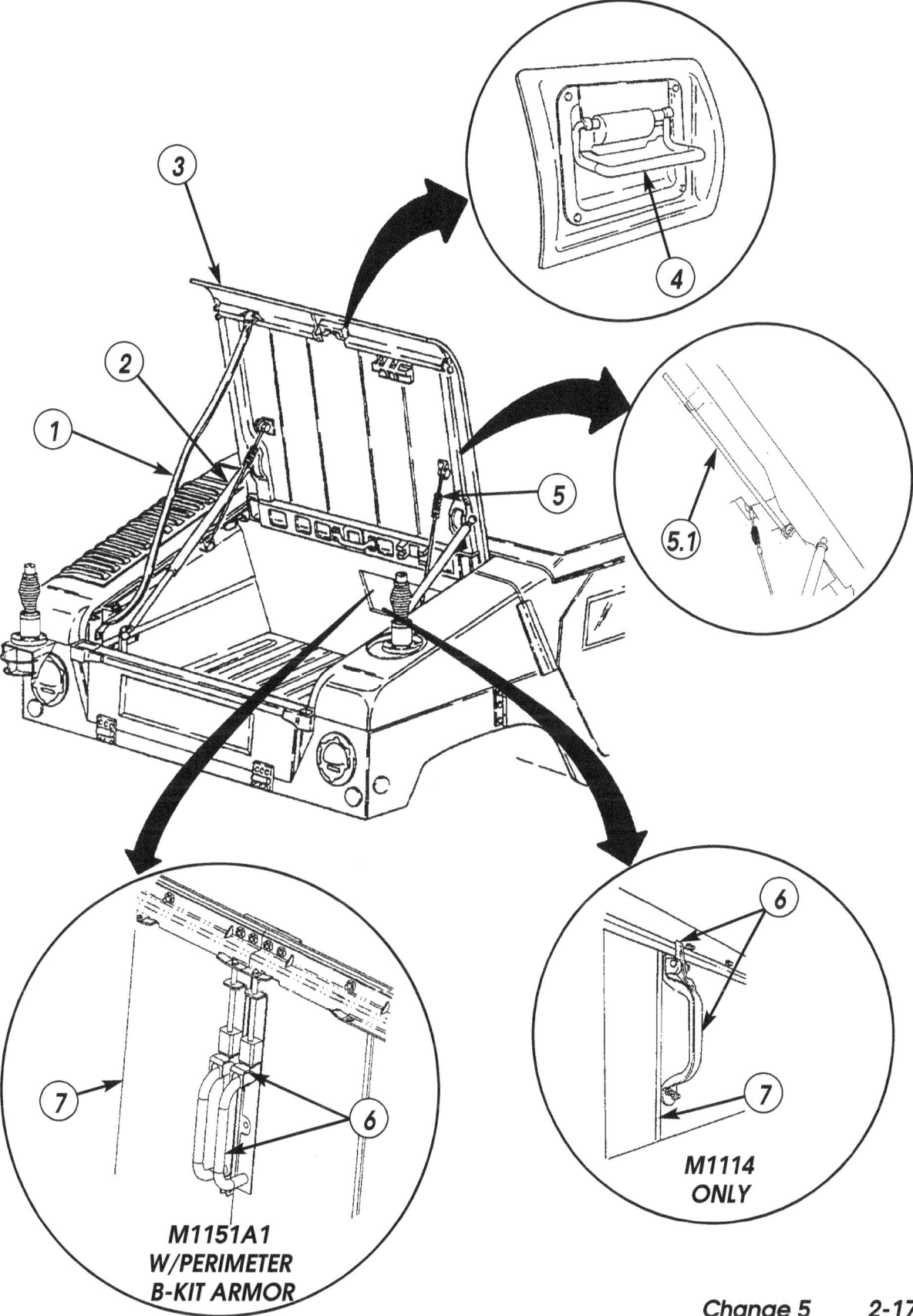

M1151A1
W/PERIMETER
B-KIT ARMOR

M1114
ONLY

2-2. CONTROLS, INDICATORS, AND EQUIPMENT (Cont'd)

■ **d. Up-Armored (M1114) and Armament Carrier (M1151/M1151A1) Equipment.**

KEY   *ITEM AND FUNCTION*

1   *Gunner's backrest* provides back support for a gunner positioned in weapon station.

2   *Weapon station hatch cover* provides sealed protective covering for roof opening when weapons are not mounted to weapon station.

3   *Hatch cover handle* is grab handle to assist gunner in opening and closing station cover.

NOTE
Items 4 and 5 apply to M1114 vehicles only.

4   *Rods* attached to cover serve to secure cover in the open position.

5   *Catch blocks* serve to support rods on weapon station while in the open position.

6   *Gunner's sling* serves as seat rest or restraint for a gunner positioned in weapon station.

7   *Hatch cover securing latches* (three each) secure cover to weapon station.

8   *Weapon station brake handle* locks the weapon station at the gunner's desired azimuth. Handle is placed in the down position for locking.

NOTE
Some M1114 vehicles are equipped with a newly configured turret brake assembly.

9   *Armament mount* is mounting bracket for weapon adapter.

10   *Universal weapons adapter pin assembly* secures weapon adapter to the armament mount.

11   *Universal weapons adapter* provides mounting base for the MK19 automatic grenade launcher, M2, caliber .50 machine gun; M60, 7.62 mm machine gun; M240B, 7.62 mm machine gun; and M249, 5.56 mm Squad Assault Weapon (SAW).

12   *Weapon station* serves as rotating mounting platform for weapon components; can be continuously rotated 360°.

NOTE
■ Items 12.1 through 12.3 apply to M1151 and M1151A1 vehicles only.

12.1 *Hatch Cover Retaining Latch* secures the latch catch for retaining station cover in the open position.

12.2 *Hatch Cover Retaining Catch* is the connecting point for retaining latch.

12.3 *Turret Positioning Handle* provides positive right-hand grip to rotate weapon station.

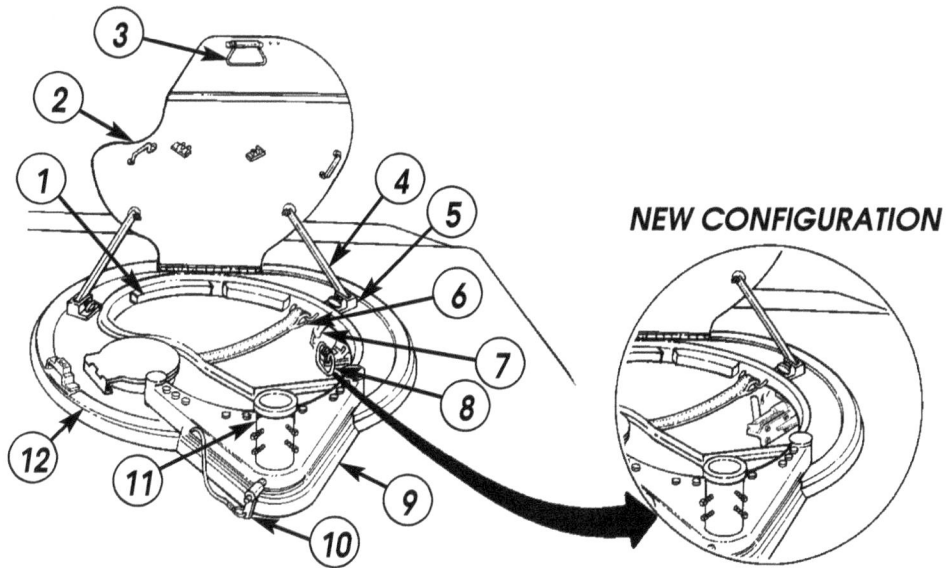

NEW CONFIGURATION

M1114

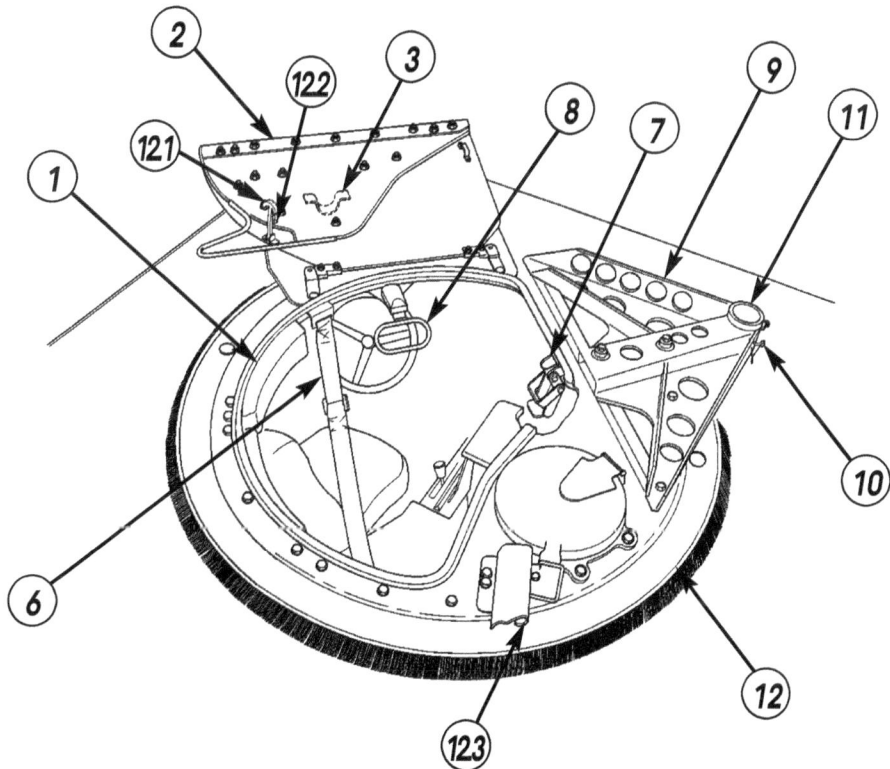

M1151/M1151A1

KEY	ITEM AND FUNCTION

13   *Traversing Unit (TU) mount adapter* adapts TU to the TU stowage pedestal or weapon station pedestal mount.

14   *Traversing unit stowage pedestal* provides mounting base for TU mount adapter.

15   *Lower traversing unit mount adapter clamp* secures TU mount adapter to the stowage pedestal or weapon station pedestal mount.

16   *Upper traversing unit mount adapter clamp* secures TU to TU mount adapter.

17   *Camouflage screen stowage straps* (four each) secure camouflage screen and support to tailgate.

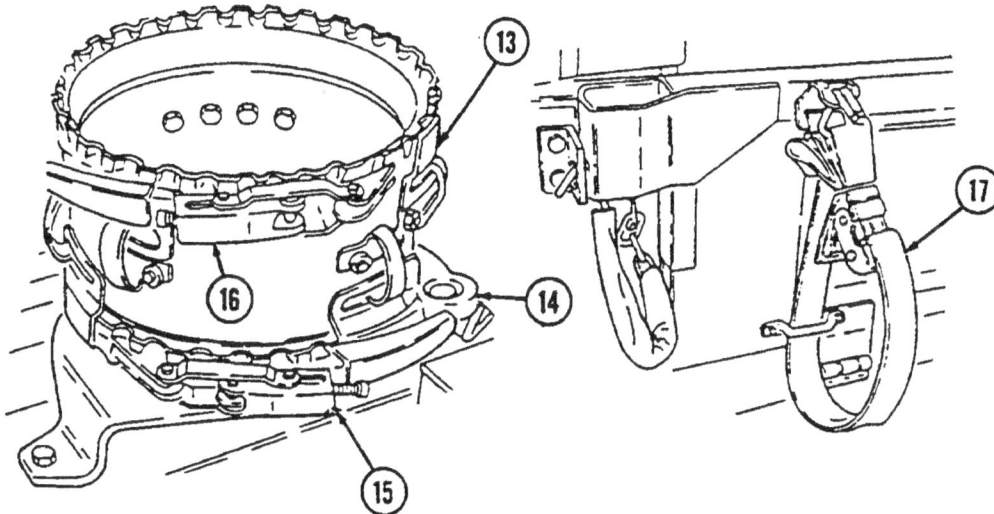

2-2. CONTROLS, INDICATORS, AND EQUIPMENT (Cont'd)

KEY   ITEM AND FUNCTION

1   *Rear seat stowage compartment net* secures extra gear and ammo in crew compartment.

2   *Hatch stowate net* secures three duffel bags to vehicle hatch.

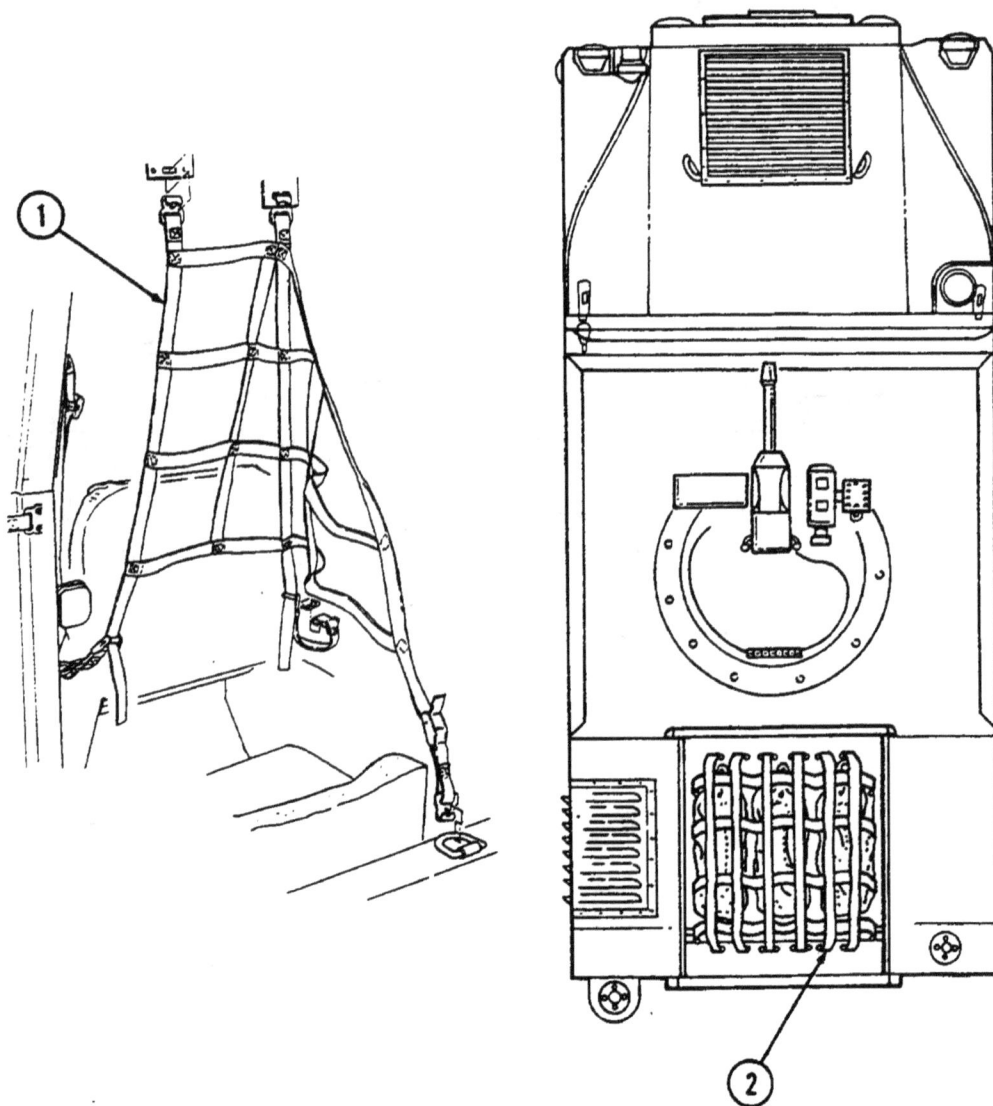

KEY   ITEM AND FUNCTION

3   *Water can footman loop and strap* secure water can to bracket.

4   *Water can bracket* provides stowage base for five-gallon water can.

5   *Fuel can footman loops and straps* secure two fuel cans to fuel can stowage brackets.

6   *Fuel can stowage brackets* permit stowage of two fuel cans.

7   *Stowage compartment net* secures items stowed in vehicle cargo area.

2-2. CONTROLS, INDICATORS, AND EQUIPMENT (Cont'd)

KEY   ITEM AND FUNCTION

1   *Ammo box racks* provides stowage base for three 40 mm ammo boxes.

2   *Ammo box footman loops and straps* (two each) secure three 40 mm ammo boxes to ammo box rack.

3   *Ammo box racks* serve as stowage base for two caliber .50 ammo boxes.

4   *Ammo box footman loops and straps* (two each) secure caliber .50 ammo boxes.

5   *Gunner platform* provides a non-slip platform for gunner.

NOTE
Items 6 through 8 apply to the adjustable gunner platform only.

6   *Adjustable gunner platform locking pins* (two each) secure gunner platform to selected height risers.

7   *Adjustable gunner platform locking lugs and holes* (two each) secure gunner platform to fully lowered position when locking pin is installed.

8   *Adjustable gunner platform risers* provide support and height adjustment for gunner platform.

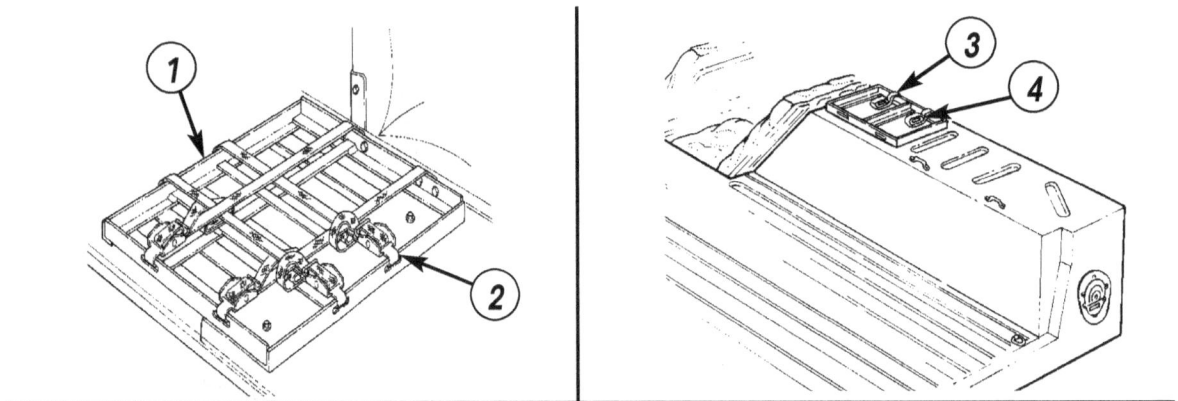

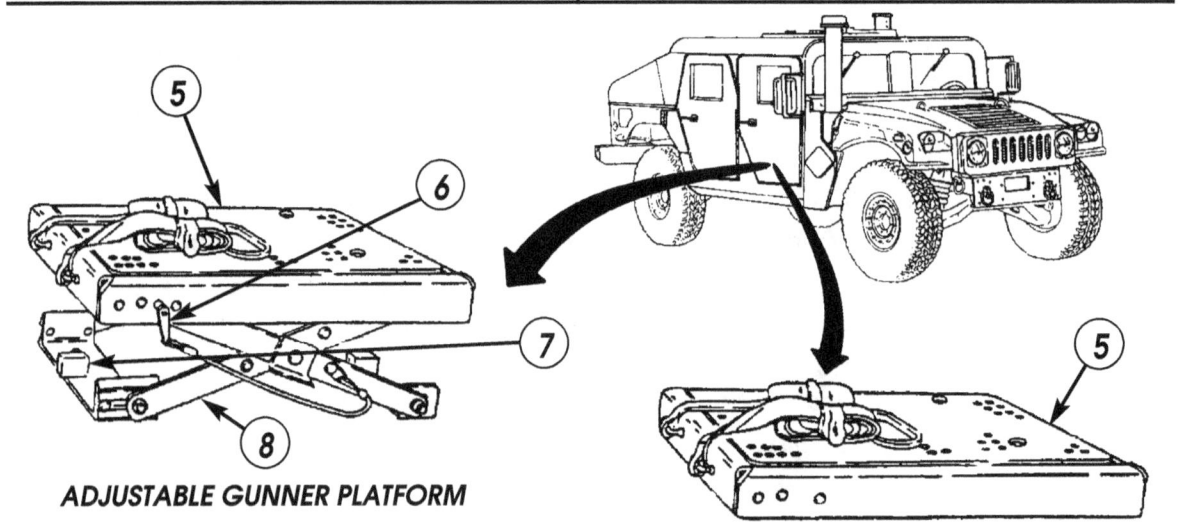

ADJUSTABLE GUNNER PLATFORM

STATIONARY GUNNER PLATFORM

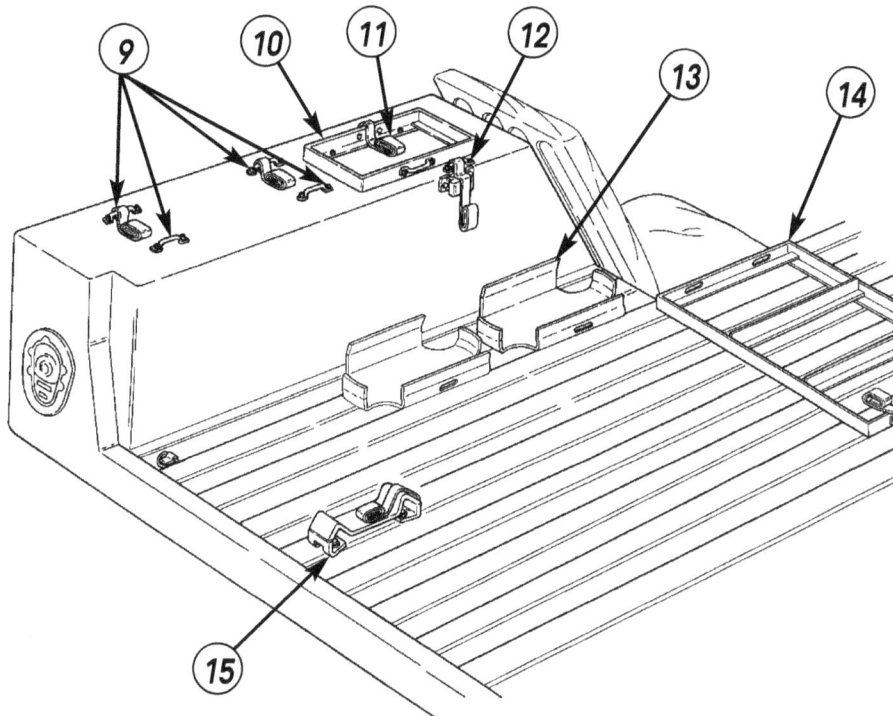

KEY   ITEM AND FUNCTION

NOTE
Items 9 through 15 apply to M1151 and M1151A1 vehicles only.

9   *Spare barrel and cleaning kit footman loops and straps* (two each) secure spare barrel and cleaning kit for M2, caliber .50 machine gun.

10   *Ammo box rack* provides stowage base for caliber .50 ammo box.

11   *Ammo box footman loop and strap* secure caliber .50 ammo box to ammo box rack.

12   *Water can footman loop and strap* secure 5 gal. water can to water can bracket.

13   *Water can bracket* provides stowage base for 5 gal. water can.

14   *Ammo box rack* provides stowage base for three 40 mm ammo boxes.

15   *Tripod mounting brackets and straps* (two each) provide stowage for M3 tripod.

2-2. CONTROLS, INDICATORS, AND EQUIPMENT (Cont'd)

KEY   ITEM AND FUNCTION

NOTE
Items 1 through 7 apply to M1151 and M1151A1 vehicles only.

1   *Ammo box rack* serves as stowage base for two caliber .50 ammo boxes.

2   *Ammo box footman loops and straps* (two each) secure two caliber .50 ammo boxes to ammo box rack.

3   *Night sight case stowage brackets* (two each) provide mounting for night sight case.

4   *Night sight case footman loops and straps* (two each) secure night sight case to night sight stowage brackets.

5   *Water can bracket* provides stowage base for 5 gal. water can.

6   *Antiskid strips* (nine each) prevent cargo from sliding around in the cargo area.

7   *Stowage compartment net* secures cargo to the cargo floor and limits movement of cargo during vehicle operation.

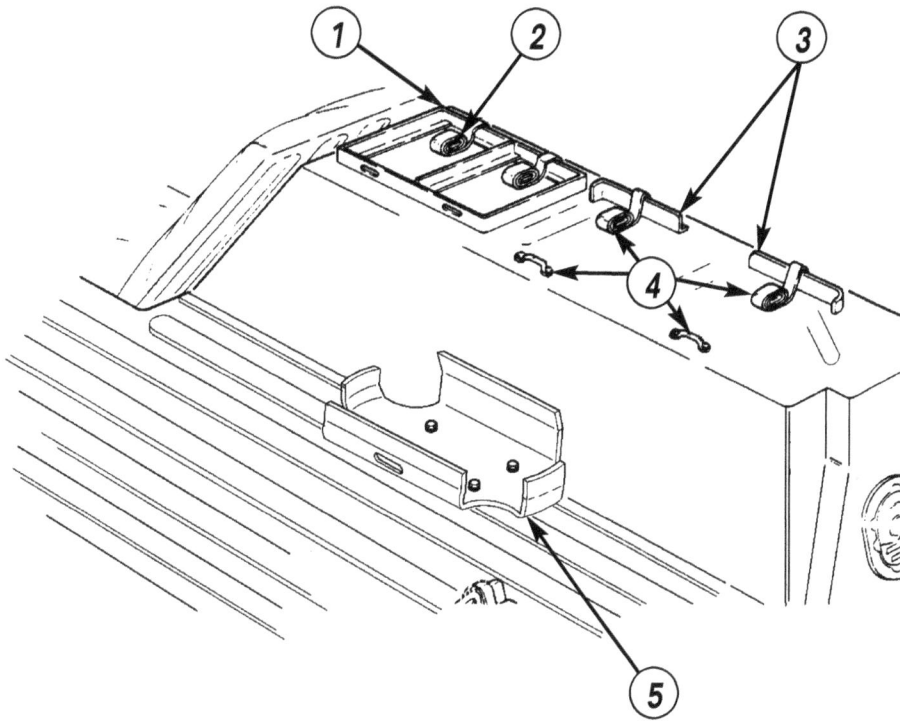

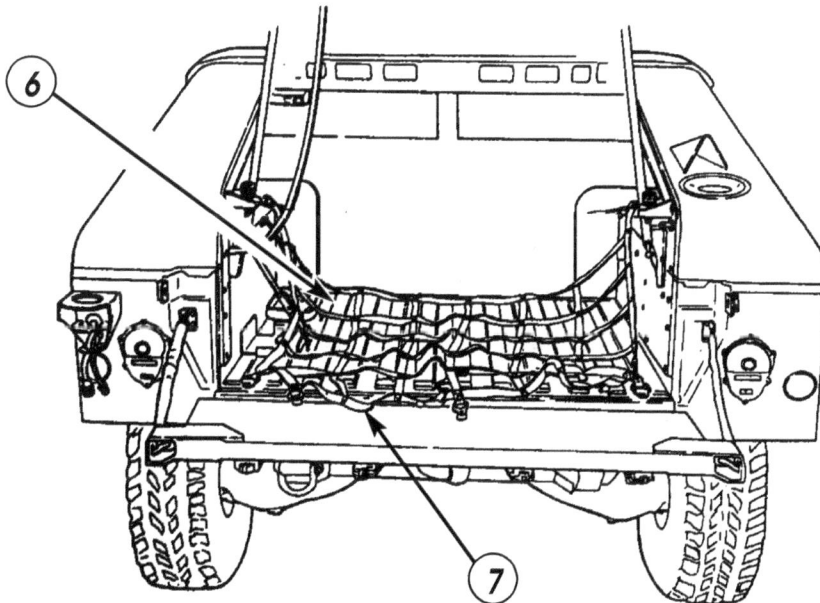

e. **S250 Shelter Carrier Equipment.**

KEY   ITEM AND FUNCTION

CAUTION

Shelter carriers are specifically designed to be operated with the S250 shelter installed. However they can be driven safely without the shelter or an equivalent payload of 1,500 lb (681 kg) for short distances (e.g., to and from maintenance, or from the railhead when being delivered), but this should not be done often or for long distances. Driving for long distances without the shelter installed, or equivalent payload of 1,500 lb (681 kg) evenly distributed in center of cargo area, will cause damage to equipment.

1   *Shelter reinforcement brackets* (four each) secure S250 shelter carrier to vehicle body.

2   *Rear suspension tiedown kit* is used to compress rear suspension to obtain an overall vehicle height of 102 in. (259 cm).

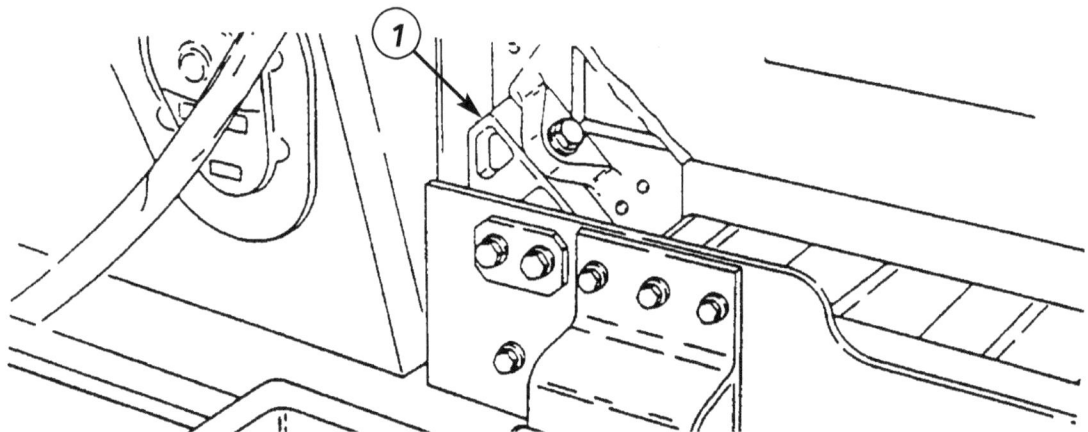

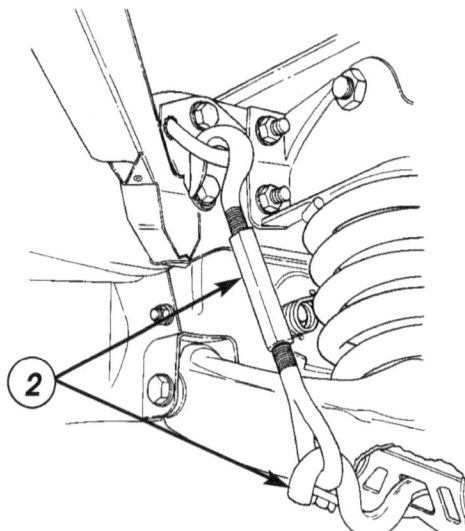

f. Troop/Cargo Winterization Kit Equipment.

KEY   ITEM AND FUNCTION

1   *Heater control assembly* operates heater and controls blower motor speed.

2   *Heater guard assembly* protects personnel from accidental contact with hot heater components.

3   *Heater* provides a flow of heated air to interior of troop/cargo enclosure assembly.

4   *Heater deflector* directs heated air from heater to all areas of troop/cargo enclosure assembly.

5   *Circuit breaker* protects heater assembly from damage caused by electrical overloading.The circuit breaker is waterproof and automatically resets.

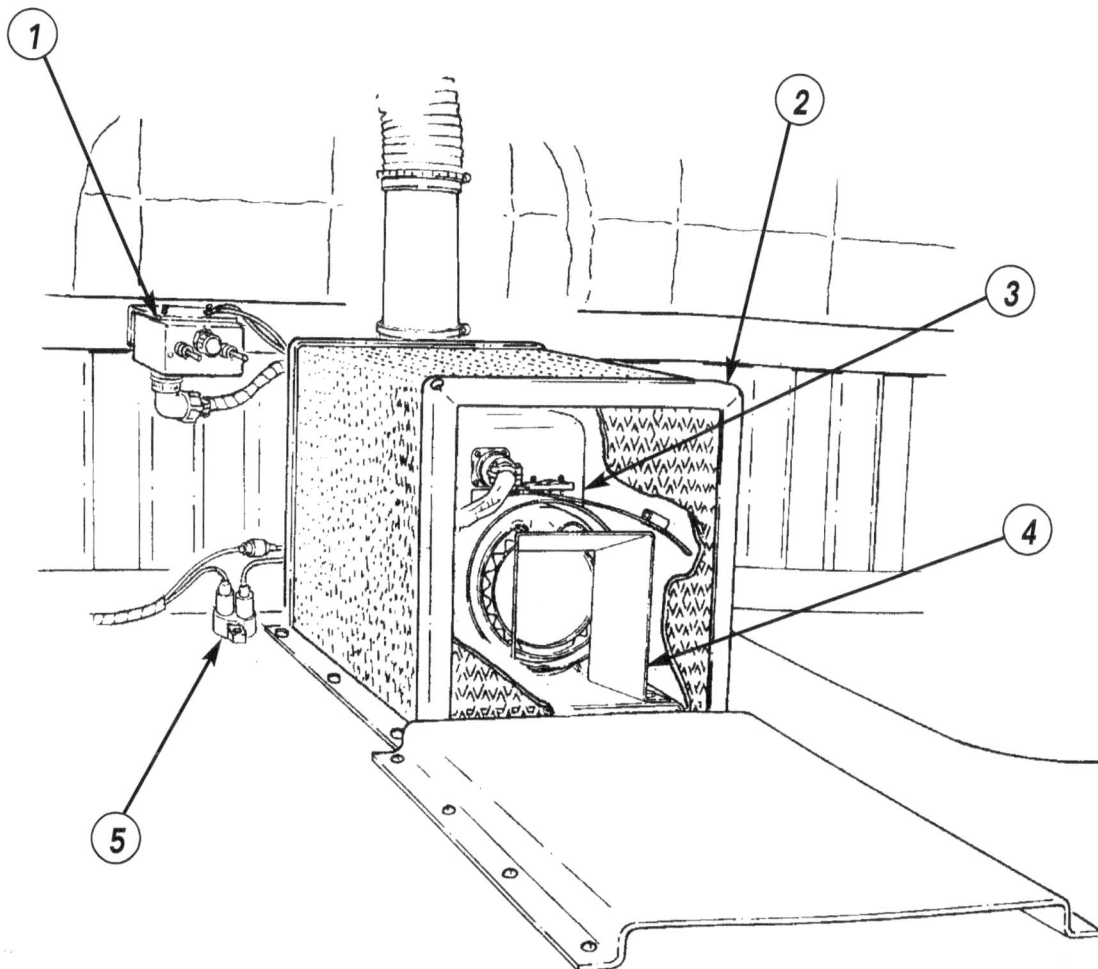

g. **TOW ITAS Carrier Equipment.**

WARNING

Never open one end of the cargo shell door without first ensuring that the opposite end is securely closed. Not doing so may cause both ends to open at the same time causing damage to equipment, mission abort, or injury to personnel.

NOTE

This manual identifies HMMWV equipment which permits the mounting and operation of the Improved Target Acquisition System, M41 (ITAS). Specific instructions for employing the Improved Target Acquisition System, M41 (ITAS) on the HMMWV are covered in FM 3-22.32.

KEY	ITEM AND FUNCTION
1	*Cargo shell door* is a double-actuating door which pivots at either end. When door forward latch is released from inside vehicle, door opens rearward and functions TOW ITAS M41 loader's door to facilitate mounting of TOW ITAS launcher and missile loading. When door rear latch is released, door opens forward and permits access to cargo area from rear of vehicle for stowing TOW ITAS launcher and equipment or ground mounting TOW ITAS launcher.
2	*Tripod legs mounting straps* secure tripod to mounting bracket.
3	*Tripod legs mounting bracket* provides stowage for tripod.
4	*Launch tube mounting straps* (two each) located on left and right side of missile rack secure TOW ITAS launch tube to missile rack tier.
5	*Tier mounting straps* (three each) located on left and right sides of missile rack secure three missile rounds to second tier of rack.
6	*Missile rack tier locking pins* secure tier of missile rack to base of missile rack.
7	*Missile rack support braces* (two each) support rack tier and pivot to outside to allow easy access to missiles for stowage or reloading.
8	*Missile stowage rack* provides stowage for six TOW ITAS missile rounds. It consists of a base and a first tier. Tiers pivot upward to facilitate missile stowage and quick access to missiles during reload operations. Missiles stowed between rack base and first tier are held in without straps. Missiles mounted on top of the first tier are secured with straps located left and right sides of rack.
9	*Missile guidance set (LBB) battery stowage box* provides stowage for the LBB battery.

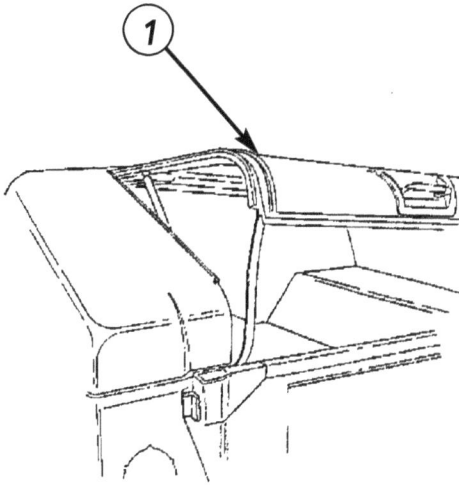

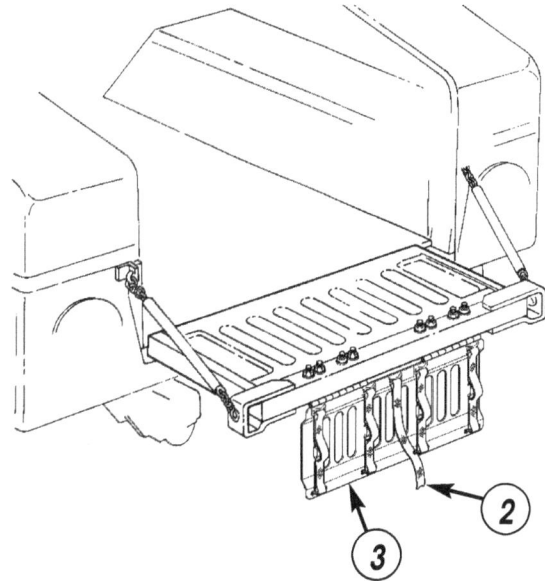

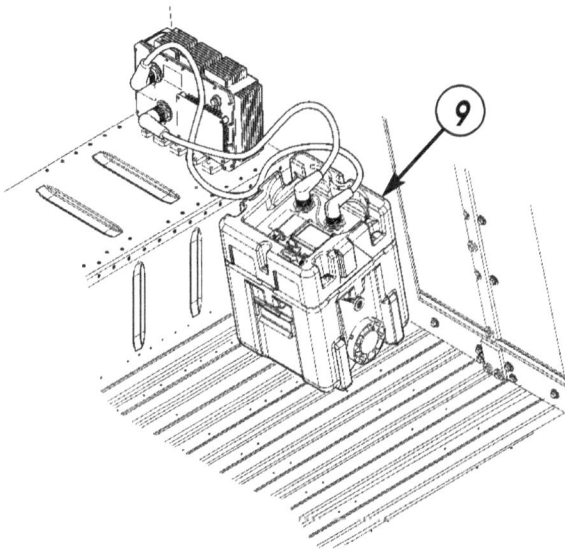

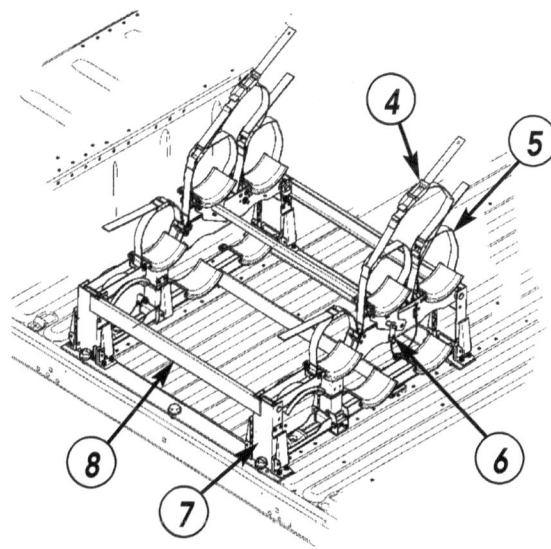

KEY ITEM AND FUNCTION

10 *Missile guidance set retaining straps* (four each) secure MGS to MGS weapon station mounting plate.

11 *Missile guidance set weapon station mounting plate* is mounting point for the MGS when the TOW ITAS launcher is assembled on the weapon station.

12 *Weapon station pedestal mount cover* protects pedestal mount opening when TOW ITAS launcher is not mounted.

13 *Weapon station pedestal mount* is mounting point for TU adapter bracket and TOW ITAS launcher.

14. *Weapon station hatch cover* provides sealed protective covering for roof opening when TOW ITAS is not mounted to weapon station.

15 *Weapon station* serves as rotating mounting platform for TOW ITAS components during the launcher mode of operation. It can be continuously rotated 360° without vehicle power conditioner (VPC) cables connected.

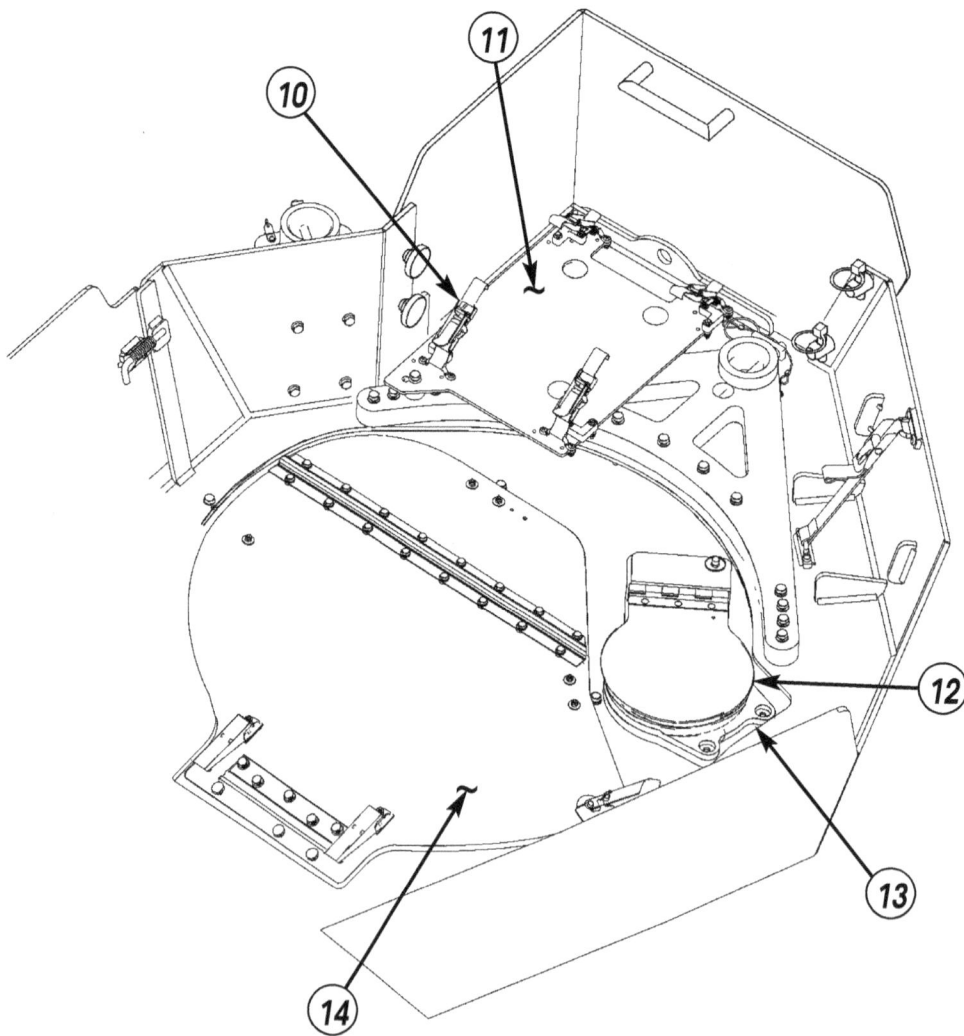

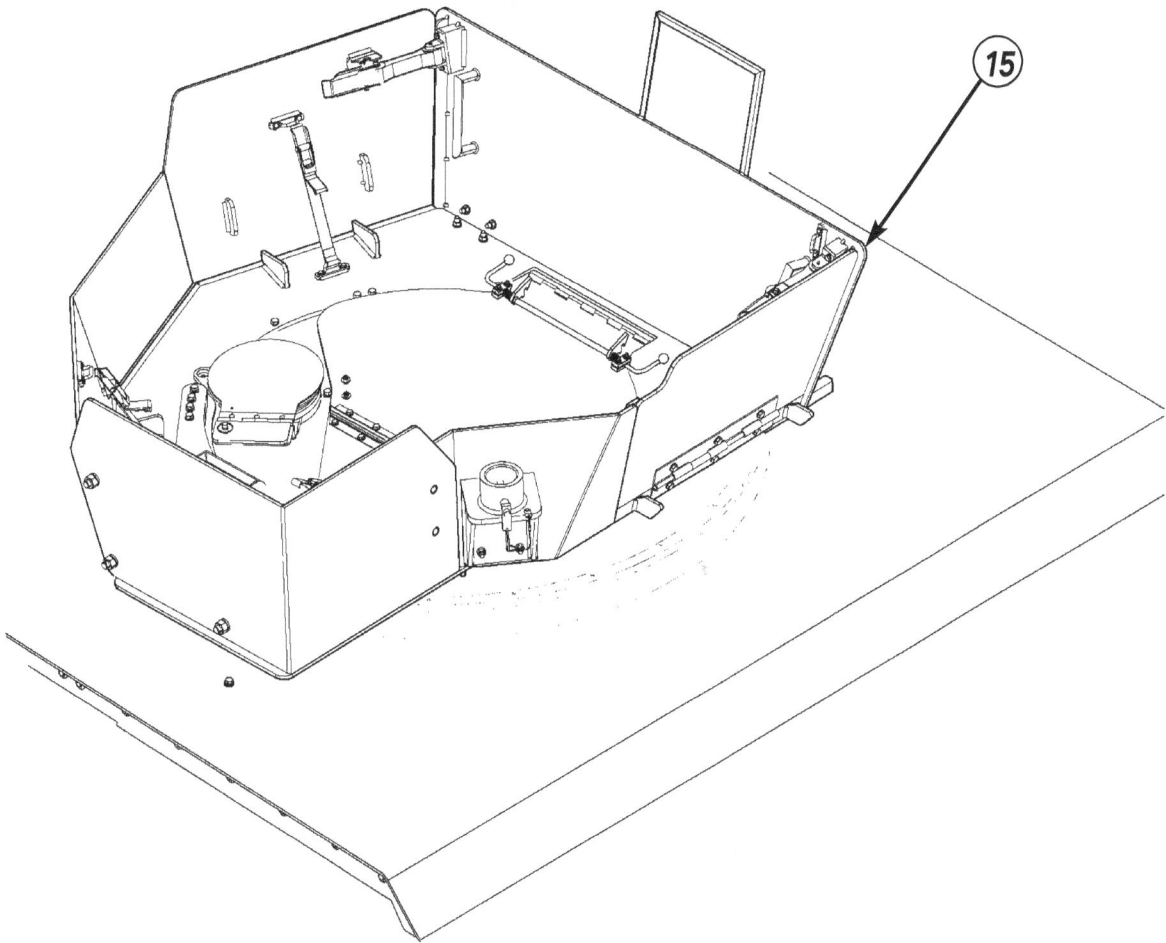

Section II. PREVENTIVE MAINTENANCE CHECKS AND SERVICES (PMCS)

2-3. GENERAL

NOTE

The army has an agreement with the door manufacturer to issue free patch kits to repair armor doors that exhibit cracks originating from the window area (M1114 doors only). If you find cracks within 1.25 in. (31.75 mm) from either side or 6.25 in. (158.75 mm) below the window opening, notify unit maintenance.

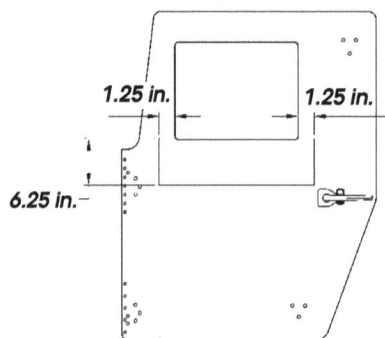

A permanent record of the services, repairs, and modifications made to these vehicles must be recorded. See DA Pam 750-8 for a list of the forms and records required and how to complete them.

2-4. PREVENTIVE MAINTENANCE CHECKS AND SERVICES REFERENCE INDEX

2-5. CLEANING INSTRUCTIONS

a. Cleaning. An after operation service performed by the operator/crew to keep the vehicle in a state of readiness. Facilities and material available to operators for vehicle cleaning can vary greatly in differing operating conditions. However, vehicles must be maintained in as clean a condition as available cleaning equipment, materials, and tactical situations permit.

WARNING

- Drycleaning solvent is flammable and will not be used near an open flame. A fire extinguisher will be kept nearby when the solvent is used. Use only in well-ventilated places. Failure to do this may result in injury to personnel and/or damage to equipment.
- Protective gloves, clothing, and/or respiratory equipment must be worn whenever caustic, toxic, or flammable cleaning solutions are used. Failure to do this may result in injury to personnel and/or damage to equipment.

CAUTION

- Do not allow cleaning compounds to come into contact with rubber, leather, vinyl, or canvas materials. Damage to equipment will result.
- Do not use compressed air when cleaning vehicle interiors. Damage to equipment will result.
- Do not allow water to enter air cleaner assembly air intake weathercap. Damage to engine will occur.

b. Cleaning Instructions for Ballistic Glass.

CAUTION

- Do not clean interior surfaces of ballistic glass by any other means than specified below.
- Do not use a scraper or other objects with sharp edges that may scratch the inside surfaces of ballistic glass.
- Do not apply stickers, labels, solvents, abrasive materials, or cleaners to ballistic glass.

   (1) Remove dust and loose abrasive particles using clean, filtered air at 20 psi (138 kPa) maximum.

   (2) Wash with mild detergent and warm water. Dry using a clean, soft, lint-free cloth.

   (3) Remove stubborn marks and stains using a clean, soft, lint-free cloth and equal parts of isopropyl alcohol or ethanol and water.

   (4) Repeat step 2.

NOTE

Clean windshield wipers of debris on a regular basis to ensure proper vision.

  c. Deleted.

  d. General Guidelines. Table 2-1 provides a general guide of cleaning materials used in removing contaminants from various parts of the vehicle.

2-5. CLEANING INSTRUCTIONS (Cont'd)

Table 2-1. General Cleaning Instructions.

Surface	Cleaning Materials Used to Remove		
	Oil/Grease	Salt/Mud/Dust/Debris	Surface Rust/Corrosion
Body	Detergent; water; rags.	Soapy water; soft brush; damp and dry rags.	Corrosion-removing compound; bristle brush; dry rags; lubricating oil.*
Vehicle Interior (Metals)	Detergent; damp and dry rags.	Damp and dry rags.	Corrosion-removing compound; bristle brush; dry rags; lubricating oil.*
Glass	Window cleaning compound; dry rags.	Window cleaning compound; dry rags.	Not applicable.
Ballistic Glass	Detergent; soapy water; dry rags.	Detergent; soapy water; dry rags.	Not applicable.
Plastic Windows	Soapy water; cream cleaner; dry rags.***	Soapy water; cream cleaner; dry rags.***	Not applicable.
Vehicle Interior (Seats and Straps)	Water; damp and dry rags.	Soapy water; damp and dry rags.	Not applicable.
Frame	Detergent rinsed with water; dry rags.	Soapy water; damp and dry rags.	Corrosion-removing compound; wire brush; dry rags; lubricating oil.*
Engine and Transmission	Drycleaning solvent; water; rags.	Soapy water; soft wire brush; damp and dry rags.	Bristle brush; warm water; dry rags.
Radiator	Not applicable.	Low pressure water or air; soapy water; damp and dry rags.	Not applicable.
Oil Cooler	Not applicable.	Low pressure water or air; soapy water; damp and dry rags.**	Not applicable.
Master Cylinder	Detergent; rinsed with soapy water; dry rags.	Soapy water; damp and dry rags.	Not applicable.
Rubber Insulation	Damp and dry rags.	Damp and dry rags.	Not applicable.
Tires	Soapy water; damp rags.	Soapy water; damp rags.	Not applicable.
Wood	Detergent; water; damp and dry rags.	Low pressure water; soapy water; damp and dry rags.	Not applicable.

* After cleaning, apply light grade of lubricating oil to all unprotected surfaces to prevent continued rust.
** If more space is needed to clean mud and debris between the oil cooler and radiator, refer to unit maintenance.
   Operator may perform this cleaning procedure under the supervision of unit level maintenance.
***After cleaning window zippers, apply zipper lubricant (appendix D, item 19).

2-6. PREVENTIVE MAINTENANCE CHECKS AND SERVICES (PMCS)

a. Designated Intervals.

NOTE

Designated intervals are performed under usual operating conditions. PMCS intervals must be performed more frequently when operating under unusual conditions.

(1) BEFORE checks and services of PREVENTIVE MAINTENANCE must be performed prior to placing vehicle or its components in operation.

(2) DURING checks and services of PREVENTIVE MAINTENANCE must be performed while the vehicle and/or its components/systems are in operation.

(3) AFTER checks and services of PREVENTIVE MAINTENANCE are performed upon completion of mission.

(4) WEEKLY checks and services of PREVENTIVE MAINTENANCE are performed once every 7 days.

(5) MONTHLY checks and services of PREVENTIVE MAINTENANCE are performed once every 30 days.

b. Procedures.

(1) For troubleshooting malfunctions, refer to table 3-1 or notify your supervisor.

(2) Use DA Form 2404 or DA Form 5988-E (automated) and report malfunctions to unit maintenance at once.

(3) Tools included with vehicle are to be used when making PREVENTIVE MAINTENANCE checks and services. Wiping cloths are needed to remove dirt or grease.

(4) Refer to appropriate TMs for PMCS requirements on mounted systems (i.e., missiles systems, radios, etc.).

c. Troublespots.

NOTE

Dirt, grease, oil, and debris may cover up a serious problem. Clean as you check. Following precautions printed on container, use drycleaning solvent (SD-3) on all metal surfaces. On rubber or plastic material, use soap and water.

(1) Check all bolts, nuts, and screws. If loose, bent, broken, or missing, either tighten or report conditions to unit maintenance.

(2) Look for loose or chipped paint, and rust or cracks at welds. Remove rust and loose paint, and spot-paint as required. If a cracked weld is found, report situation to unit maintenance.

(3) Inspect electrical wires and connectors for cracked or broken insulation. Also look for bare wires and loose or broken connections. Tighten loose connections. Report other problems to unit maintenance.

(4) Check hoses and fluid lines for wear, damage, and leaks. Ensure clamps and fittings are tight. (Refer to para. 2-7 for information on leaks.)

(5) Check hinges for security and operation.

(6) Check data, caution, and warning plates for security and legibility.

2-6. PREVENTIVE MAINTENANCE CHECKS AND SERVICES (PMCS) (Cont'd)

    d. Not Ready/Available. If a vehicle is not able to perform the mission, equipment will be reported as not ready/available. Refer to DA Pam 750-8.

    e. Correct Assembly or Stowage. Check each component for installation as an assembly, that it is in the right place, and has no missing parts.

2-7. FLUID LEAKAGE

Wetness around seals, gaskets, fittings, or connections indicates leakage.A stain also denotes leakage. If a fitting or connector is loose, tighten it. If broken or defective, report it. Use the following as a guide:

    a. Class I. Leakage indicated by wetness or discoloration, but not great enough to form drops.

    b. Class II. Leakage great enough to form drops, but not enough to cause drops to drip from item being checked/inspected.

    c. Class III. Leakage great enough to form drops that fall from the item being checked/inspected.

CAUTION

Operation is allowable with class I or II leakage except for brake system.Any brake fluid leakage must be reported.WHEN IN DOUBT, NOTIFY YOUR SUPERVISOR.When operating with class I or II leaks, check fluid levels more frequently. Class III leaks must be reported immediately to your supervisor or to unit maintenance. Failure to do this may result in damage to vehicle and/or components.

2-8. LUBRICATION REQUIREMENTS

For lubrication requirements and procedures, refer to appendix G.

Table 2-2. Preventive Maintenance Checks and Services.

Item No.	Interval	Location Item to Check/ Service	Crewmember Procedure	Not Fully Mission Capable If:
1	Before	Left Front, Side Exterior	**WARNING** Always remember the WARNINGS, CAUTIONS, and NOTES before operating this vehicle and prior to PMCS. **NOTE** Perform your before, after, and weekly checks if: **a.** You are the assigned driver but have not operated the vehicle since the last weekly inspection. **b.** You are operating the vehicle for the first time. **c.** See separate manual for TOW ITAS PMCS. DRIVER **CAUTION** If leaks are detected in the area of the transfer case oil cooler, do not attempt to tighten retaining nuts; internal damage to the transfer case oil cooler may result. Notify unit maintenance. **NOTE** If leakage is detected, further investigation is needed to determine the location and cause of the leak. **a.** Visually check underneath vehicle for any evidence of fluid leakage.	**a.** Any brake fluid leaks; class III leak of oil, fuel, or coolant.

Table 2-2. Preventive Maintenance Checks and Services (Cont'd).

Item No.	Interval	Location Item to Check/Service	Crewmember Procedure	Not Fully Mission Capable If:
1	Before	Left Front, Side Exterior (Cont'd)	**b.** Visually check front and left side of vehicle for obvious damage that would impair operation. DRIVER	**b.** Any damage prevents operation.
2	Before	Left Side Tires	**WARNING** Operating a vehicle with a tire in an underinflated condition or with a questionable defect may lead to premature tire failure and may cause equipment damage and injury or death to personnel. **NOTE** The radial tire is a bidirectional tire and the tread may be positioned in either direction. Visually check tires for underinflation and defects. DRIVER	Tire missing, deflated, or unserviceable.
3	Before	Rear Exterior	**NOTE** If leakage is detected, further investigation is needed to determine the location and cause of the leak. **a.** Visually check underneath vehicle for evidence of fluid leakage. **b.** Visually check rear of vehicle for obvious damage that would impair operation. **c.** Inspect bumper supports for cracks before towing trailer.	**a.** Any brake fluid leaks; class III leak of oil, fuel, or coolant. **b.** Any damage prevents operation. **c.** Any damage that would prevent operation.

Table 2-2. Preventive Maintenance Checks and Services (Cont'd).

Item No.	Interval	Location Item to Check/ Service	Crewmember Procedure	Not Fully Mission Capable If:
4	Before	Right Front, Side Exterior	DRIVER **NOTE** If leakage is detected, further investigation is needed to determine the location and cause of the leak. **a.** Visually check underneath vehicle for evidence of fluid leakage. **b.** Visually check front and right side of vehicle for obvious damage that would impair operation.	**a.** Any brake fluid leaks; class III leak of oil, fuel, or coolant. **b.** Any damage prevents operation.
5	Before	Right Side Tires	DRIVER **WARNING** Operating a vehicle with a tire in an underinflated condition or with a questionable defect may lead to premature tire failure and may cause equipment damage and injury or death to personnel. **NOTE** The radial tire is a bidirectional tire and the tread may be positioned in either direction. Visually check tires for under-inflation and defects.	Tire missing, deflated, or unserviceable.

Table 2-2. Preventive Maintenance Checks and Services (Cont'd).

Item No.	Interval	Location Item to Check/ Service	Crewmember Procedure	Not Fully Mission Capable If:
6	Before	Front	DRIVER **NOTE** If leakage is detected, investigation is needed to determine the location and cause of the leak. **a.** Visually check front of vehicle for obvious damage that would impair operation. **b.** Visually check underneath vehicle for evidence of fluid leakage.	**a.** Any damage prevents operation. **b.** Any brake leaks; class III leak of oil, fuel, or coolant.
6.1	Before	Power Steering Reservoir (P/N RCSK 18330)	DRIVER **CAUTION** • Do not permit dirt, dust, or grit to enter power steering reservoir. Damage to power steering system will result if power steering fluid becomes contaminated. • Do not overfill power steering reservoir. Damage to power steering system will result.	

Table 2-2. Preventive Maintenance Checks and Services (Cont'd).

Item No.	Interval	Location Item to Check/ Service	Crewmember Procedure	Not Fully Mission Capable If:
6.1	Before	Power Steering Reservoir (P/N RCSK 18330) (Cont'd)	 HOT COLD ADD Check fluid in power steering reservoir (para. 3-20). Fluid should be between HOT and COLD marks. Add fluid if level is below COLD mark. DRIVER	
6.2	Before	Power Steering Reservoir (P/N 94252A)	**CAUTION** • Do not permit dirt, dust, or grit to enter power steering reservoir. Damage to power steering system will result if power steering fluid becomes contaminated. • Do not overfill power steering reservoir. Damage to power steering system will result.	

Table 2-2. Preventive Maintenance Checks and Services (Cont'd).

Item No.	Interval	Location Item to Check/ Service	Crewmember Procedure	Not Fully Mission Capable If:
6.2	Before	Power Steering Reservoir (P/N 94252A) (Cont'd)	Check fluid in power steering reservoir (para. 3-20.1). Fluid should be between HOT and COLD marks. Add fluid if level is below COLD mark.	
7	Before	Serpentine Drivebelt and Pulleys	DRIVER **a.** Visually check drive and idler pulleys for evidence of excessive wear or misalignment. **b.** Check if serpentine drivebelt is missing, broken, cracked, frayed, loose, misaligned, or split.	**a.** Pulleys worn, broken, or misaligned. **b.** Serpentine drivebelt is missing or broken. Drivebelt fiber has more than one crack 1/8 in. (3.2 mm) in depth or 50%, or frays more than 2-in. (5.1-cm) long. Drivebelt is loose or misaligned (off one or more grooves on any pulley).

Table 2-2. *Preventive Maintenance Checks and Services (Cont'd)*.

Item No.	Interval	Location Item to Check/ Service	Crewmember Procedure	Not Fully Mission Capable If:
8	Before	Cooling System	DRIVER **WARNING** If engine has been recently operated, do not remove radiator cap to check coolant level. Cooling system is under pressure, and escaping steam or coolant can cause burns. **CAUTION** • Type 1, ethylene glycol (green), and Type 2, propylene glycol (purple), should never be mixed due to their difference in toxic properties. Failure to comply may result in damage to equipment. • Using antifreeze without mixing it with water can cause high operating temperatures, blockage of cooling system passages, and damage to water pump seals. **NOTE** Type 1 antifreeze is an ethylene glycol based coolant, green in color. Type 1 can be added to factory-filled pink coolant. When it becomes necessary to flush factory coolant, Type 1, ethylene glycol, will be used. When mixing Type 1 antifreeze with water, distilled water is recommended. Tap water should only be used in emergency situations. Check coolant level in coolant tank. Level should be at or above the FULL COLD line. Add coolant if below the FULL line.	

Table 2-2. *Preventive Maintenance Checks and Services (Cont'd).*

Item No.	Interval	Location Item to Check/ Service	Crewmember Procedure	Not Fully Mission Capable If:
8.1	Before	Doors	<u>DRIVER (M1114 ONLY)</u> Check door for proper operation of door lock.	Door does not lock.
8.2	Before	Doors (Frag 5/ Armored)	<u>DRIVER (M1114, M1151A1, M1152A1, M1165A1, M1167)</u> **a.** Check door for proper operation. **b.** Check door latch for proper operation. **c.** Check combat lock for proper operation.	**a.** Excessive force is required to operate. **b.** Door does not latch. **c.** Combat lock does not lock.
9	Before	Seat and Seatbelts	<u>DRIVER</u> **NOTE** Vehicle operation with inoperative seatbelts may violate AR 385-55. **a.** Check all seatbelts for security, damage, and operation of buckle and clasp ends. **b.** Check operation of seat adjusting mechanism (driver's seat only).	 **b.** Seat adjustment lock broken or missing.
10	Before	Fire Extinguisher	<u>DRIVER</u> **a.** Check for missing or damaged fire extinguisher. **b.** Check gauge for proper pressure of about 150 psi (1,034 kPa). **c.** Check for damaged or missing seal.	**a.** Fire extinguisher missing or damaged. **b.** Pressure gauge needle in recharge area. **c.** Seal broken or missing.

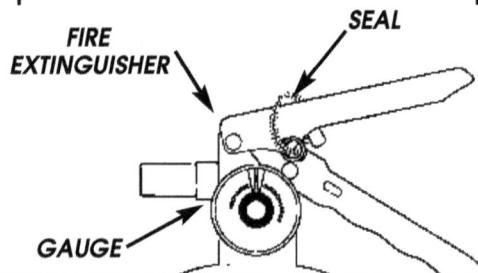

FIRE EXTINGUISHER

SEAL

GAUGE

Table 2-2. Preventive Maintenance Checks and Services (Cont'd).

Item No.	Interval	Location Item to Check/ Service	Crewmember Procedure	Not Fully Mission Capable If:
10.1	Before	Automatic Fire Extinguisher System 	DRIVER **NOTE** • Smoking in a vehicle with Automatic Fire Extinguishing System (AFES) installed may activate the sensor and cause system to discharge. • Large, high intensity spotlights may activate the sensor and cause system to discharge. **CAUTION** Do not handle the fire extinguisher unless the anti-recoil plug is installed in the valve outlet port and manual lever lock pin is installed in the lever lock holes. a. Check that pressure gauges on both fire extinguishers read at or above minimum pressure on extinguisher label using the chart below.	a. Pressure gauge reads below pressure shown on label.

TEMP. °F			-40	-22	-4	14	32	50	68	86	104	122	140	158
TEMP. °C			-40	-30	-20	-10	0	10	20	30	40	50	60	70
P min, PSIG			585	620	655	690	730	770	810	850	905	955	1005	1060

WARNING

Do not remove manual lever lock pin. Injury to personnel may happen if fire suppression system is accidently discharged.

Table 2-2. *Preventive Maintenance Checks and Services (Cont'd)*.

Item No.	Interval	Location Item to Check/ Service	Crewmember Procedure	Not Fully Mission Capable If:
			CAUTION	
			• Ensure that stowed items do not interfere with fire suppression system operation. Do not stow any items around the valve outlet port nozzles that may negate the extinguishers disbursement radius in the case of a fire. Failure to do so could prevent proper fire suppression system operation.	
			• Ensure optical fire sensors are kept clean. If optical fire sensors are not kept clean, it could decrease the system ability to quickly and correctly identify a fire and activate the automatic fire extinguishing system.	
			b. Check system wiring harness and sensors for presence, security and damage.	**b.** Wiring harness or sensor box is disconnected, improperly installed, damaged or missing.
			c. Check for loose, improperly installed, or missing fire extinguisher, manual lever lock pin or tube assembly.	**c.** Fire extinguisher, manual lever lock pin or tube assembly missing.

Table 2-2. *Preventive Maintenance Checks and Services (Cont'd).*

Item No.	Interval	Location / Item to Check/ Service	Crewmember Procedure	Not Fully Mission Capable If:
			d. Check that anti-recoil plugs are properly stowed.	
			e. Turn rotary switch to RUN position, check control module LED indicator for a continuous ON green light to see if system is fully functional.	**e.** Green LED indictor light is off, steady blink, slow blink, or double blink.
			f. Check fire sensors LED for continuous green light and cleanliness.	**f.** LED light does not come on, is blinking, or covered with dust or any foreign object.
11	Before	Gear Shifter Lever	**a.** Check transmission shift lever operation. Shift transmission lever through all operating ranges. Lever should move freely through all range positions.	**a.** Lever inoperable or binds between range detents.
			b. Check transfer shift lever operation. With transmission in N, shift transfer lever through all range positions. Lever should move freely through all range positions.	**b.** Lever inoperable or does not engage in all ranges with engine not running.

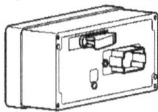

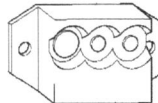

Table 2-2. Preventive Maintenance Checks and Services (Cont'd).

Item No.	Interval	Location Item to Check/ Service	Crewmember Procedure	Not Fully Mission Capable If:
12	Before	Instrument Panel	**WARNING** If gauges, instruments, or instrument lights are inoperable or not within ranges described in these checks, immediately shut off engine and notify your supervisor or unit maintenance personnel. Continued operation of vehicle may result in injury to personnel or damage to equipment. **NOTE** If engine is warm, wait-to-start light may not come on. During cranking or after starting, light may go on and off a few times. **a.** Check wait-to-start light and brake warning light.Turn rotary switch to RUN.Wait-to-start and brake warning light should come on.	**a.** Wait-to-start light does not come on when engine is cold, or wait light stays on continually. Brake warning light does not come on.

Table 2-2. Preventive Maintenance Checks and Services (Cont'd).

Item No.	Interval	Location — Item to Check/ Service	Crewmember Procedure	Not Fully Mission Capable If:
12	Before	Instrument Panel (Cont'd)	**b.** Start engine and check the following: **c.** Engine oil pressure gauge. **d.** Voltmeter. **e.** Air restriction gauge. **f.** Brake warning light should go off when hand brake is released. **g.** Check fuel gauge. **h.** Check coolant temperature gauge.	**b.** Engine will not start. **c.** Oil pressure is less than 6 psi (41 kPa) at idle or 15 psi (103 kPa) under load. **d.** Voltmeter needle stays in yellow or red range. **e.** Air restriction indicator reaches red zone. **f.** Brake warning light stays on after hand brake is released or comes on while driving. **h.** Coolant temperature gauge inoperative or reads greater than 250°F (120°C) and/or overheat lamp illuminates.

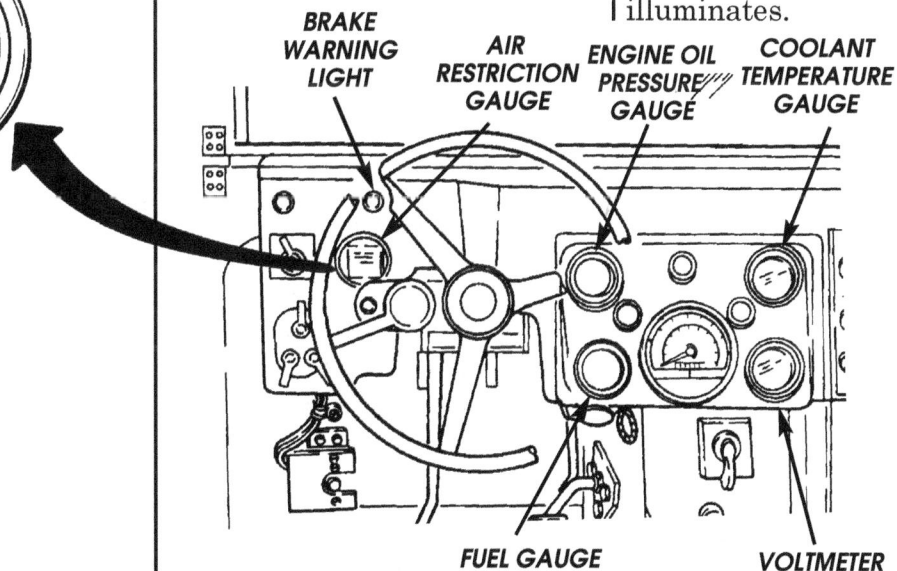

AIR FILTER RESTRICTION GAUGE PUSH TO RESET

BRAKE WARNING LIGHT — AIR RESTRICTION GAUGE — ENGINE OIL PRESSURE GAUGE — COOLANT TEMPERATURE GAUGE

FUEL GAUGE — VOLTMETER

Table 2-2. Preventive Maintenance Checks and Services (Cont'd).

Item No.	Interval	Location Item to Check/ Service	<u>Crewmember</u> Procedure	Not Fully Mission Capable If:
13	Before	Steering	<u>DRIVER</u> Check steering wheel for operation. With engine running, turn steering wheel from left to right. Steering wheel should move freely.	Steering wheel inoperable or binds.
14	Before	Brakes	<u>DRIVER</u> **NOTE** Engine must be warmed up and idling, transmission in Ⓓ (overdrive), transfer in H (high), and parking brake released to perform the following check. **a.** Check brake pedal travel. With vehicle at idle, transfer in H, and transmission in Ⓓ, allow vehicle to move forward. As vehicle moves, slowly depress brake pedal. Pedal should travel 1 to 1-1/2 in. (2.5 to 3.8 cm) before brakes take hold. After brakes take hold, pedal may exceed the 1 to 1-1/2 in. (2.5 to 3.8 cm) travel. This is normal. **b.** Check parking brake. With parking brake fully applied, transmission in Ⓓ or R, and transfer in H, vehicle should not move. **c.** Check parking brake lever safety mechanism to ensure that it latches when parking brake is applied.	**a.** Brakes will not stop the vehicle. **b.** Parking brake inoperable or unable to hold vehicle. **c.** Parking brake lever safety mechanism is not functioning properly.

Table 2-2. Preventive Maintenance Checks and Services (Cont'd).

Item No.	Interval	Location Item to Check/ Service	Crewmember Procedure	Not Fully Mission Capable If:
15	Before	Weapon Station	DRIVER (M1114, M1151, M1151A1, AND M1167 ONLY) **NOTE** Weapon station binding should be checked with weapon system or equivalent weight applied to turret. Refer to appropriate system TM to determine weight of weapon system. **a.** Check weapon station for binding by rotating 360° in both directions at least five times. **b.** Check armament mounting plate and bearing sleeve for security of mounting and obvious damage that would impair operation.	**a.** Weapon station binds. **b.** Armament weapons required for missions and mounting plate or bearing sleeve missing, or any damage that will prevent or impair mounting of armament weapons.
16	During	Controls and Indicators	DRIVER **a.** Monitor fuel gauge. **b.** Monitor engine oil pressure gauge. **c.** Monitor coolant temperature gauge.	**b.** Engine oil pressure gauge reads less than approximately 30 psi (207 kPa) under normal driving conditions or less than 10 psi (69 kPa) at idle. **c.** Coolant temperature gauge reads greater than 250°F (120°C) and/or overheat lamp illuminates.

Table 2-2. Preventive Maintenance Checks and Services (Cont'd).

Item No.	Interval	Location Item to Check/Service	Crewmember Procedure	Not Fully Mission Capable If:
15.1	Before	Traversing Unit	DRIVER (M1114, M1151A1, M1167 ONLY) **NOTE** If equipped with an operational Battery Powered Motorized Traversing Unit (BPMTU), a not fully mission capable Manual Traversing Unit (MTU) does not deadline the weapon system. **a.** Check Manual Traversing Unit (MTU) gear linkage assembly for bent, broken or missing linkage. **b.** Inspect MTU support bracket for loose or missing linkage. **c.** Inspect traversing handle for cracks or loose screws and quick release pin for damage or missing. **d.** Check disengage handle for proper operation. **e.** Check operation of MTU while engaged with weapon station by rotating 360 degrees in both directions at least five times.	**a.** Gear linkage assembly bent, broken, or missing linkage. **b.** Loose or missing mounting hardware. **c.** Handle missing or unserviceable, or quick release pin damaged or missing. **d.** Check disengage handle for proper operation. **e.** MTU is inoperable.
15.2	Before	Battery Powered Motorized Traversing Unit (BPMTU)	DRIVER (M1114, M1151A1 ONLY) **NOTE** If equipped with an operational MTU, a not fully mission capable BPMTU does not deadline the weapon system. **WARNING** Push in emergency stop switch to the "OFF" position before joystick or other components are removed or replaced. Failure to do so may cause injury to personnel or damage to equipment.	

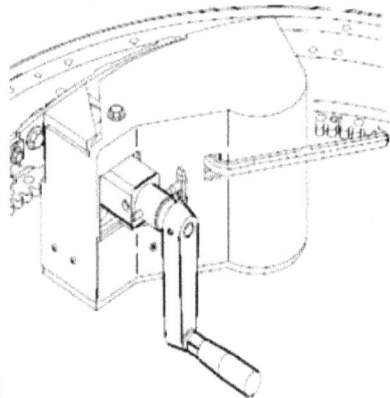

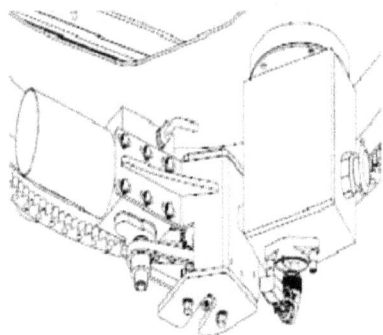

Table 2-2. Preventive Maintenance Checks and Services (Cont'd).

Item No.	Interval	Location Item to Check/Service	Crewmember Procedure	Not Fully Mission Capable If:
15.2	Before	Battery Powered Motorized Traversing Unit (cont'd)	**a.** Inspect power cables for exposed wiring, breaks, frayed insulation, loose or damaged connectors and loose, damaged, or missing mounting hardware.	**a.** Power cable broken, frayed, or damaged. Mounting hardware loose, damaged, or missing.
			NOTE Emergency stop switch must be in the "ON" position to charge batteries.	
			b. Check control box emergency stop switch, battery status indicator button, and LED display light for proper operation.	**b.** Control box emergency stop switch inoperable.
			NOTE Turret has no brake when motor engagement control is in the "NEUTRAL" (vertical) position.	
			c. Check operation of drive assembly motor engagement control, and chain for proper adjustment.	**c.** Drive assembly motor engagement control will not engage.
			CAUTION Turn emergency stop switch to the "OFF" position prior to plugging in the joystick cable. Failure to comply may result in damage to equipment.	
			d. Check joystick operation. Ensure joystick moves freely, and ensure housing is sealed properly.	**d.** Joystick inoperable or housing is not properly sealed.
			NOTE Allowing batteries to fully discharge greatly reduces life span and rechargability of battery. To prolong battery life, recharge when single bar flashes on the battery status indicator.	

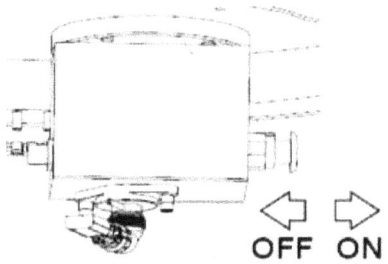

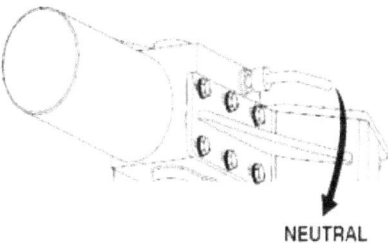

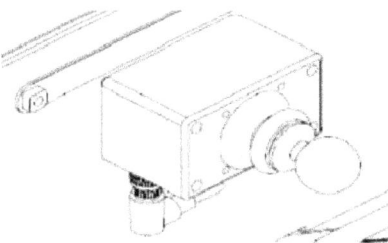

Table 2-2. Preventive Maintenance Checks and Services (Cont'd).

Item No.	Interval	Location Item to Check/ Service	Crewmember Procedure	Not Fully Mission Capable If:
16	During	Controls and Indicators (Cont'd)	**d.** Monitor air restriction gauge. **e.** Monitor voltmeter. **f.** Monitor brake warning light. **g.** Check speedometer operation. <u>DRIVER</u>	**d.** Air restriction gauge indicates restriction in the air cleaner **e.** Voltmeter indicates a loss of voltage. **f.** If light stays lit, brakes may not function properly. **g.** Speedometer needle does not move, jerks unevenly during sustained speeds, or appears stuck.
17	During	Brakes	Check brakes for pulling or grabbing. <u>DRIVER</u>	Brakes pull or grab.
18	During	Steering	Be alert for vibration, excessive sway, leaning to one side, or unstable handling. Check steering response for unusual free play, binding, or shimmy. <u>DRIVER</u>	Handling is unstable; turning is difficult or free play, binding, or shimmy detected.
18.1	During	Accelerator Pedal	Check response to accelerator feed. Check for sticking or binding pedal. <u>DRIVER</u>	Pedal sticking or binding.
19	During	Power-train	Be alert for unusual noises or vibrations from engine, transmission, transfer, differentials, propeller shafts, axle shafts, or wheels. <u>DRIVER</u>	Unusual noise or vibration detected.
20	During	Trans-mission	Check transmission for proper operation.	Transmission slips or will not shift.

Table 2-2. Preventive Maintenance Checks and Services (Cont'd).

Item No.	Interval	Location Item to Check/ Service	Crewmember Procedure	Not Fully Mission Capable If:
21	During	Air Conditioner	ALL MODELS EXCEPT M1113 **NOTE** Perform the following inspection only if the air conditioner is required for climatic conditions. Turn air conditioner on. Refer to para. 2-35 for M1114 and para.2-35.1 for M1151/M1151A1/ M1152/M1152A1/M1165/M1165A1/ M1167.Wait 5 minutes to allow temperature to stabilize. Check vents for cool air. If air is not cooler than ambient temperature, notify your supervisor and record on DA Form 2404.	Climatic conditions require air-condition-ing and A/C is inoperable, or outlet duct air is not cooler than ambient temperature.
21.1	During	De-Icer	DRIVER (M1114 ONLY) **NOTE** Perform the following inspection only if the de-icer is required for climatic conditions. Turn de-icer on. Refer to para. 2-36. Check to see if de-icer functions by removing steam, frost, or ice from windshield.	De-icer does not operate and mission requires de-icer.

21.2	After	Gear Shifter Lever	DRIVER
			  a. Start engine and check transmission shift lever operation. Shift transmission through all operating ranges. Lever should move freely through all range positions.
			
			  b. Check transfer shift lever operation.With transmission in N, shift transfer lever through all range positions. Lever should move freely through all range positions.
				a. Lever inoperable or binds between range detents.
				
				  b. Lever inoperable or does not engage in all ranges with engine not running.

Table 2-2. *Preventive Maintenance Checks and Services (Cont'd).*

Item No.	Interval	Location Item to Check/ Service	Crewmember Procedure	Not Fully Mission Capable If:
22	After	Trans- mission Fluid	DRIVER **CAUTION** • Do not permit dirt, dust, fluid or grit to enter transmission oil dipstick tube. Internal transmission damage will result if transmission oil becomes contaminated. • Do not overfill transmission. Damage to transmission will result. An over-full transmission can also indicate a transfer case fluid leak. Notify unit maintenance if transmission fluid is above crosshatch mark. **NOTE** • Transmission fluid level should be checked with engine running, parking brake set, transmission shift lever in P, and vehicle on level ground. Fluid level should be at crosshatch marks on dipstick. • Let vehicle idle with all accessories off for three minutes. • Engine operating temperature of 185-250°F (85-120°C) must be reached before performing AFTER checks. Check transmission fluid level. Apply brake and move shift lever through each gear range. Pause for about three seconds in each range, ending in P. If level is below the crosshatch marks, add sufficient fluid to bring the level to the crosshatch marks.	

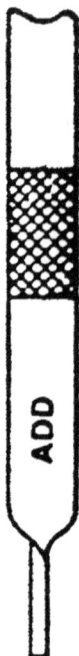

Table 2-2. *Preventive Maintenance Checks and Services (Cont'd).*

Item No.	Interval	Location Item to Check/ Service	Crewmember Procedure	Not Fully Mission Capable If:
23	After	Fuel Filter	DRIVER **WARNING** Do not perform fuel system checks, inspec-tions, or maintenance while smoking or near fire, flames, or sparks. Fuel may ignite, causing damage to vehicle and injury or death to personnel. **NOTE** A rubber hose can be attached to drainvalve to catch fuel in container before opening drainvalve. If fuel is clear, put fuel back in fuel tank. a. Check fuel for contamin-ation. With engine running, open drainvalve. Allow fuel to drain into suitable container until it runs clear. Close valve.	a. Fuel is not clear after draining 1 pt. (0.47 L).

Table 2-2. Preventive Maintenance Checks and Services (Cont'd).

Item No.	Interval	Location Item to Check/ Service	Crewmember Procedure	Not Fully Mission Capable If:
23	After	Fuel Filter (Cont'd)	**NOTE** Fuel retained in the drain-valve may drip when vehicle vibrations occur. This is normal and does not constitute a leak. Wipe drainvalve with rag until excess fuel is removed. **b.** Check for leaks. **c.** Stop engine and remove rubber hose from drainvalve, if installed.	**b.** Class III leakage evident.
24	After	Left Side Tires	DRIVER **WARNING** Operating a vehicle with a tire in an underinflated condition or with questionable defect may lead to premature tire failure and may cause equipment damage and injury or death to personnel. Visually check tires for under-inflation, cuts, gouges, cracks, or bulges. Remove all penetrating objects.	Tire deflated or otherwise unserviceable.

Table 2-2. *Preventive Maintenance Checks and Services (Cont'd).*

Item No.	Interval	Location Item to Check/ Service	Crewmember Procedure	Not Fully Mission Capable If:
25	After	Mirror (Left Side)	DRIVER **NOTE** Vehicle operation with damaged or missing outside rearview mirrors may violate AR 385-55. Check mirror for defects, cracks, and serviceability.	
26	After	Left Front, Side Exterior	DRIVER **NOTE** If leakage is detected, further investigation is needed to determine the location and cause of the leak. **a.** Visually check underneath vehicle for evidence of fluid leakage. **b.** Visually check halfshaft cv boots and ball joint boots for rips, tears, or cuts. **c.** Inspect frame crossmembers and underbody support for missing hardware, cracks, bends, and breaks. Notify unit maintenance if rust is present, but the base metal is sound. **d.** Visually check for body damage that would impair operation of vehicle.	**a.** Any brake fluid leaks; class III leak of oil, fuel, or coolant. **c.** Crossmembers or underbody support are missing any hardware, are cracked, broken, or bent or rusted-through condition is present that would affect vehicle operation. **d.** Any damage will prevent operation.

Table 2-2. Preventive Maintenance Checks and Services (Cont'd).

Item No.	Interval	Location Item to Check/ Service	Crewmember Procedure	Not Fully Mission Capable If:
27	After	Rear Exterior	DRIVER **NOTE** If leakage is detected, further investigation is needed to determine the location and cause of the leak. **a.** Visually check underneath vehicle for evidence of fluid leakage. **b.** Visually check halfshaft cv boots and ball joint boots for rips, tears, or cuts. **c.** Inspect frame crossmembers and underbody support for missing hardware, cracks, bends, and breaks. Notify unit maintenance if rust is present, but the base metal is sound. **d.** Inspect bumper or crossmember and inner braces in area around towing pintle for cracks or breaks.	**a.** Any brake fluid leaks; class III leak of oil, fuel, or coolant. **c.** Crossmembers or underbody support are missing any hardware, are cracked, broken, or bent or rusted-through condition is present that would affect vehicle operation. **d.** Bumper, crossmember or an inner brace is cracked or broken.

Table 2-2. Preventive Maintenance Checks and Services (Cont'd).

Item No.	Interval	Location Item to Check/ Service	Crewmember Procedure	Not Fully Mission Capable If:
28	After	Right Side Tires	DRIVER **WARNING** Operating a vehicle with a tire in an underinflated condition or with questionable defect may lead to premature tire failure and may cause equipment damage and injury or death to personnel. Visually check tires for underinflation, cuts, gouges, cracks, or bulges. Remove all penetrating objects.	Tire deflated or otherwise unserviceable.
29	After	Mirror (Right Side)	DRIVER **NOTE** Vehicle operation with damaged or missing outside rearview mirrors may violate AR 385-55. Check mirror for defects, cracks, and serviceability	
30	After	Right Front, Side Exterior	DRIVER **NOTE** If leakage is detected, further investigation is needed to determine the location and cause of the leak. **a.** Visually check underneath vehicle for evidence of fluid leakage. **b.** Visually check halfshaft cv boots and ball joint boots for rips, tears, or cuts.	**a.** Any brake fluid leaks; class III leak of oil, fuel, or coolant.

Table 2-2. Preventive Maintenance Checks and Services (Cont'd).

Item No.	Interval	Location Item to Check/ Service	Crewmember Procedure	Not Fully Mission Capable If:
30	After	Right Front Side, Exterior (Cont'd)	**c.** Inspect frame crossmembers and underbody support for missing hardware, cracks, bends, and breaks. Notify unit maintenance if rust is present, but the base metal is sound.	**c.** Crossmembers or underbody support are missing any hardware, are cracked, broken, or bent or rusted-through condition is present that would affect vehicle operation.
			d. Visually check front and right side of vehicle for obvious damage that would impair operation.	**d.** Any damage will prevent operation.
			DRIVER	
31	After	Engine Oil	**CAUTION** • Do not permit dirt, dust, or grit to enter engine oil dipstick tube. Internal engine damage will result if engine oil becomes contaminated. • Do not overfill engine crankcase. Damage to engine will result. Check engine oil level. Level should be between ADD and FULL. If level is below ADD, add oil to bring level between the ADD and FULL marks.	Oil appears milky.
			DRIVER	
32	After	Power Steering Lines and Fittings	**CAUTION** Notify unit maintenance if power steering system has class III leak. Loss of power assist could occur if this condition exists.	

Table 2-2. Preventive Maintenance Checks and Services (Cont'd).

Item No.	Interval	Location Item to Check/ Service	Crewmember Procedure	Not Fully Mission Capable If:
32	After	Power Steering Lines and Fittings (Cont'd)	Check power steering lines and fittings for leaks.	Class III leakage evident.
			DRIVER	
33	After	Cooling System	Inspect radiator hoses for leakage.	Class III leakage evident.
			DRIVER	
34	After	Master Cylinder	Visually check master cylinder lines for leaks and security of cover.	Any leak, or cover missing.
35	After	Lights	**CAUTION** Never set the rotary switch to RUN to check the lights. This drains the batteries and can burn out the glow plugs and control box. **NOTE** Vehicle operation with damaged or inoperable headlights may violate AR 385-55. a. Check for presence and operation of service drive, turn signal, blackout marker, marker, blackout drive, and side marker lights. b. Check operation of tail/stop-lights. Push down brake pedal approximately 1/4 in. (6.4 mm). Tail/stoplights should come on.	

Table 2-2. Preventive Maintenance Checks and Services (Cont'd).

Item No.	Interval	Location Item to Check/ Service	Crewmember Procedure	Not Fully Mission Capable If:
36	After	Horn	**NOTE** Vehicle operation with inoperative horn may violate AR 385-55. Check operation of horn if tactical situation permits. DRIVER	
37	After	Wind-shield and Wipers	**NOTE** Vehicle operation with damaged windshield may violate AR 385-55. **a.** Check windshield for damage that would impair operator's vision. **NOTE** Vehicle operation with inoperative wipers may violate AR 385-55. **b.** Check windshield wiper, blade, and washer fluid reservoir for defects, damage, and proper operation. **c.** Check windshield wiper motor for proper operation. **d.** Check washer reservoir fluid level.	**a.** Windshield is cracked, broken, or discolored (cloudy) sufficiently to impair operator's vision.

Table 2-2. *Preventive Maintenance Checks and Services (Cont'd).*

Item No.	Interval	Location Item to Check/ Service	Crewmember Procedure	Not Fully Mission Capable If:
37	After	Wind-shield and wipers (Cont'd)	**e.** Inspect inner surface of windshield glass (spall liner) for complete breaks, delamination, scratches, gouges, tape, decals, adhesives, or blurred vision.	**e.** The bond between glass and frame is separated from glass or frame. Any complete break on inner surface of glass. Any digs, gouges, or scratches on inner surface of glass. Any complete break on inner surface of glass.
			f. Inspect outer surface of windshield glass for a complete break.	**f.** Any complete break on the outer surface of windshield glass.
38	After	Light Switches	DRIVER **NOTE** Ensure all switches are in the OFF position. Failure to turn switches to the OFF position when not in use will drain the batteries. Check and ensure all switches are in the OFF position.	
38.1	Weekly	Hand Throttle	DRIVER **a.** Check hand throttle and mounting bracket for security. Check throttle release button to ensure hand throttle cable operates properly. **b.** Check hand throttle cable for corrosion, nicks, breaks, or burns.	

Table 2-2. Preventive Maintenance Checks and Services (Cont'd).

Item No.	Interval	Location Item to Check/Service	Crewmember Procedure	Not Fully Mission Capable If:
39	Weekly	Tires	DRIVER **WARNING** • Do not exceed 50 psi (345 kPa) cold radial tire inflation pressure. Overinflation of tire may result in damage to equipment and injury or death to personnel. • Load Range D valves and tires are not compatible with Load Range E wheels. Load Range E valves and tires are not compatible with Load Range D wheels. Failure to comply may result in damage to equipment and injury or death to personnel. **NOTE** • The radial tire is a bidirectional tire and the tread may be positioned in either direction. • Refer to tables 1-15 and 1-15.1 for vehicle tire pressures.	

Table 2-2. Preventive Maintenance Checks and Services (Cont'd).

Item No.	Interval	Location Item to Check/ Service	Crewmember Procedure	Not Fully Mission Capable If:
39	Weekly	Tires (Cont'd) THREAD WEAR INDICATOR (TWI) WEAR BAR TIRE TREAD DEPTH	**a.** Check tire tread depth. Tread should not be worn beyond level of wear bar (1/16 in.) (15.9 mm or less).Wear bars are molded across the tread pattern in the valley between the center rib and lugs. The Tread Wear Indicator letters (TWI) are molded on the sidewall to aid in locating the wear bar. **NOTE** • The wear bars are not evident on new or very low mileage tires.The wear bars will appear after usual use. • Some tires will show a diamond instead of TWI indicating where the wear bar is located. **b.** Check for missing or loose wheel stud nuts and lug nuts. Tighten loose lug nuts and have unit maintenance tighten stud nuts and lug nuts to proper torque. STUD NUTS LUG NUTS	**a.** Any tread is worn even to height of tread wear indicator (1/16 in. (1.59 mm) or less).Any cut, gouge, or crack that extends to the cord body or any bulges. Tires exhibit excessive inner and outer wear or balance **b.** Any wheel stud nut or lug nut is broken or missing.

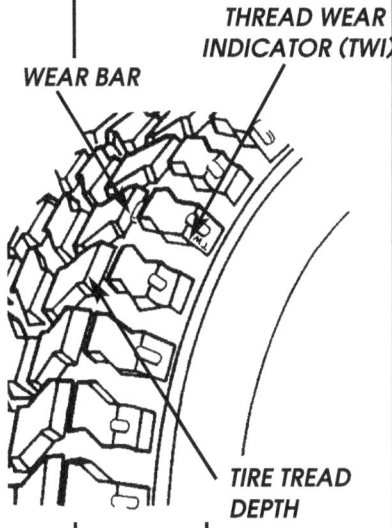

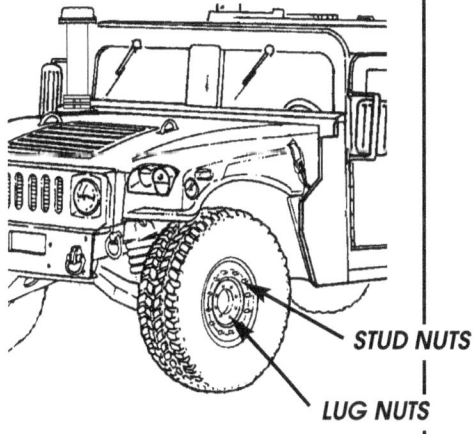

Table 2-2. *Preventive Maintenance Checks and Services (Cont'd).*

Item No.	Interval	Location Item to Check/ Service	Crewmember Procedure	Not Fully Mission Capable If:
39	Weekly	Tires (Cont'd)	**WARNING** Do not exceed 50 psi (345 kPa) cold radial tire inflation pressure. Over-inflation of tire may result in premature tire failure, damage to equipment, and injury or death to personnel. **c.** Gauge tires for correct air pressure using tire inflation gauge. Adjust as necessary. **CAUTION** The M1113 shelter carriers are specifically designed to be operated with the S250 shelter installed. They can be driven safely without the shelter installed, or with equivalent payload of 1,500 lb (681 kg), for short distances (e.g., to and from maintenance, or from the rail head when being delivered), but this should not be done often or for long distances. Driving for long distances without the shelter installed, or equivalent payload of 1,500 lb (681 kg) evenly distributed in center of cargo area, will cause damage to equipment. DRIVER	
40	Weekly	Exhaust System	Check exhaust system for security of all mounts, tightness of clamps and bolts, rusted conditions, damaged pipes, and any indication of an exhaust leak. DRIVER	Any mounts are broken, pipes are rusted through or broken, or any indication of an exhaust leak.
41	Weekly	Shock Absorbers	Visually inspect shock absorbers for leaks, damage, and security of mounting.	Class III leakage or damage is evident; mounting damaged or loose.

Table 2-2. Preventive Maintenance Checks and Services (Cont'd).

Item No.	Interval	Location Item to Check/ Service	Crewmember Procedure	Not Fully Mission Capable If:
42	Weekly	Doors and Windows	DRIVER (M1113, M1152, AND M1165 ONLY) **a.** Check operation of doors and windows. **b.** Check crew door assembly for visible cracks that would make door unserviceable or unable to secure properly. **c.** Check crew door assembly latch, hinges, and door handle for damage, looseness, or missing parts.	**b.** Visible cracks or door does not secure properly. **c.** Loose, missing, or unserviceable parts.
42.1	Weekly	Doors and Windows	DRIVER (M1114, M1151, M1151A1, M1152A1, M1165A1, AND M1167 ONLY) **a.** Inspect crew door locks for proper operation. **b.** Inspect inner surface of door glass (spall liner) for complete breaks, delamination, scratches, gouges, tape, decals, adhesives, or blurred vision. **c.** Inspect outer surface of door glass for a complete break. **d.** Inspect door armor for cracks. If cracks are found notify unit maintenance. For M1114 vehicles only, see note in para. 2-3.	
43	Weekly	Tailgate	DRIVER Check operation of tailgate. Check that tailgate latches securely and operates properly.	

Table 2-2. *Preventive Maintenance Checks and Services (Cont'd).*

Item No.	Interval	Location Item to Check/ Service	Crewmember Procedure	Not Fully Mission Capable If:
44	Weekly	Body	DRIVER (M1114, M1151, M1151A1, AND M1167 ONLY) **a.** Inspect cargo shell door for bends, warping, binding, and ease of operation. Inspect latching mechanisms for proper operation. Inspect lift cylinders for bends and security of mounting. **a.1.** Inspect retaining wire rope for damage and security of mounting. **b.** Inspect strap assembly for frays and security of mounting. **c.** Check cargo shell door for alignment as follows: (1) Insert a piece of paper between the door seal and door opening. (2) With door closed, seal should offer resistance to pulling out paper. If door seal does not offer resistance, adjustment is required. **d.** Check crew compartment armor for security of mounting. **e.** Check armor for loose or missing fasteners. Notify unit maintenance if any are found. **f.** Check armor components for cracks.	**a.** Lift cylinders or latches bent, warped, binding, or inoperative. **a.1.** Retaining wire rope is damaged, missing, or not secured. **f.** Armor components are cracked.

Table 2-2. *Preventive Maintenance Checks and Services (Cont'd).*

Item No.	Interval	Location Item to Check/ Service	Crewmember Procedure	Not Fully Mission Capable If:
			WARNING	
			Unauthorized welds and/or drilling of vehicle armor will degrade armor protection capabilities and could result in severe injury to personnel.	
			g. Inspect underbody armor for missing retainer plates or screws.	**g.** Vehicle missing a retainer plate or more than one screw.
			h. Inspect underbody plates for any deformation (buckling, collapsing, or bending).	**h.** Deformation is visually detected.
			i. Check operation of C-partition door and ensure it will lock in all three positions.	**i.** C-partition door will not close and lock in all three positions, or inoperative.
44.1	Weekly	Geared Fan Drive (Serial Numbers 3000000 and Above Only)	Check for missing grease fittings.	One or more grease fittings missing.

Table 2-2. *Preventive Maintenance Checks and Services (Cont'd).*

Item No.	Interval	Location Item to Check/ Service	Crewmember Procedure	Not Fully Mission Capable If:
45	Weekly	Vehicular Heater	DRIVER (Vehicle w/Vehicular Winterization Kit) **a.** Check heater and heater controls for proper operation. **b.** Check fuel lines and fittings for leaks, cracks, or breaks. **c.** Check electrical cables and connections for security, or frayed or broken wires. **d.** Check heater exhaust pipe for damage, security of mounting, and missing components. **e.** Check fuel filter for leaks or damage.	**a.** Heater inoperable and mission requires heater. **b.** Class III fuel leakage is evident and mission requires heater. **c.** Wires frayed or broken. **d.** Heater exhaust damaged or components missing. **e.** Class III leak evident.
46	Weekly	Air Cleaner	DRIVER **WARNING** If NBC exposure is suspected, all air filter media should be handled by personnel wearing protective equipment. Consult your unit NBC officer or NBC NCO for appropriate handling or disposal instructions. Check air cleaner weathercap, air cleaner assembly, air intake hose, and air horn for security of mounting and damage.	Damage to air cleaner weathercap, body, air intake hose, or mounting allows unfiltered air to enter the engine.

Table 2-2. *Preventive Maintenance Checks and Services (Cont'd).*

Item No.	Interval	Location Item to Check/ Service	Crewmember Procedure	Not Fully Mission Capable If:
47	Weekly	Alternator Brackets	DRIVER Visually check power steering and alternator brackets for cracks, damage, or loose bolts.	Bracket is cracked or bolts damaged or loose.
48	Weekly	Cooling System	DRIVER **a.** Check fan and fan pulley for damage.	**a.** Fan blade or pulley is bent, broken, cracked, or loose.
			b. Check radiator for leaks, clogged or damaged fins, and loose or damaged hoses to and from the engine.	**b.** Class III leakage evident.
			c. Check support mounts, side brackets, and side bracket weldments on radiator for missing hardware, damage, or broken welds.	**c.** Support mounts broken, damaged, or missing hardware. Side brackets damaged or two or more weldments broken, allowing movement of radiator.

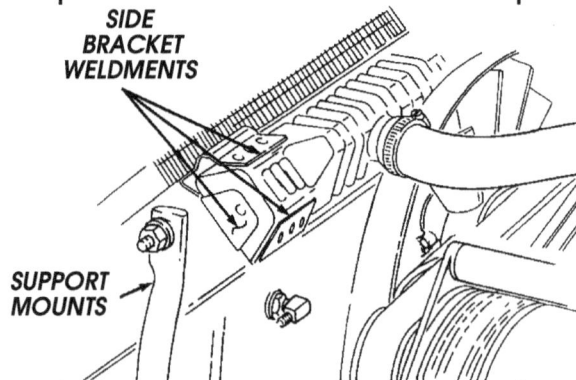

SIDE BRACKET WELDMENTS

SUPPORT MOUNTS

Table 2-2. *Preventive Maintenance Checks and Services (Cont'd).*

Item No.	Interval	Location Item to Check/ Service	Crewmember Procedure	Not Fully Mission Capable If:
48	Weekly	Cooling System (Cont'd)	**d.** Deleted **e.** Check fan shroud for damage. **f.** Check engine oil cooler and hoses for damage and leaks. <u>DRIVER</u>	**d.** Deleted **e.** Fan shroud broken cracked, or loose. **f.** Class III leak evident.
49	Weekly	Batteries	**WARNING** • Do not smoke, have open flames, or make sparks around batteries, especially if the caps are off. Batteries can explode and may cause injury or death. • Remove all jewelry such as rings, dog tags, bracelets, etc. If jewelry contacts battery terminal, a direct short may result, causing severe injury to personnel, or damage to equipment. **a.** Remove companion seat and check batteries for damaged casing, terminal posts, and security of mounting. Notify unit maintenance if any defects are found.	**a.** One or more batteries missing, unserviceable, or leaking; terminal or cables loose, corroded, or holddowns not secure.
49.1	Weekly	Battery Powered Motorized Traversing Unit (BPMTU) Batteries	**a.** Check the BPMTU batteries for loose or corroded connections, damaged casing, terminal posts, cables, and loose mounting.	**a.** One or more batteries missing, unserviceable, or leaking; terminal or cables loose, corroded, or holddowns not secure.

Table 2-2. Preventive Maintenance Checks and Services (Cont'd).

Item No.	Interval	Location Item to Check/ Service	Crewmember Procedure	Not Fully Mission Capable If:
49	Weekly	Batteries (Cont'd)	**b.** Electrolyte should be filled to the level/split ring in the battery filler opening (vent). If fluid is low, fill with distilled water to split ring (appendix D, item 7). If fluid is gassing (boiling), notify unit maintenance. **NOTE** Water in battery box can be caused by debris plugging battery box drain holes. If water is present, clean debris from battery box drain holes. **c.** Check battery box for corrosion or water on bottom of battery tray.	
50	Weekly	Weapon Station	<u>DRIVER</u> **a.** Inspect weapon station hatch and hinge for bends, cracks, warped, or damaged areas. **b.** Inspect brake for ease of operation. **c.** Inspect gunner's sling for tears, frays, or damaged hook.	**a.** Hatch or hinges inoperable. **b.** Brake does not operate. **c.** Sling is torn, shows wear, or hook is damaged.
51	Weekly	Tiedowns	<u>DRIVER</u> **a.** Inspect stored equipment footman loops for presence and security of mounting. Inspect straps for tears or frays. **b.** Inspect stowage brackets, footman loops, and tiedowns for security of mounting, damage, and missing components.	

Table 2-2. *Preventive Maintenance Checks and Services (Cont'd).*

Item No.	Interval	Location Item to Check/ Service	Crewmember Procedure	Not Fully Mission Capable If:
51	Weekly	Tiedowns (Cont'd)	**c.** Inspect all tiedown strap assemblies for proper operation, frays, damage, cleanliness, and security of mounting.	
			DRIVER	
52	Weekly	Tow Pintle	Check pintle hook for looseness, damaged locking mechanism, and presence of cotter pin.	
			DRIVER (M1114 ONLY)	
53	Weekly	Environment Control System	**a.** Check security of windshield de-icer control box mounting.	
			NOTE Insufficient cooling could be a result of loss of R134a refrigerant. This is a gas, therefore leaks cannot be detected. If leaks in lines or fittings are suspected, the vehicle is to be considered non-mission capable. Notify your supervisor.	
			b. Check Heating, Ventilation, and Air Conditioning (HVAC) vents, and mounting hardware for damage, leaks, missing components, and security of mounting.	**b.** Leaks in lines or fittings.
			c. Inspect exposed wiring harnesses for breaks, frayed insulation, loose or damaged connectors, and loose, damaged, or missing mounting hardware.	**c.** Wiring harness broken, frayed, or damaged. Mounting hardware missing.

Table 2-2. Preventive Maintenance Checks and Services (Cont'd).

Item No.	Interval	Location Item to Check/ Service	Crewmember Procedure	Not Fully Mission Capable If:
54	Weekly	Parking Brake	DRIVER Check combination service/ parking brake assemblies; inspect parking brake for obstruction of the actuating lever or broken or missing spring.	Actuating lever or spring is broken or missing.
54.1	Weekly	Steering Gear Mounting	DRIVER With vehicle running, observe steering gear mounting area for movement while assistant turns steering wheel left and right.	Any movement observed in steering gear mounting area.
55	Weekly	Windshield Washer	DRIVER **a.** Visually check windshield washer reservoir for damage. **b.** Check windshield washer fluid level.	
55.1	Weekly	Rear Troop/ Cargo Heater	DRIVER (Vehicle w/Troop/Cargo Winterization Kit) **a.** Check heater and heater controls for proper operation. **b.** Check fuel lines and fittings for leaks, cracks, or breaks. **c.** Check electrical cables and connections for security and frayed or broken wires. **d.** Check heater exhaust pipe for damage, security of mounting, and missing components. **e.** Check fuel filter for leaks or damage.	**a.** Heater inoperable and mission requires heater. **b.** Class III fuel leakage is evident and mission requires heater. **c.** Frayed or broken wires. **d.** Heater exhaust damaged or components missing. **e.** Class III leak evident.

Table 2-2. Preventive Maintenance Checks and Services (Cont'd).

Item No.	Interval	Location Item to Check/ Service	Crewmember Procedure	Not Fully Mission Capable If:
55.2	Weekly	Deep Water Fording Kit (if installed)	DRIVER Inspect vent lines and connectors for security, cracks, and deterioration.	Vent lines cracked, plugged, or missing.
55.3	Weekly	TOW ITAS Missile Rack	DRIVER **a.** Inspect TOW ITAS missile rack locking pins and support braces for presence and ease of operation. **b.** Inspect straps for tears and frays.	a. Rack will not stow six TOW ITAS missiles.
55.4	Weekly	Floorboard	Check for presence of TU adapter.	TU adapter missing or damaged.
56	Monthly	Corrosion	DRIVER Visually inspect vehicle for indication of corrosion or cracks and/or breaks.	Any corroded-through condition, cracks or breaks that would affect vehicle operation.
57	Monthly	Tailgate	DRIVER Check tailgate for corroded-through condition and/or damage. If tailgate does not latch securely or is damaged, notify unit maintenance.	Any corroded-through condition, or damage that would affect vehicle operation.
58	Monthly	Winch	DRIVER **a.** Check winch controls for proper operation. **b.** Check winch cable for kinks, frays, or breaks. **c.** Inspect tree saver strap for cuts and abrasions. If red safety thread is visible in main body of strap, notify unit maintenance for replacement of strap.	

Table 2-2. *Preventive Maintenance Checks and Services (Cont'd).*

Item No.	Interval	Location Item to Check/ Service	Crewmember Procedure	Not Fully Mission Capable If:
58 (cont'd)	Monthly	Winch	**d.** If it is known the strap has been overloaded, notify unit maintenance for replacement of strap. **e.** Wrap winch cable (para. 2-28.1). DRIVER	
59	Monthly	Zippers	**a.** Check canvas top and door zippers for corrosion and/or damage. **b.** Clean zippers with toothbrush.Apply interlock lubricant (appendix D, item 19) to canvas top, door zippers, and threads that hold zipper in place. DRIVER	
60	Monthly	TOW ITAS Power Cable	Inspect TOW ITAS power cable at the point where it exits the battery box. Chafing of the nylon braid that covers the cable is acceptable. If wire inside the cable is exposed, notify unit maintenance.	Wire inside the cable is exposed.

Section III. OPERATION UNDER USUAL CONDITIONS

2-9. GENERAL

This section provides instructions for vehicle operations under moderate temperature, humidity, and terrain conditions. For vehicle operation under unusual conditions, refer to section IV of this chapter.

2-10. OPERATION UNDER USUAL CONDITIONS REFERENCE INDEX

2-10. OPERATION UNDER USUAL CONDITIONS REFERENCE INDEX (Cont'd)

2-11. BREAK-IN SERVICE

CAUTION

Do not tow trailer during the first 500 mi (805) km) of operation. Damage to equipment may occur.

Break-in precautions are no longer required during the first 500 mi (805 km) of operation, with the exception of trailer towing.

2-12. STARTING THE ENGINE

WARNING

- The automatic transmission has a PARK position. Never use the shift lever in place of the parking brake. Set the parking brake. Make sure the transmission shift lever is in the P (park) position and transfer case shift lever is NOT in the N (neutral) position. Damage to equipment and injury to personnel may occur if these instructions are not followed.

- Chock blocks will be used when parking a vehicle with inoperative parking brakes, operating in extreme cold conditions, parking on inclines, or whenever and wherever maintenance is being performed. Failure to do so may result in injury to personnel or damage to equipment.

- Hearing protection is required for driver and passengers when engine is running. Noise levels produced by these vehicles exceed 85 dBA, which may cause injury to personnel.

   a. Ensure transmission shift lever (3) is in P (park) position and transfer case shift lever (4) is NOT in N (neutral) position.

NOTE

To apply parking brake, grasp handle firmly and pull upward until handle is locked in a straight-up position.

a.1. Ensure parking brake (5) is applied.

   b. Adjust driver's seat (para. 2-16).

WARNING

Right rearview mirror may restrict visibility.

c. Adjust left and right rearview mirrors. Ensure both mirrors provide a clear view (para. 3-24).

d. Ensure all windows are clean. If not, clean windows before attempting to move vehicle (table 2-1).

WARNING

Ensure all slack from seatbelt is removed. Seatbelt retracts and will lock only during sudden stops or impact. Injury to personnel will result if an accident occurs and seatbelt is not used or adjusted properly.

NOTE

Fasten unused seatbelts to protect the belt ends from damage or dirt contamination.

e. Fasten and adjust seatbelt (para. 2-18).

CAUTION

- Do not leave rotary switch in RUN position for extended periods of time. Glow plugs will continue to cycle and batteries will discharge leading to a no-start condition.

- Glare from the sun may make it difficult to tell if the wait-to-start lamp assembly is illuminated. If this occurs, shade the light with your hands to ensure that the wait-to-start lamp assembly goes out before attempting to start the vehicle.

- Starting the engine before the wait-to-start lamp goes out can result in starting problems.

e.1. Ensure all electrical switches (lights, wipers, and blower motor) in vehicle are turned to OFF.

f. Turn rotary switch (1) to RUN and wait until wait-to-start lamp assembly (2) goes out. Transmission indicator lamp (2.1) will be illuminated.

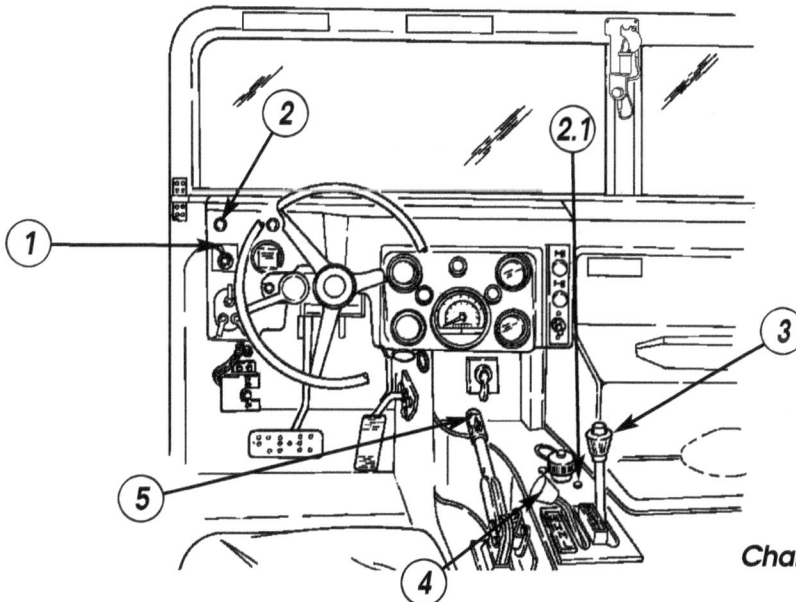

2-12. STARTING THE ENGINE (Cont'd)

CAUTION

- If ambient temperature is above 0°F (-18°C), do not operate starter continuously for more than 20 seconds; wait 10 to 15 seconds between periods of starter operation. Failure to do this may result in damage to the starter.

- If any instrument reading is not normal, stop engine. Failure to do this may result in damage to the engine. Refer to troubleshooting, table 3-1.

- If engine does not start, leave rotary switch in RUN position and wait 10 to 15 seconds before trying to restart. Failure to do this may result in damage to glow plugs or starter.

- If you accidentally turn the rotary switch to ENG STOP position after an unsuccessful attempt to start, wait 90 seconds before trying to restart. Failure to do this may result in damage to glow plugs.

   g. Place rotary switch (1) to START. Release lever after engine starts. Lever will return automatically to RUN. Transmission indicator lamp (5.1) goes out when engine starts.

NOTE

- Before the engine reaches operating temperature, the wait-to-start lamp may flicker and a clicking noise may be heard. This is due to glow plug relay cycling and is a normal condition.

- If the transmission lamp stays illuminated, notify maintenance. Transmission damage can result if cause of light illumination is not determined.

   h. Allow engine to warm up for approximately one minute and ensure instruments indicate the following:

   (1) Oil pressure gauge (3) should register above approximately 6 psi (41 kPa) with engine at idle.

   (2) Voltmeter (5) should register in green area.

   (3) Fuel gauge (6) should indicate fuel level in fuel tank.

   (4) Air restriction gauge (2) should not register within the red zone.

   i. Stop engine if any of the following conditions occur:

   (1) Excessive engine vibration.

   (2) Oil pressure does not register or suddenly drops to or less than approximately 6 psi (41 kPa) with engine at idle.

   (3) Air restriction gauge (2) is within the red zone.

   j. If engine overheating occurs:

   (1) Park vehicle, allow engine to idle.

   (2) Observe coolant temperature gauge (4) for steady cooling.

CAUTION

Stop engine if coolant temperature gauge suddenly increases beyond approximately 250°F (120°C), or damage to engine will result.

(3) If engine coolant temperature continues to increase, or does not decrease, as indicated by coolant temperature gauge (4), stop engine. Perform troubleshooting procedures in table 3-1.

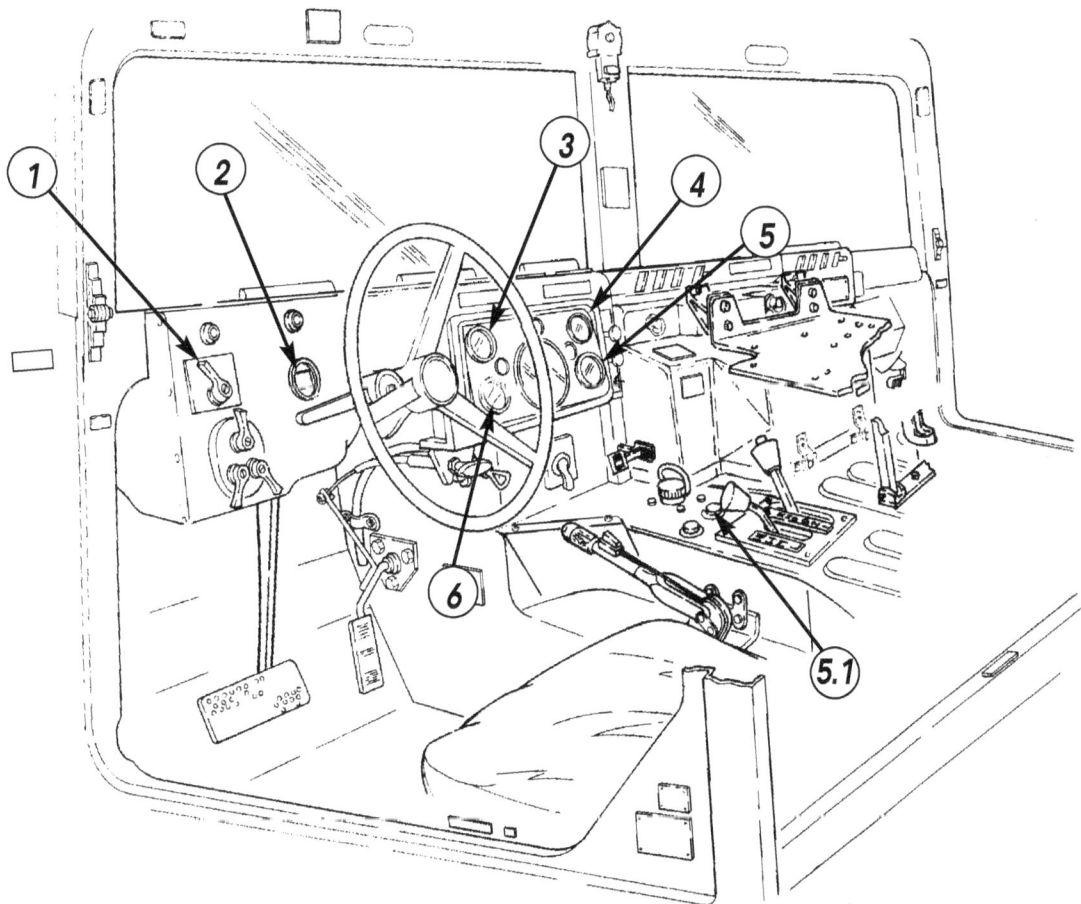

2-13. PLACING VEHICLE IN MOTION

WARNING

- This vehicle has been designed to operate safely and efficiently within the limits specified in this TM. Operation beyond these limits are prohibited IAW AR 750-1 without written approval from the Commander, U.S.Army Tank-automotive and Armaments Command,ATTN:AMSTA-CM-S,Warren, MI 48397-5000.
- Communications shelters AN/GRC-122 and AN/GRC-142 RATT may overload truck by up to 500 lbs (227 kg). Use caution when driving to avoid damage to equipment or injury to personnel.

CAUTION

Do not tow trailer during the first 500 mi (805 km) of operation. Damage to equipment may occur.

NOTE

- The following procedures apply to a vehicle being driven in good weather on high traction surfaces where little or no wheel slippage is evident. For operating the vehicle under unusual conditions; i.e., unusual terrain, cold weather, ice, snow, dusty or sandy areas, mud or rain, refer to section IV.
- Before you operate your vehicle, be sure to perform the Preventive Maintenance Checks and Services (PMCS) shown in table 2-2.

a. Be sure all auxiliary equipment and tools are stored for travel.

CAUTION

- Vehicle must be stopped, and transmission shift lever placed in P (park) or N (neutral), before transfer case can be shifted. Failure to do this will result in damage to drivetrain.
- Do not place transfer case shift lever in H/L (high/lock range) or L (low range) on high traction surfaces where little or no wheel slippage is evident, particularly when encountering sharp, continuous turns. Failure to operate the vehicle with transfer case in H (high range) on high traction surfaces, particularly when encountering sharp, continuous turns, can damage drivetrain.

b. With transmission shift lever (2) in P, select H transfer case driving gear range using transfer case shift lever (3).This range is for normal driving in good weather or on high traction surfaces.

c. Start engine (para. 2-12).

d. Set vehicle light switch (7) (para. 2-15).

CAUTION

Ensure parking brake is released completely before operating the vehicle. Failure to do so may cause damage to equipment.

NOTE

- To release parking brake, grasp handle firmly and push forward until handle is seated in its most forward position.
- Parking brake handle has a safety release button which must be depressed to release the parking brake.
- If covering warning lamp with tape, put a pinhole in the tape to see when the light is on.

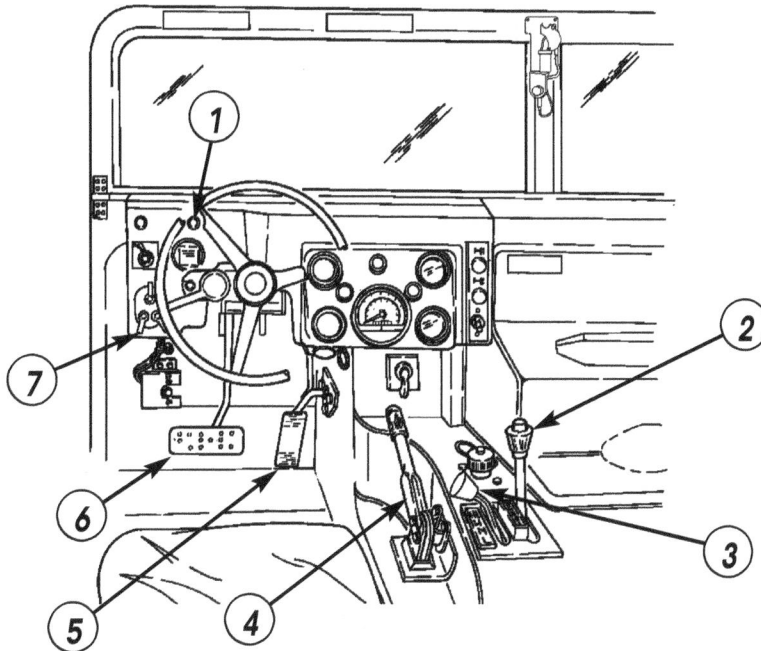

e. Depress service brake pedal (6). Depress parking brake release button and release parking brake lever (4).The brake warning lamp assembly (1) should go out.

f. Place transmission shift lever (2) in (D)(overdrive) for normal driving.

g. Release service brake pedal (6), and depress accelerator pedal (5).Accelerate at a safe, steady speed.

h. Upshift or downshift transmission shift lever (2) when road and/or traffic conditions change.

WARNING

- Use extreme caution when transporting personnel.Although certain design characteristics of the vehicle, such as vehicle width, ground clearance, independent suspension, etc., provide improved capabilities, accidents can still happen.

- Operators are reminded to observe basic safe driving techniques/skills when operating the vehicle, especially when transporting personnel.Vehicle speed must be reduced consistent with weather and road/terrain conditions. Obstacles such as stumps and boulders must be avoided. Failure to use basic safe driving techniques/skills may result in loss of control and an accident or rollover resulting in injury or death to personnel and damage to equipment. Since the troop/cargo area has minimal overhead protections and does not have seatbelts, personnel seated here are at greater risk of serious injury.

NOTE

To help judge clearance more accurately, guide rods can be used at the discretion of the unit commander.

i. Deleted.

2-13. PLACING VEHICLE IN MOTION (Cont'd)

j. Operators are reminded to observe basic safe driving techniques/skills when operating the vehicle, especially when transporting personnel.

2-14. STOPPING THE VEHICLE AND ENGINE

a. Release accelerator pedal (5).

b. Depress service brake pedal (6) to bring vehicle to a gradual stop.

WARNING

• Never use the transmission shift lever in place of the parking brake. Set the parking brake. Make sure the transmission shift lever is in the P (park) position and transfer case shift lever is NOT in the N (neutral) position. Damage to equipment and injury to personnel may occur if these instructions are not followed.

• Chock blocks shall be used when parking a vehicle with inoperative parking brakes, in extreme cold conditions, parking on inclines, or whenever and wherever maintenance is being performed. Failure to do this may result in injury to personnel or damage to equipment.

NOTE

To apply parking brake, grasp handle firmly, and pull upward until handle is locked in a straight-up position.

c. Once vehicle is completely stopped, apply parking brake lever (4).

d. Move transmission shift lever (3) to P (park).

e. Turn light switch (7) to OFF.

f. Place rotary switch (1) to ENG STOP.

NOTE

If there is engine run-on, pinch the fuel return line with your fingers or a pair of pliers to stop the engine. Notify unit maintenance.

g. Lock steering wheel with cable (2), and chock wheels if tactical situation permits.

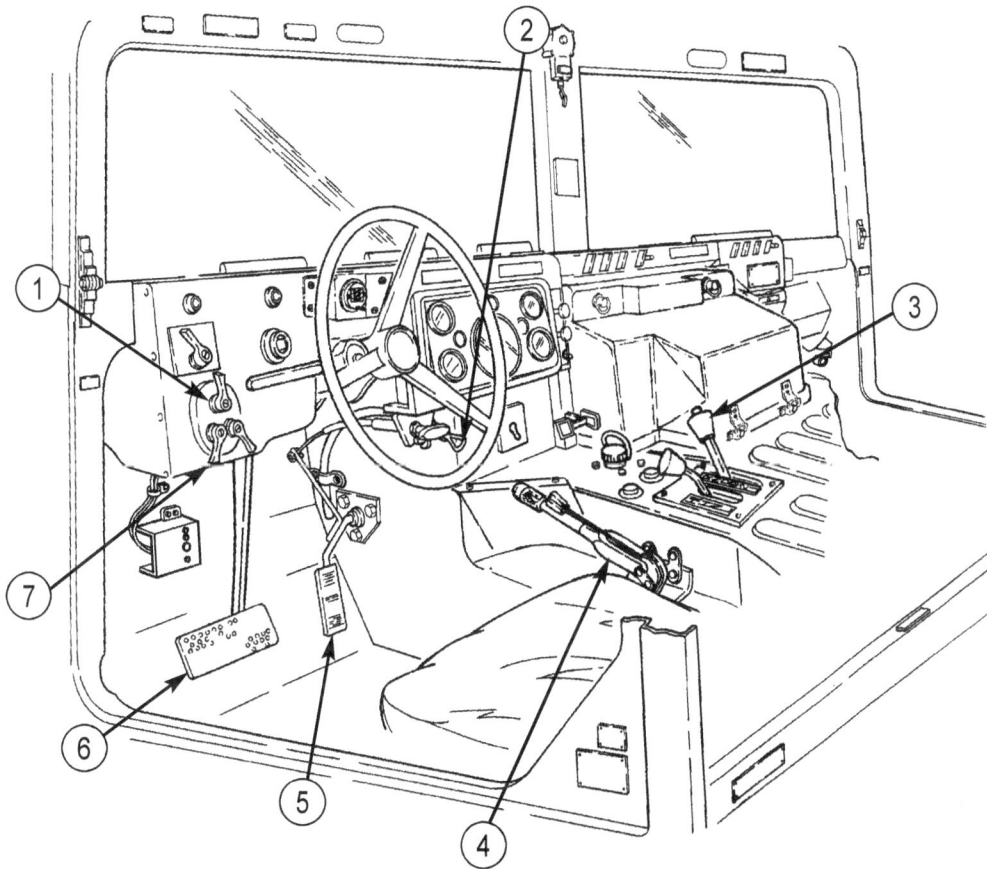

2-15. OPERATION OF VEHICLE LIGHT SWITCH

a. To illuminate instrument panel:

　(1) Lift unlock lever (2) to UNLOCK and hold in position.

　(2) Turn selector switch lever (1) to any ON position except B.O. MARKER.

　(3) Turn auxiliary lever (3) to DIM or PANEL BRT (bright).

b. For normal daylight driving, turn selector lever (1) to STOPLIGHT.

c. For night driving, turn selector lever (1) to SERVICE DRIVE. The headlight dimmer foot switch is depressed to provide high beam service light operation. The indicator light (4) on the instrument panel should illuminate when high beams are operating.

d. In blackout operations:

　(1) When driving vehicle, turn selector lever (1) to B.O. DRIVE.

　(2) When vehicle is parked, turn selector lever (1) to B.O. MARKER.

e. To illuminate parked vehicle at night (if tactical situation permits):

　(1) Keep selector lever (1) in SERVICE DRIVE.

　(2) Turn auxiliary lever (3) to PARK.

f. Move turn signal lever up for right turns, and down for left turns.

NOTE

- When activated, the warning flashers override operation of the brake lights.
- The headlights cannot be turned on when auxiliary lever (3) is in the PARK position.

g. For hazard warning lights (blinking lights):

　(1) Turn selector lever (1) to SERVICE DRIVE or STOPLIGHT.

　(2) Pull warning hazard tab out and move turn signal lever up to lock lever in place.

　(3) To deactivate, move turn signal lever back to NEUTRAL.

h. For horn operation, turn selector lever (1) to either STOPLIGHT or SERVICE DRIVE.

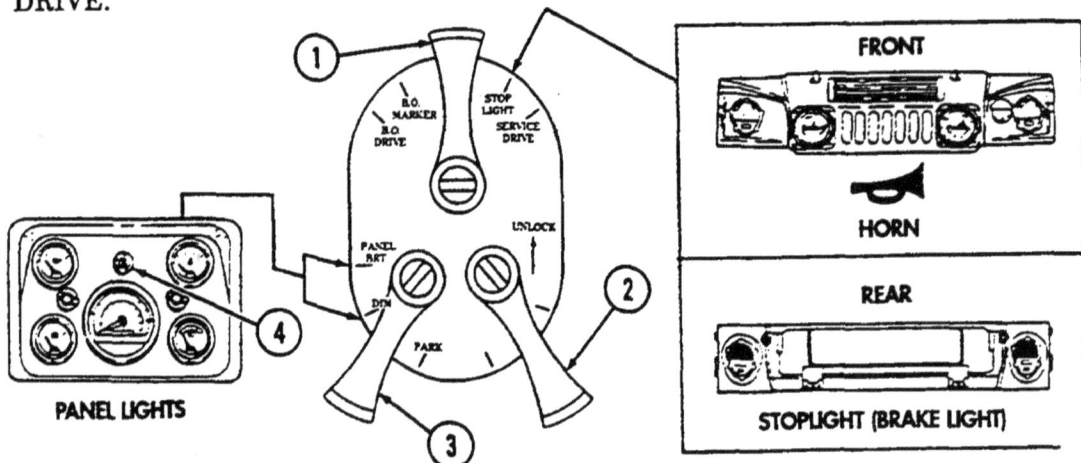

PANEL LIGHTS

FRONT

HORN

REAR

STOPLIGHT (BRAKE LIGHT)

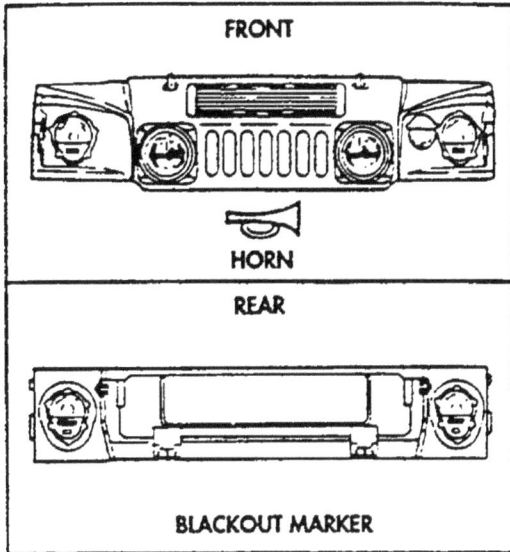

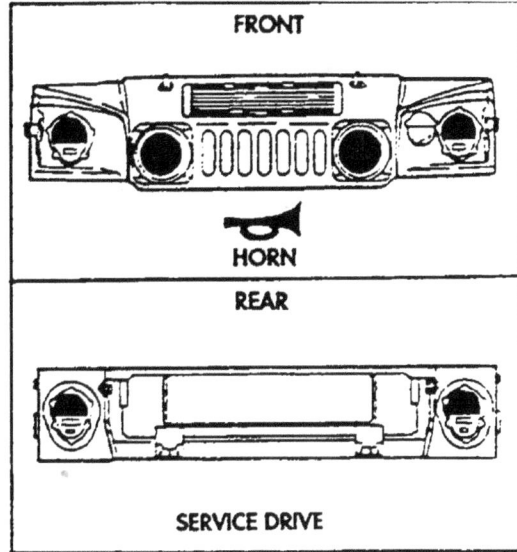

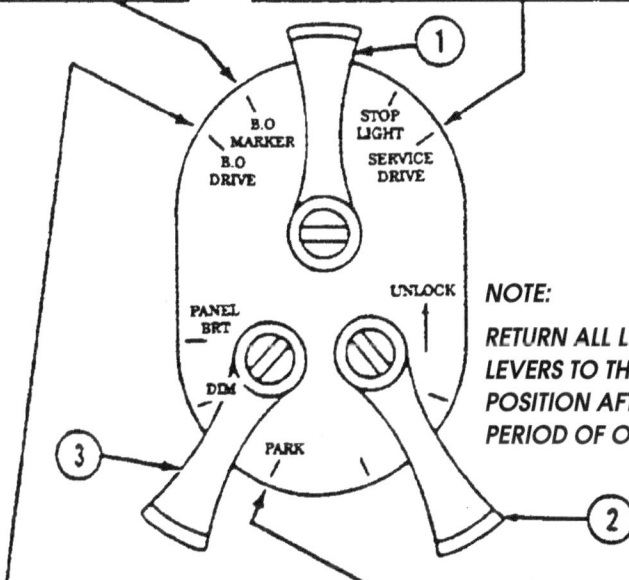

FRONT

HORN

REAR

BLACKOUT MARKER

FRONT

HORN

REAR

SERVICE DRIVE

B.O MARKER
B.O DRIVE

STOP LIGHT
SERVICE DRIVE

PANEL BRT

UNLOCK

DIM

PARK

NOTE:

TO MOVE SELECTOR SWITCH LEVER FROM OFF TO ANY ON POSITION EXCEPT B.O. MARKER, UNLOCK SWITCH LEVER MUST BE LIFTED TO UNLOCK.

NOTE:

RETURN ALL LIGHT SWITCH LEVERS TO THE OFF POSITION AFTER EACH PERIOD OF OPERATION.

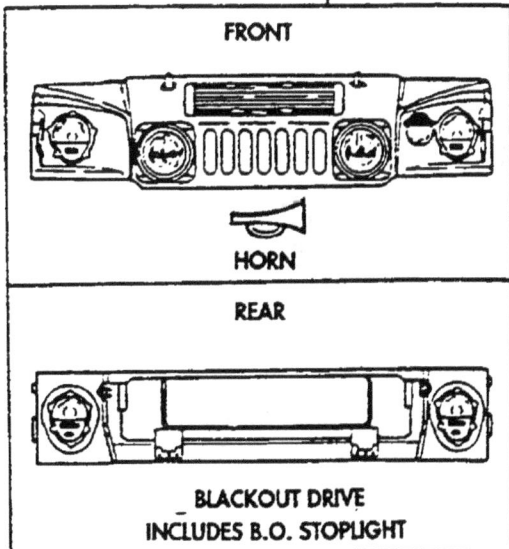

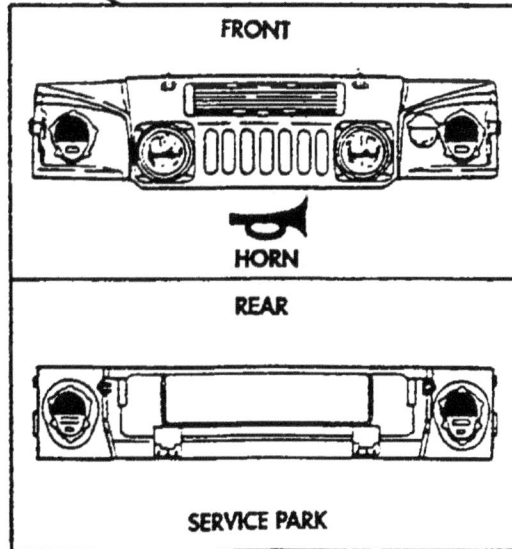

FRONT

HORN

REAR

BLACKOUT DRIVE
INCLUDES B.O. STOPLIGHT

FRONT

HORN

REAR

SERVICE PARK

2-16. ADJUSTING DRIVER'S SEAT

a. Adjusting Seat Horizontally.

(1) Pull seat slide lever (2) upward and slide seat (3) to desired position.

(2) Release seat slide lever (2) to lock seat (3) into position.

b. Adjusting Seat Vertically.

(1) To lower seat (3), pull seat height adjustment lever (1) upward and allow body weight to lower seat (3).

(2) To raise seat (3), pull seat height adjustment lever (1) upward and remove body weight from seat (3).

(3) Release seat adjustment lever (1) to lock seat (3) in desired position.

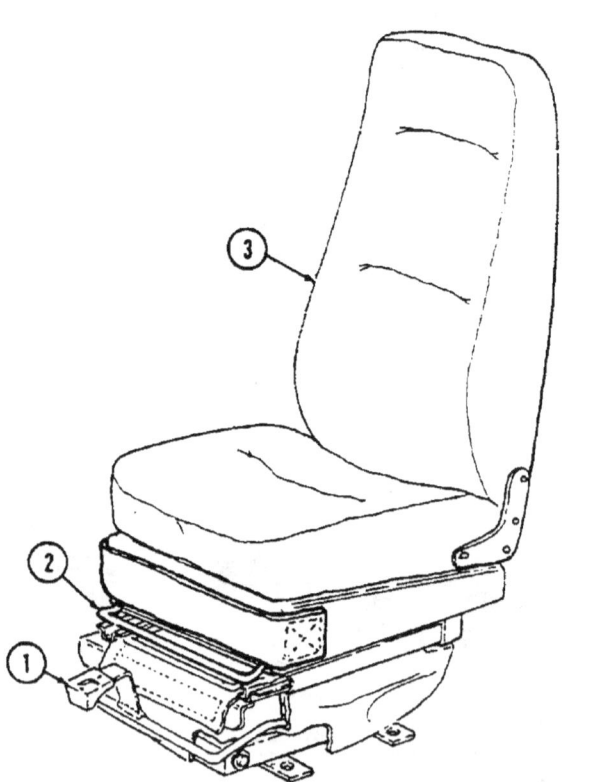

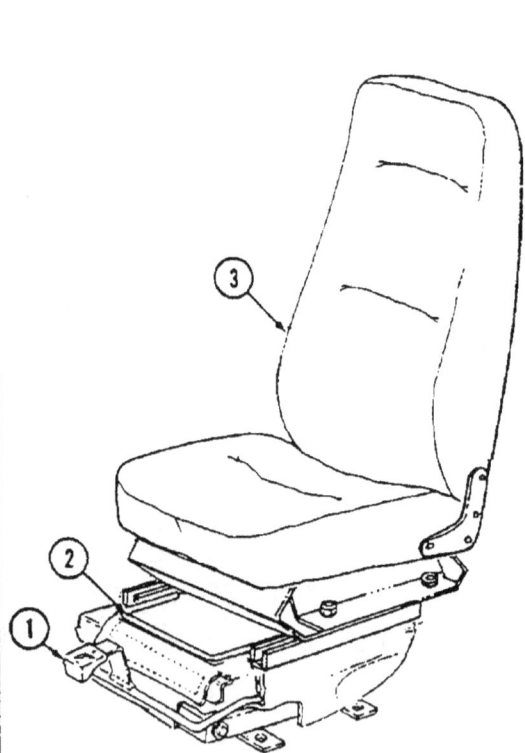

M1113   **ALL MODELS EXCEPT M1113**

2-17. COMPANION SEAT ASSEMBLY AND BATTERY BOX COVER REPLACEMENT

a. Removal.

(1) Release two latches (3) from strikers (4) securing companion seat assembly (1) and battery box cover (5) to battery box (2).

(2) Lift and pull companion seat assembly (1) and battery box cover (5) forward and remove from battery box (2).

b. Installation.

(1) Install companion seat assembly (1) and battery box cover (5) on battery box (2).

(2) Secure seat assembly (1) and battery box cover (5) to battery box (2) by connecting two latches (3) to strikers (4).

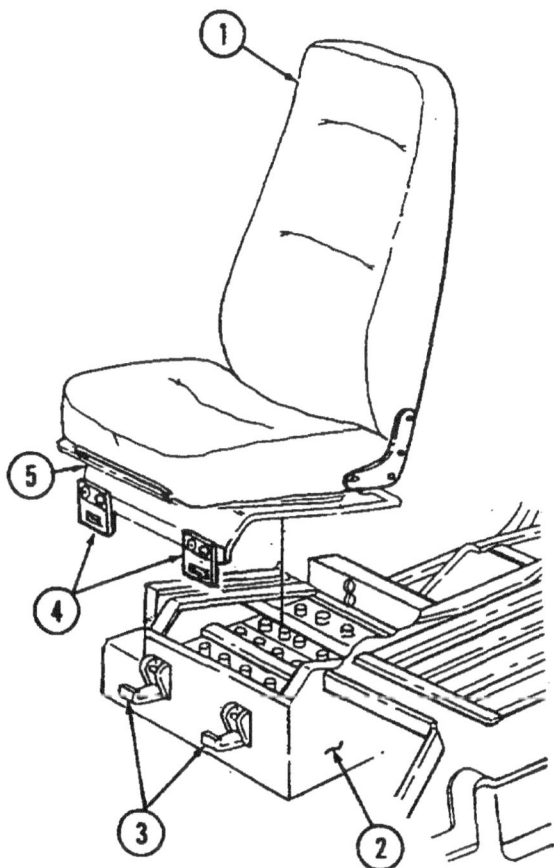

ALL MODELS EXCEPT M1114   M1114

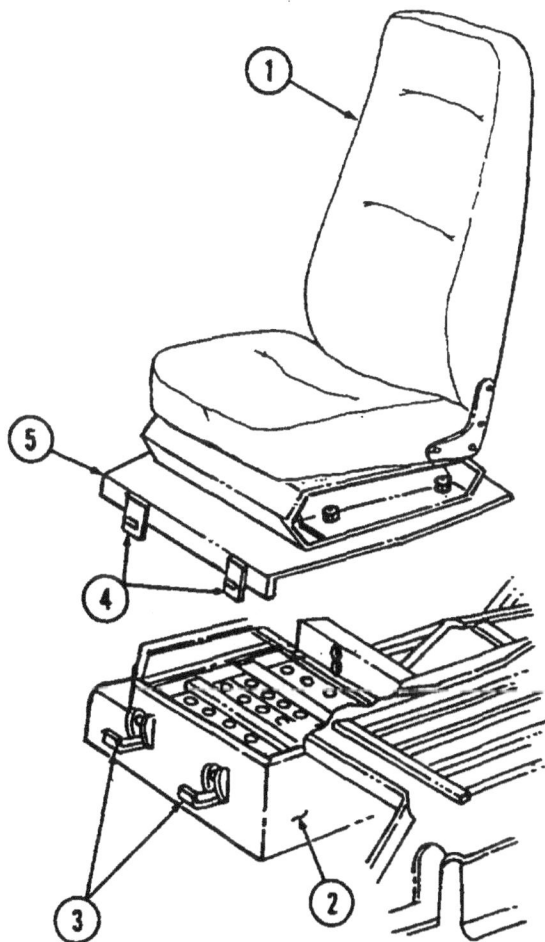

2-18. SEATBELT OPERATION – THREE-POINT SYSTEM

WARNING

Ensure all slack from seatbelt is removed. Seatbelt retracts and will lock only during sudden stops and/or impact. Injury and/or death to personnel will result if an accident occurs and seatbelt is not in use or adjusted properly.

a. Seatbelt Fastening:
Pull shoulder harness and seatbelt (3) across body and fasten latch plate (2) to belt buckle (1).

b. Seatbelt Adjustment:
Pull shoulder harness strap through latch plate (2) and remove slack from seatbelt (3). Seatbelt (3) must fit snugly across operator's and passenger's hips, shoulders, and chests.

c. Seatbelt Unfastening:
Push release button (4) on belt buckle (1) to release seatbelt (3).

2-18.1 SEATBELT OPERATION – IMPROVED PERSONAL RESTRAINT SYSTEM

WARNING

• Ensure all slack from shoulder strap and lap strap is removed. The straps will lock only during sudden stops and/or impact.

• Lap strap and shoulder strap must be worn together. Injury and/or death to personnel may result if shoulder strap is worn without the lap strap or lap strap is worn without the shoulder strap.

NOTE

If strap is allowed to retract suddenly, it is possible for the retractor to lock with the webbing fully retracted in which case it may not automatically unlock. To unlock, pull slowly on the strap in order to pack the webbing and allow the strap to extend an inch. Allow the strap to retract slowly allowing the retractor to unlock.

a. **Shoulder Strap/Lap Fastening:**
   (1) Pull shoulder strap (1) across body and fasten latch plate (9) to upper slot (8) of buckle assembly (5).
   (2) Pull lap strap (2) across body and fasten latch plate (3) to lower slot (4) of buckle assembly (5).

b. **Shoulder Strap/Lap Strap Unfastening:**
   Pull up on buckle assembly latch (6) to release both the shoulder strap (1) and lap strap (2).

c. **Buckle Assembly Adjustment:**
   (1) Lift the buckle adjuster knob (7) observing it is springloaded and moves freely.
   (2) Ensure adjuster can move freely throughout its full range.
   (3) Ensure adjuster knob (7) can lock into each of the adjustment positions.

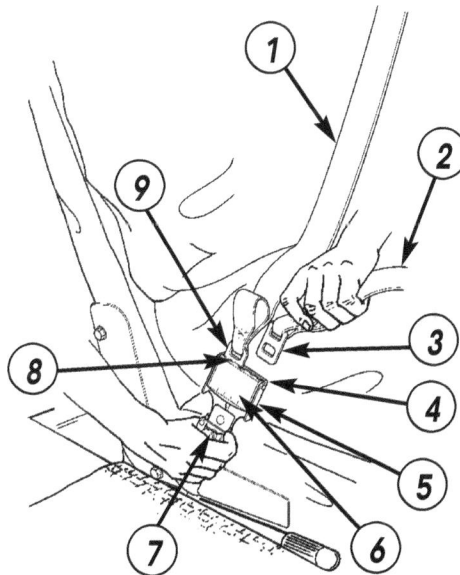

2-19. DEFROSTER OPERATION (M1113)

a. Start engine (para. 2-12).

b. Push fresh air intake lever (3) forward to close grille.

c. Turn heater fan switch (4) to desired setting, HIGH/LOW.

d. Position heater control knob (2) to desired setting.

e. Push in defroster control knob (1).

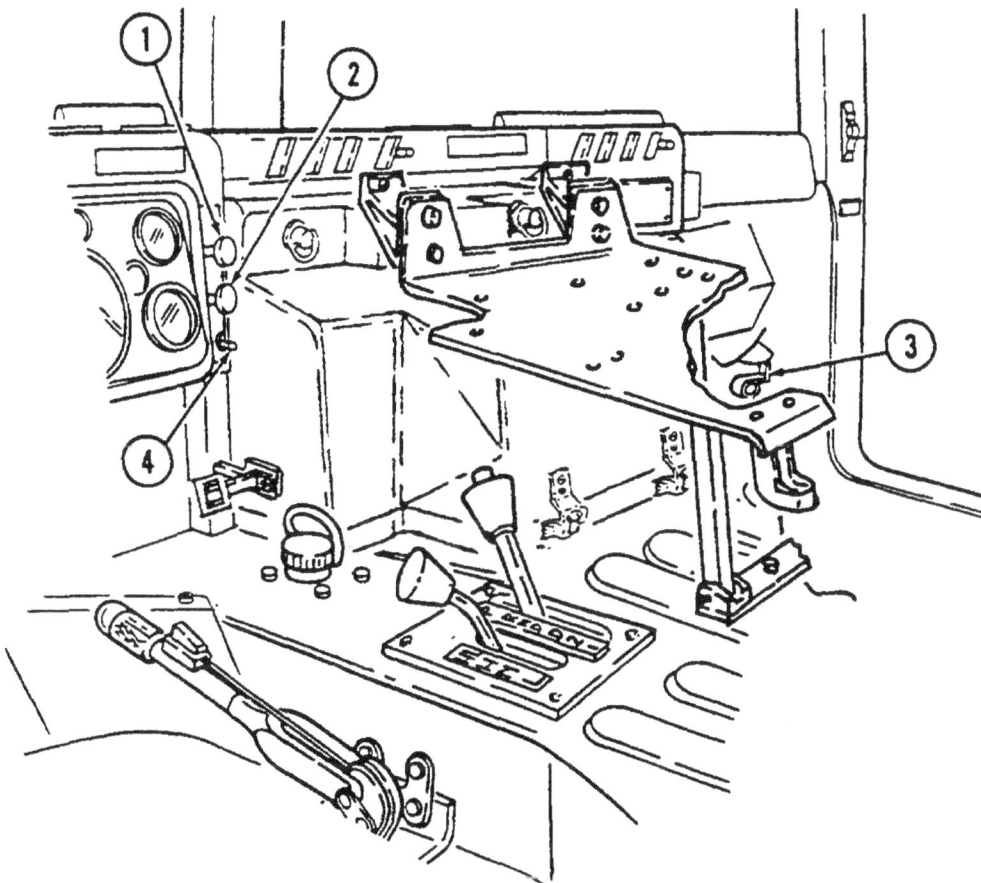

2-20. DEFROSTER OPERATION (M1114)

a. Operation.

   (1) Start engine (para. 2-12).

   (2) Slide defroster register (1) to the right to open defroster.

   (3) Pull fresh air control knob (2) in to close vents (5).

   (4) Position heater control knob (3) to desired setting.

   (5) Flip heater fan switch (4) up to turn on heater.

b. After Operation.

   (1) Flip heater fan switch (4) down to turn off heater.

   (2) Slide defroster register (1) to the left to close defroster.

   (3) Stop engine (para. 2-14).

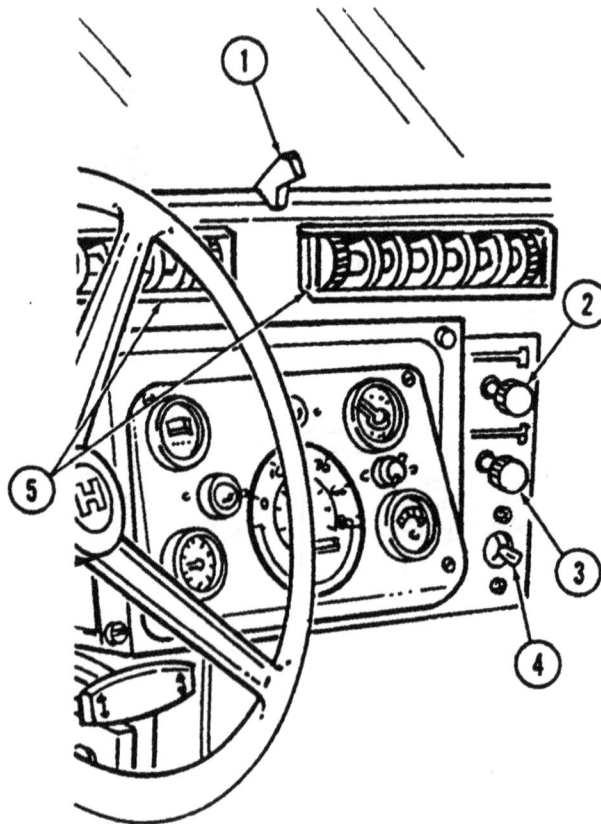

2-20.1. DEFROSTER OPERATION (M1151, M1151A1, M1152, M1152A1, M1165, M1165A1, M1167)

a. Operation.

(1) Start engine (para. 2-12).

(2) Slide defroster register (2) to the right to open defroster.

(3) Adjust vents (1) to closed position.

(4) Flip fan switch (4) to desired setting, HIGH/LOW.

(5) Flip A/C / HEAT switch (3) down to turn on heater.

b. After Operation.

(1) Flip fan switch (4) to OFF position.

(2) Slide defroster register (2) to the left to close defroster.

(3) Stop engine (para. 2-14).

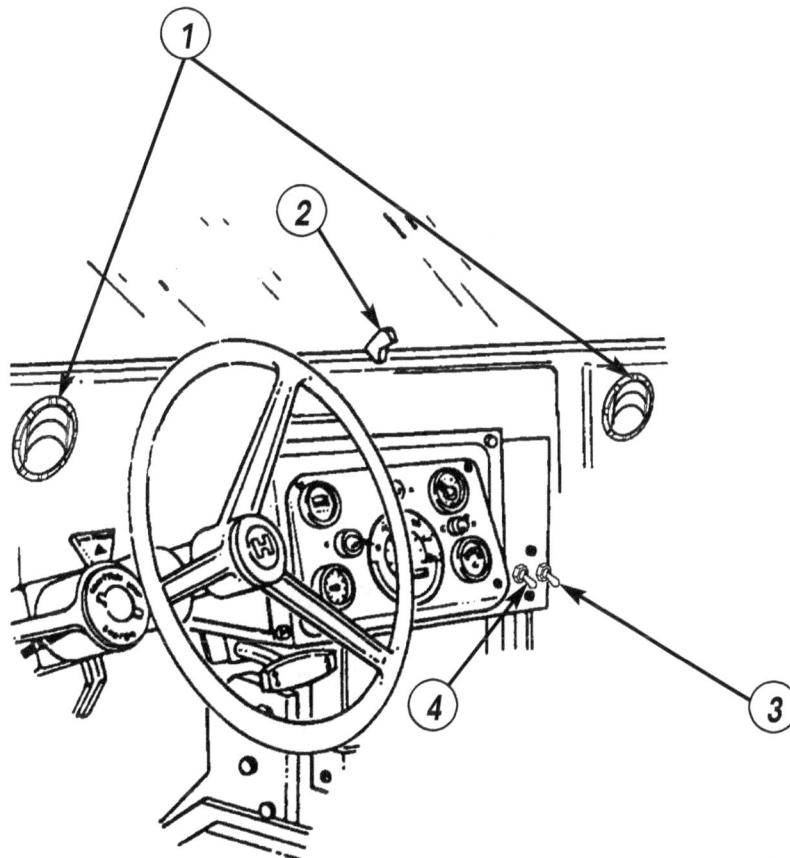

2-21. HEATER OPERATION (M1113, M1114)

a. Start engine (para. 2-12).

NOTE

Perform step b for M1113 models only.

b. Push fresh air intake lever (3) forward to close grille.

c. Turn heater fan switch (4) to desired setting, HIGH/LOW.

NOTE

For maximum heat, pull heater control knob all the way out.

d. Position heater control knob (2) to desired setting.

e. To shut off outside air, pull out control knob (1).

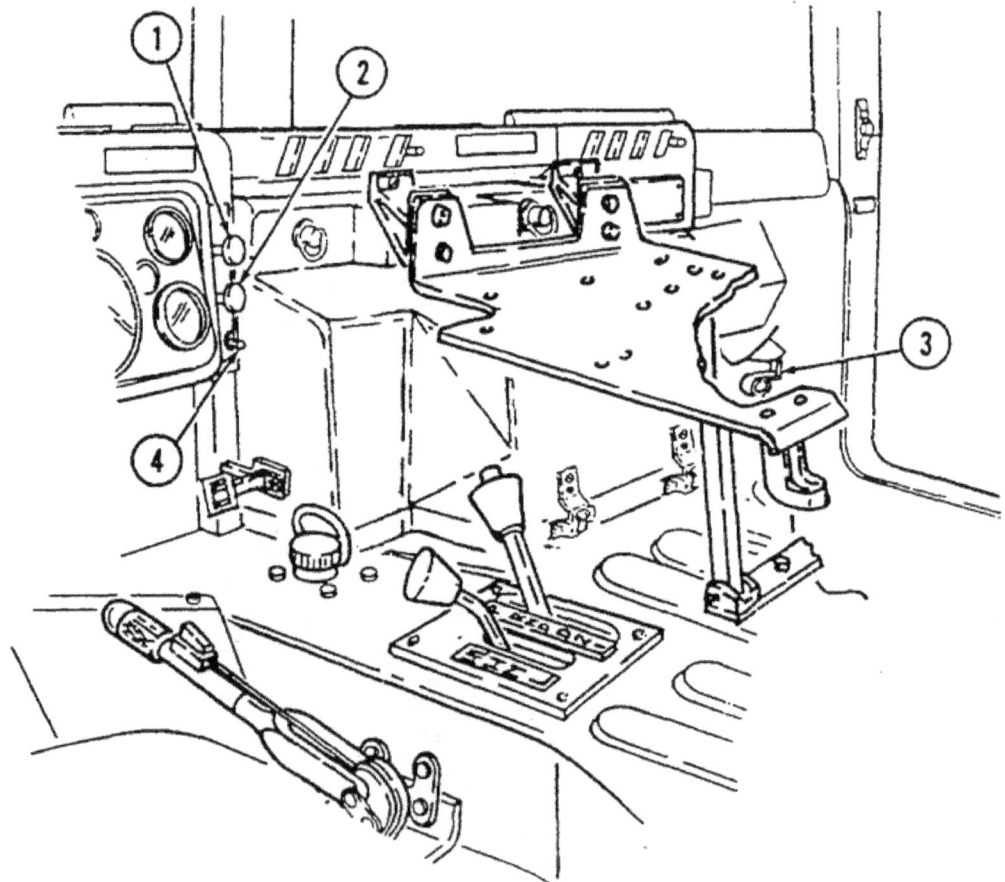

2-21.1. REAR HEATER ASSEMBLY OPERATION (M1114)

a. Start engine (para. 2-12).

b. Pull valve (2) to UP position to close vent (1).

c. Push valve (2) to DOWN position to open vent (1).

NOTE

Pull valve all the way down for maximum heat.

d. Position valve (2) to desired setting.

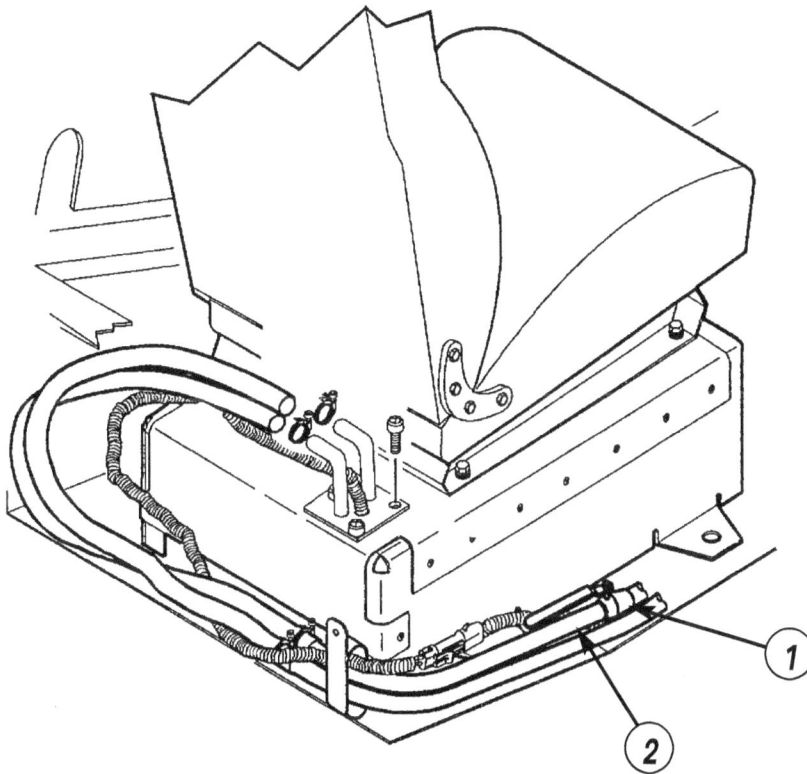

2-21.2. HEATER OPERATION (M1151, M1151A1, M1152, M1152A1, M1165, M1165A1, M1167)

a. Operation.

(1) Start engine (para. 2-12).

(2) Flip A/C / HEAT switch (1) down to turn on heater.

(3) Flip fan switch (2) to desired setting, HIGH/LOW, to activate heated air output.

b. After Operation.

(1) Flip fan switch (2) to OFF position.

(2) Stop engine (para. 2-14).

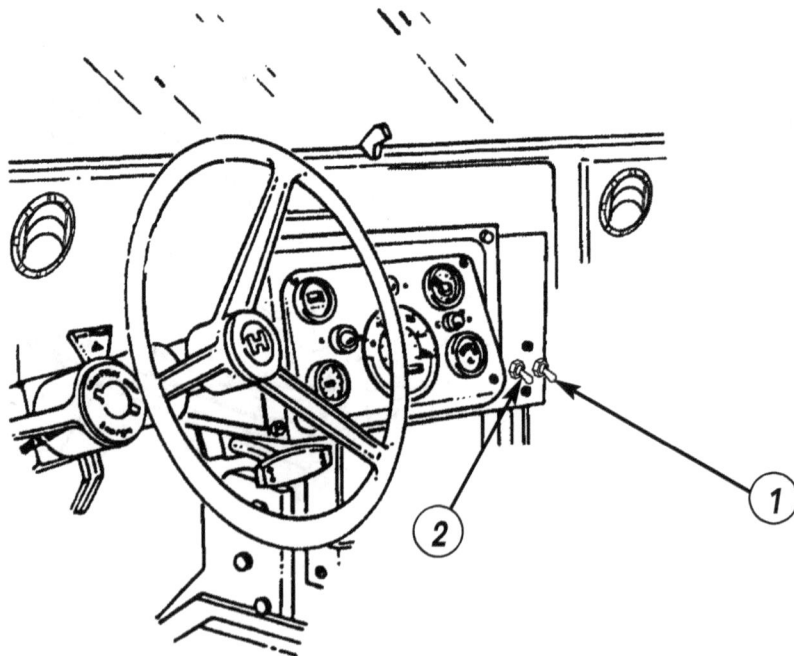

2-22. TAILGATE OPERATION

a. **Lowering Tailgate.**

WARNING

Do not use tow pintle as a step when entering or exiting vehicle cargo area. Using tow pintle as a step may result in injury to personnel.

CAUTION

The tailgate should not be lowered farther than the length of tailgate chains.Tailgate chains should always be used to support tailgate when open. Do not allow tailgate to slam against lifting shackles. Damage to equipment may occur.

(1) Remove two tailgate chain hooks (2) securing tailgate (1) to rear of vehicle body.

(2) Lower tailgate (1) and secure with two tailgate chain hooks (2).

b. **Raising Tailgate.**

(1) Raise two tailgate chain hooks (2) and raise tailgate (1).

(2) Secure tailgate (1) to rear of vehicle body with two tailgate chain hooks (2).

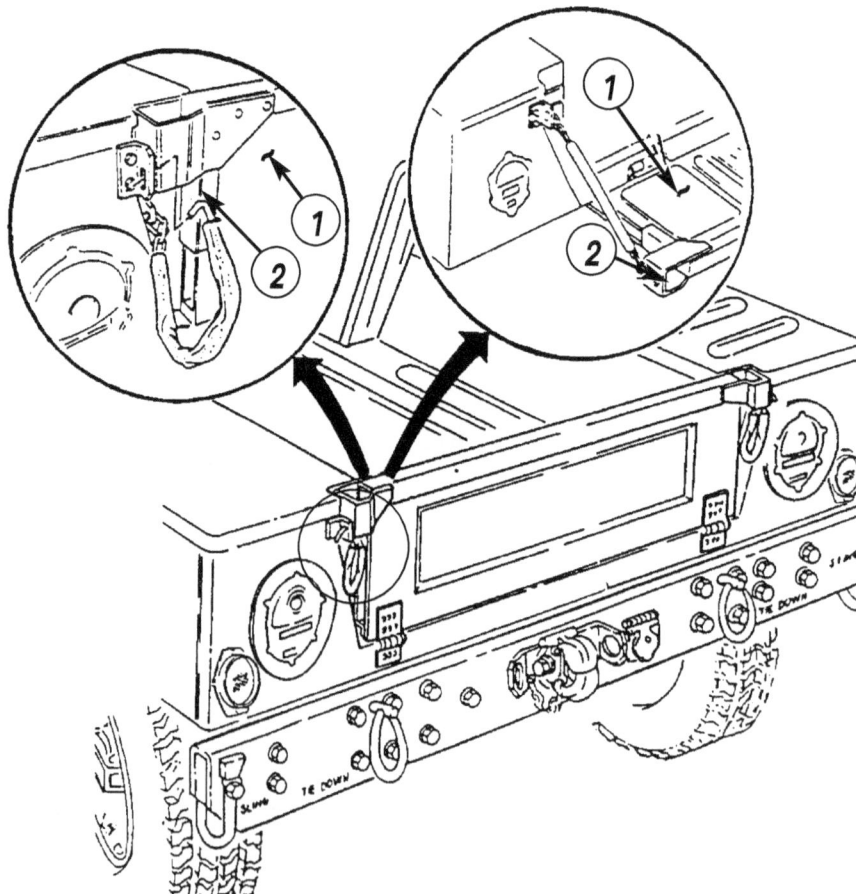

2-22.1. OPTIONAL 12-VOLT AUXILIARY POWER OUTLET OPERATION

a. Operating 12-Volt Auxiliary Power Outlet.

(1) Remove screw-on cover (2) from 12-volt auxiliary power outlet (1).

WARNING

Use only 12-volt accessories with compatible plug ends when connecting to 12-volt auxiliary power outlet. Damage to equipment or serious injury to personnel may result if non-compatible accessories are connected to 12-volt auxiliary power outlet.

CAUTION

Use a twisting motion when connecting accessories to 12-volt auxiliary power outlet. Forcefully pushing accessory plugs onto 12-volt auxiliary power outlet may cause damage to equipment.

(2) Connect 12-volt accessory plug to 12-volt auxiliary power outlet (1).

CAUTION

Do not use 12-volt accessories for long periods of time while vehicle engine is off.This will cause premature battery failure and will prevent vehicle starting.

(3) When finished using 12-volt accessory, disconnect 12-volt accessory plug from 12-volt auxiliary power outlet (1) and install 12-volt auxiliary screw-on cover (2).

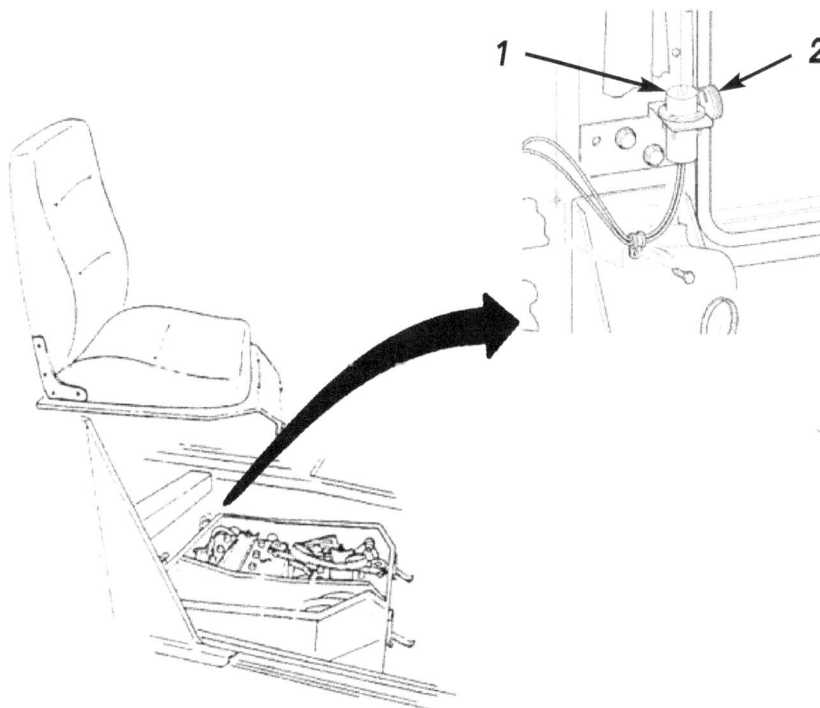

2-23. SLAVE STARTING OPERATION

a. Position slaving vehicle and disabled vehicle close enough for cable hookup.

b. Stop slaving vehicle engine.

c. Remove cover from slave receptacle of disabled vehicle and slaving vehicle.

WARNING

Ensure all battery cables in disabled vehicle are properly connected before connecting slave cable. Damage to batteries, cables, or serious injury to personnel may result from improperly connected batteries.

CAUTION

Use a twisting motion when installing slave cable to the receptacle. Forcefully pushing the cable onto the receptacle may cause damage to the receptacle mount.

NOTE

Ensure all electrical switches in both vehicles are turned off.

d. Connect slave cable to the slave receptacle of both vehicles.

e. Start slaving vehicle engine.

f. Start disabled vehicle engine.

CAUTION

Use a twisting motion when disconnecting slave cable from the receptacle. Forcefully pulling the cable from the receptacle may cause damage to the receptacle mount.

g. After engine starts, disconnect slave cable from both vehicles.

NOTE

For ease of removal, apply hand cleaner (appendix D, item 16) on the inside of the cover before installing receptacle covers.

h. Install receptacle covers on both vehicles.

2-24. MAX TOOL KIT STOWAGE (M1113) OPERATION

a. Tool Removal.

(1) Disconnect three strap assemblies (2) in right footwell area (1) securing ax (4), max tool carrying bag (5), and jack and tools bag (3). Remove ax (4) and max tool carrying bag (5).

(2) Remove the following attachments from max tool carrying bag (5):

- Reversible rake/hoe attachment (12)
- Rake/hoe thumbscrew attachment (11)
- Shovel attachment (6)
- Six safety locking pins (8)
- Broad pick attachment (10)
- Pick attachment (9)
- Mattock attachment (7)

NOTE

- Using the ax and attachments, the max tool kit can be incorporated into seven basic hand tools.
- Ax blade must be covered with sheath before using kit.
- Read all safety and assembly instructions enclosed with max tool kit before using kit.

b. Reversible Rake/Hoe Attachment.

(1) Connect rake/hoe attachment (12) (with either rake or hoe in the working position) into socket (13) on end of ax (4).

NOTE

Thumbscrew must be seated tightly in rake/hoe attachment and socket of ax. Check thumbscrew often to be sure it does not work loose.

(2) Install thumbscrew (11) on rake/hoe attachment (12) and in socket (13) of ax (4). Tighten thumbscrew (11).

(3) To remove rake/hoe attachment (12), remove thumbscrew (11) from rake/hoe attachment (12) and ax (4).

c. Shovel Attachment.

NOTE

Shovel attachment is shown. Broad pick, pick, and mattock attachments are attached basically the same.

(1) Connect shovel attachment (6) into socket (13) on end of ax (4).

(2) Insert safety locking pin (8) into hole in end of taper on shovel attachment (6).

(3) To remove shovel attachment (6), remove safety locking pin (8) from shovel attachment (6) and ax (4).

d. Tool Installation.

(1) Replace attachments in the max tool carrying bag (5).

(2) Install max tool carrying bag (5) and jack and tools bag (3) with three strap assemblies (2) in right footwell area (1) of vehicle.

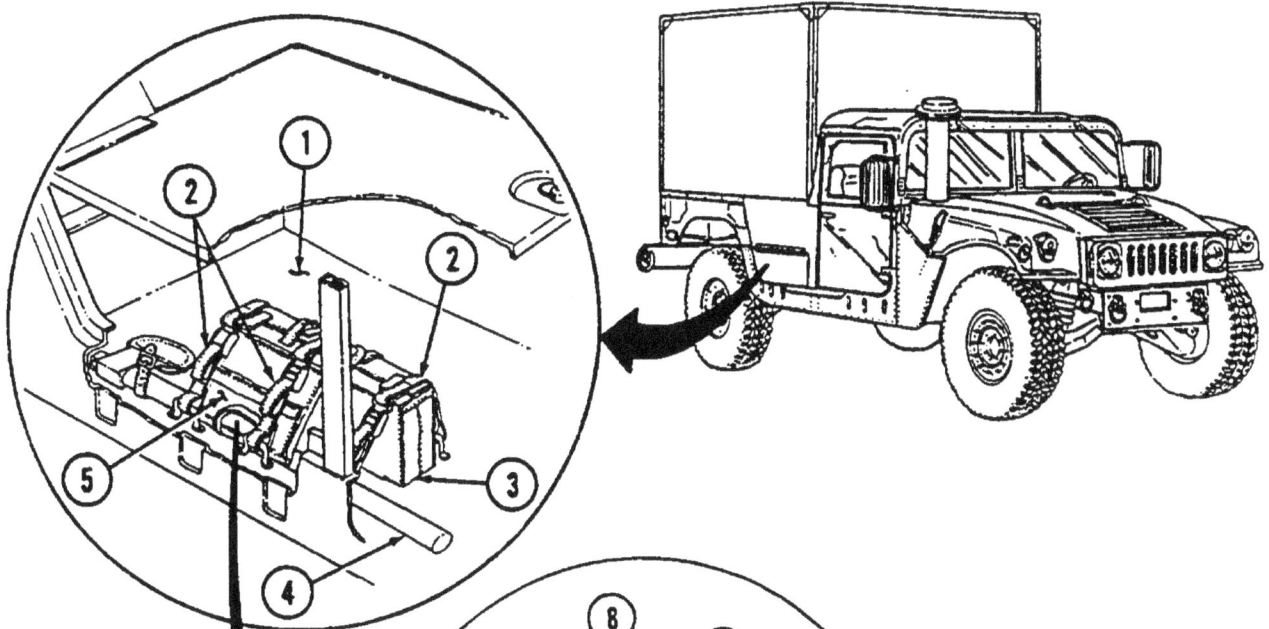

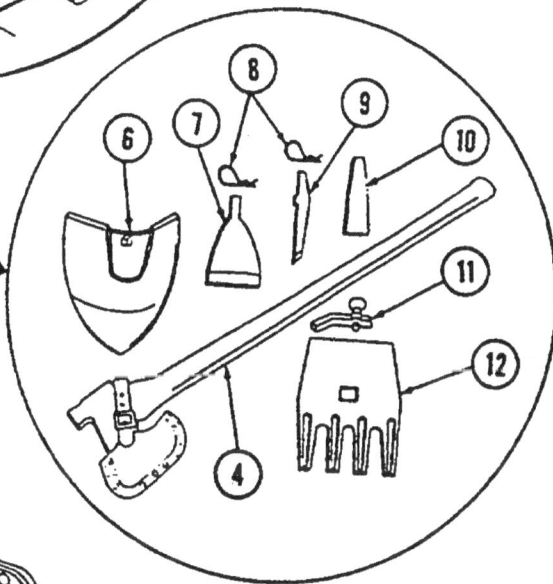

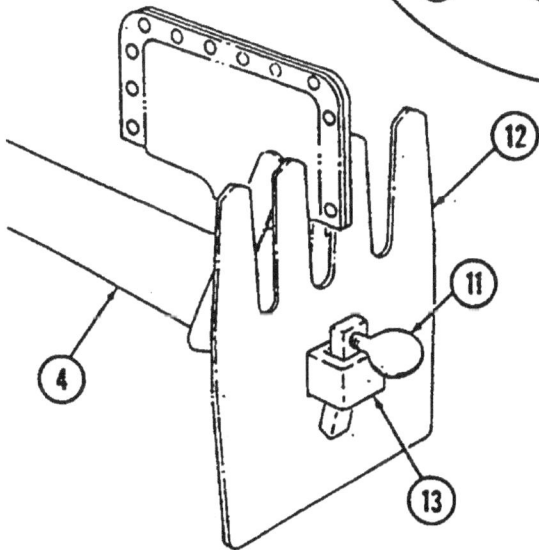

REVERSIBLE RAKE/HOE ATTACHMENT

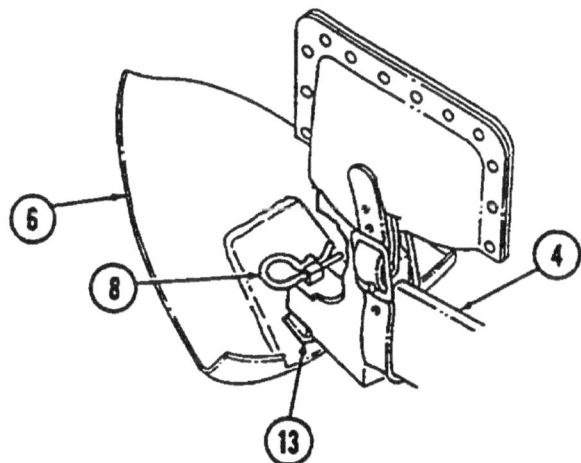

SHOVEL ATTACHMENT

2-24.1. MAX TOOL KIT STOWAGE (M1152, M1152A1, M1165, M1165A1) OPERATION

NOTE

Stowage location on M1165 and M1165A1 vehicles is in left wheelwell.
Stowage location on M1152 and M1152A1 vehicles is in right wheelwell.
The following procedure is for M1152 and M1152A1 vehicles.

a. Tool Removal.

(1) Disconnect two strap assemblies (4) in right footwell (3) and remove ax (1) and max tool kit case (2).

(2) Remove the following attachments from max tool kit case (2):

- Reversible rake/hoe attachment (11)
- Rake/hoe thumbscrew attachment (10)
- Shovel attachment (5)
- Six safety locking pins (7)
- Broad pick attachment (9)
- Pick attachment (8)
- Mattock attachment (6)

NOTE

- Using the ax and attachments, the max tool kit can be incorporated into seven basic hand tools.

- Ax blade must be covered with sheath before using kit.

- Read all safety and assembly instructions enclosed with max tool

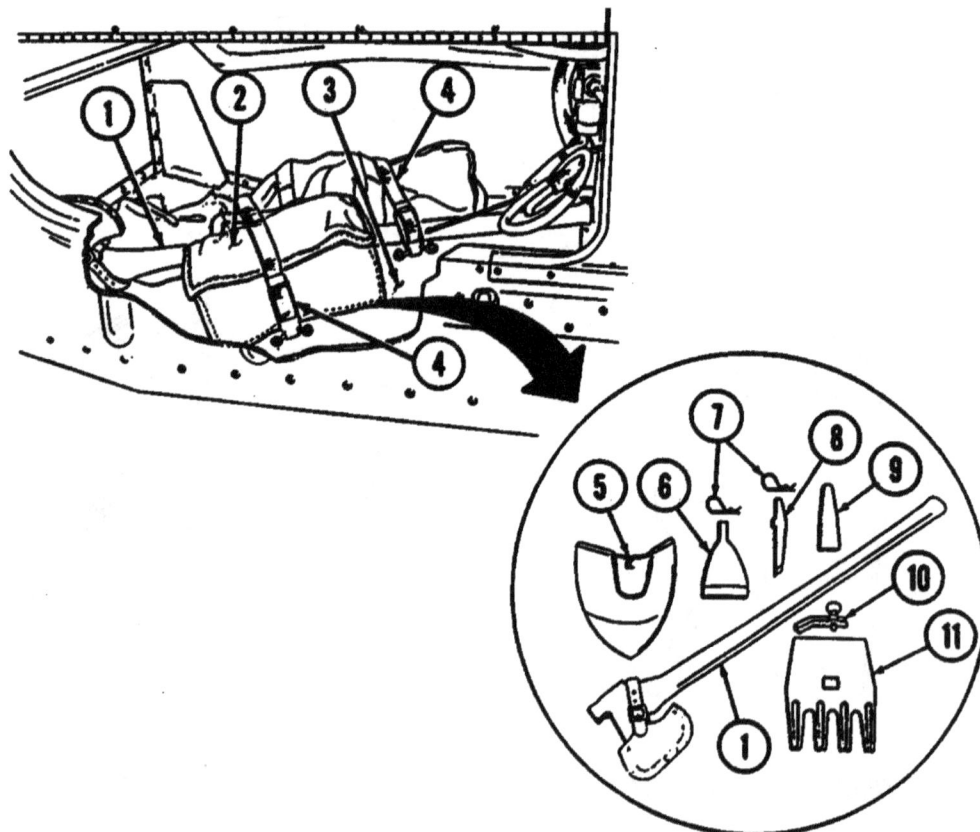

b. Reversible Rake/Hoe Attachment.

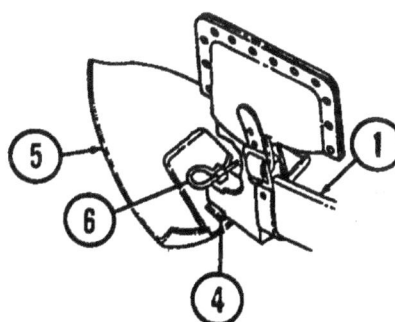

(1) Connect rake/hoe attachment (2) (with either rake or hoe in the working position) into socket (4) on end of ax (1).

NOTE

Thumbscrew must be seated tightly in rake/hoe attachment and socket of ax. Check thumbscrew often to be sure it does not work loose.

(2) Install thumbscrew (3) on rake/hoe attachment (2) and in socket (4) of ax (1). Tighten thumbscrew (3).

(3) To remove rake/hoe attachment (2), remove thumbscrew (3) from rake/hoe attachment (2) and ax (1).

c. Shovel Attachment.

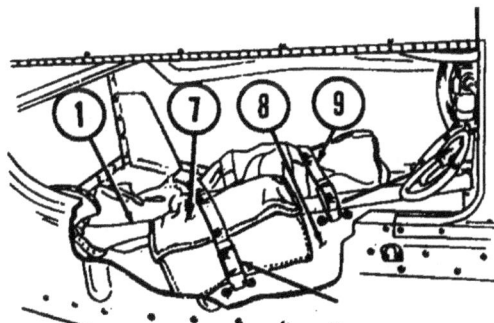

NOTE

Shovel attachment is shown. Broad pick, pick, and mattock attachments are attached basically the same.

(1) Connect shovel attachment (5) into socket (4) on end of ax (1).

(2) Insert safety locking pin (6) into hole in end of taper on shovel attachment (5).

(3) To remove shovel attachment (5), remove safety locking pin (8) from shovel attachment (5) and ax (1).

d. Tool Installation.

(1) Replace attachments in the max tool kit case (7).

(2) Install max tool kit case (7) and ax (1) with two strap assemblies (9) in right footwell (8) of vehicle.

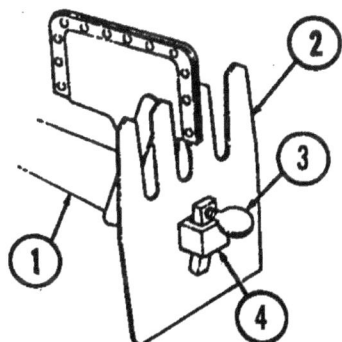

REVERSIBLE RAKE/HOE
ATTACHMENT

SHOVEL ATTACHMENT

2-25. MAX TOOL KIT STOWAGE (M1114, M1151, M1151A1, M1167) OPERATION

a. Tool Removal.

    (1) Open cargo shell door (para. 2-32).

    (2) Lower tailgate (para. 2-22).

    (3) Disconnect four strap assemblies (2) securing ax (3) and max tool carrying bag (1) to tailgate (4). Remove ax (3) and max tool carrying bag (1).

    (4) Remove the following attachments from max tool carrying bag (1):

- Reversible rake/hoe attachment (11)
- Rake/hoe thumbscrew attachment (10)
- Shovel attachment (5)
- Six safety locking pins (7)
- Broad pick attachment (9)
- Pick attachment (8)
- Mattock attachment (6)

NOTE

- Using the ax and attachments, the max tool kit can be incorporated into seven basic hand tools.
- Ax blade must be covered with sheath before using kit.
- Read all safety and assembly instructions enclosed with max tool kit before using kit.

b. Reversible Rake/Hoe Attachment.

    (1) Connect rake/hoe attachment (11) (with either rake or hoe in the working position) into socket (12) on end of ax (3).

NOTE

Thumbscrew must be seated tightly in rake/hoe attachment and socket of ax. Check thumbscrew often to be sure it does not work loose.

    (2) Install thumbscrew (10) on rake/hoe attachment (11) and in socket (12) of ax (3). Tighten thumbscrew (10).

    (3) To remove rake/hoe attachment (11), remove thumbscrew (10) from rake/hoe attachment (11) and ax (3).

c. Shovel Attachment.

NOTE

Shovel attachment is shown. Broad pick, pick, and mattock attachments are attached basically the same.

    (1) Connect shovel attachment (5) into socket (12) on end of ax (3).

    (2) Insert safety locking pin (7) into hole in end of taper on shovel attachment (5).

    (3) To remove shovel attachment (5), remove safety locking pin (7) from shovel attachment (5) and ax (3).

d. Tool Installation.

    (1) Replace attachments in max tool carrying bag (1).

    (2) Install max tool carrying bag (1) and ax (3) on tailgate (4) with four strap assemblies (2).

    (3) Raise tailgate (para. 2-22).

    (4) Close cargo shell door (para. 2-32).

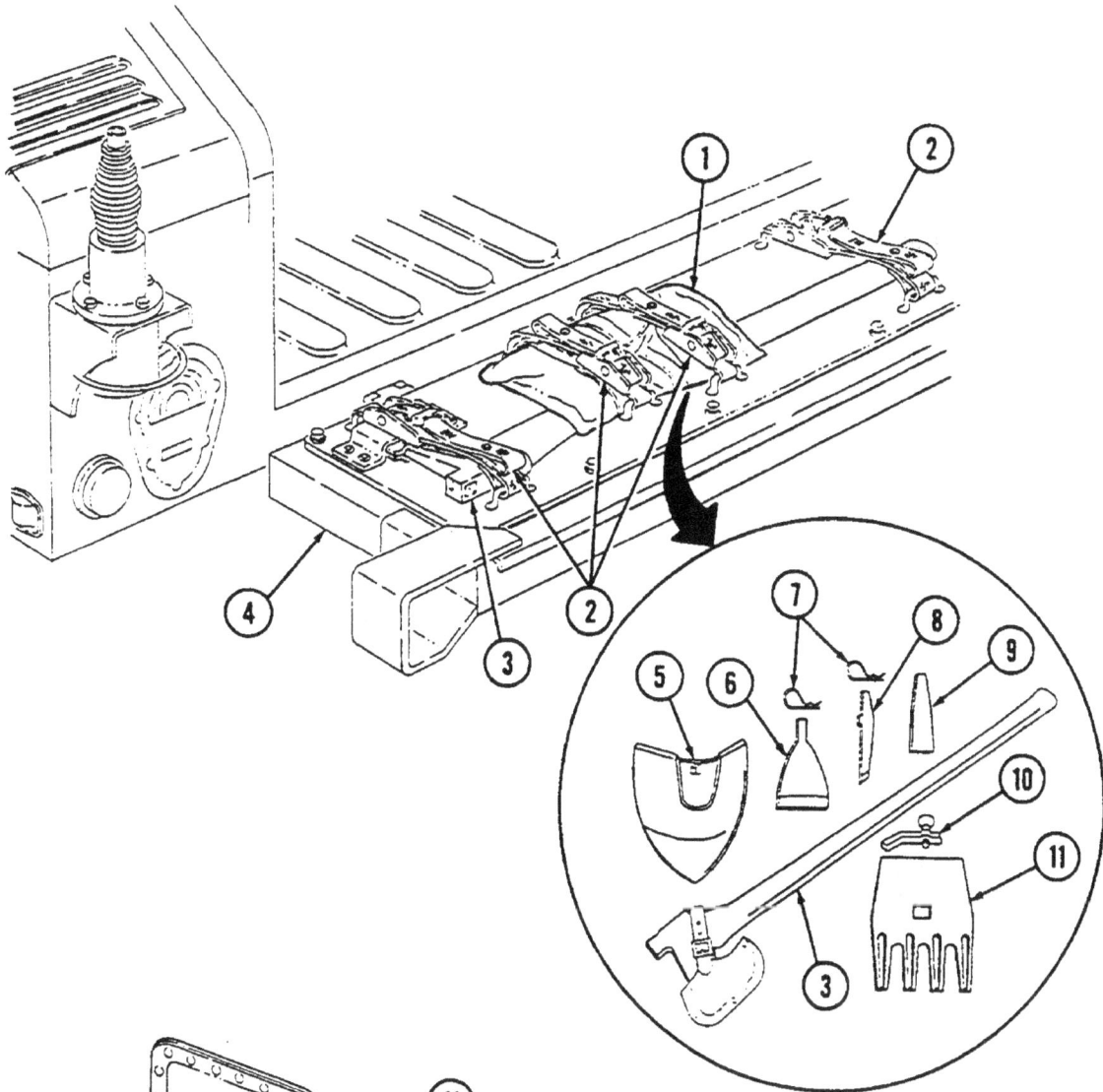

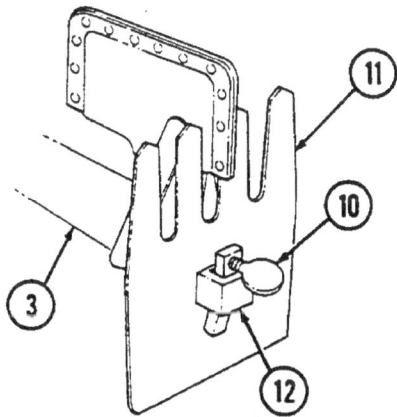

**REVERSIBLE RAKE/HOE
ATTACHMENT**

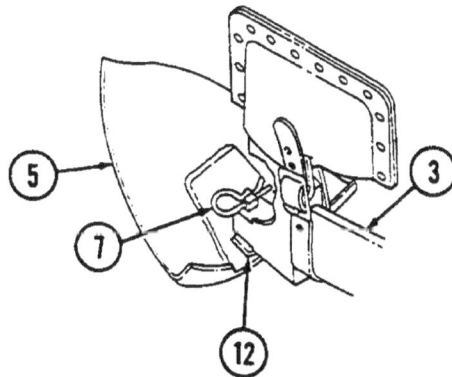

SHOVEL ATTACHMENT

2-26. TOWING OPERATION

CAUTION

- Before initiating vehicle recovery, operator should be familiar with basic vehicle recovery techniques and precautions. Refer to FM 20-22, Vehicle Recovery Operations.
- Do not exceed a towing speed of 30 mph (48 kph), or a towing distance of 50 mi (48 km), without first removing the front propeller shaft and/or rear propeller shaft, as specified in table 2-3. Failure to remove the necessary propeller shafts may result in damage to the transmission and/or transfer case.
- Do not exceed 15 mph (24 kph) towing speed when towing a shelter vehicle. Failure to comply may cause damage to equipment.
- Avoid sharp turns and U-turns when towing. Failure to comply may cause damage to equipment.

NOTE

- Towing operations for ECV vehicles are basically the same.
- If propeller shafts are to be removed, notify unit maintenance personnel.
- Towing pintle (front bumper) provides improved driver control when moving trailers in hard-to-maneuver areas and during aircraft loading operations.

Table 2-3. Towing Operations.

Vehicle Towing Mode	Prop Shafts
Rear wheels up	Front off
Front wheels up	Rear off
Four wheels on ground	Front and rear off

a. **Towing Vehicle from Front (Four Wheels on Ground with Like Vehicle).**

CAUTION

- Always use a towbar when towing vehicle. Failure to do so may cause damage to equipment.
- Place transmission and transfer case in N (neutral) (para. 2-13) prior to towing a HMMWV vehicle from the front by a like vehicle at GVW with all wheels on the ground. Failure to comply may cause damage to one or both vehicles.

(1) Attach towbar (2) to the towing shackle brackets (4) of the vehicle to be towed and to the pintle hook (1) of the towing vehicle.

(2) Attach safety chain (5) to vehicle frames directly behind bumper (3) or bumperette (6). Let safety chain (5) dip to about 1 ft (30.5 cm) from the ground.

(3) Place transmission and transfer case shift levers in N (neutral) (para.2-13).

(4) Turn on hazard warning lights on both towing and disabled vehicles (para. 2-13).

(5) Depress parking brake release button and release parking brake lever (para. 2-15).

(6) Proceed with towing operation.The vehicle is capable of towing a vehicle of similar weight, fully loaded, for a distance of 50 mi (80 km).

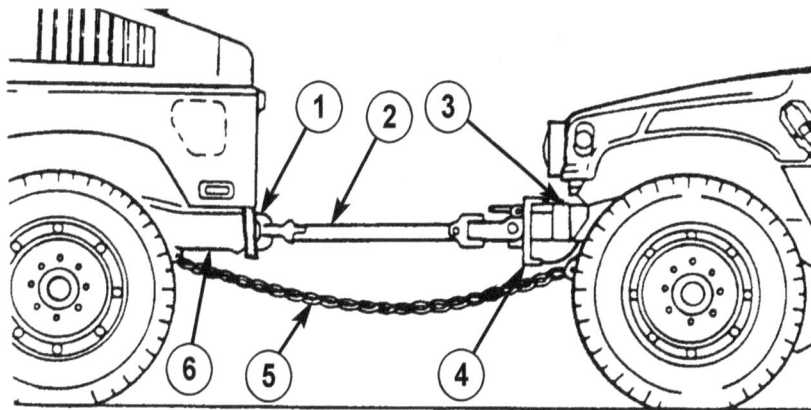

b. Towing Vehicle from Front (Front Wheels Up).

NOTE

- Towing operations for ECV vehicles are basically the same.

- Ensure that towbar is connected to towing brackets as shown in illustration.

(1) Attach towbars (2) to brackets (10) and to wrecker towing pintle.

CAUTION

- Ensure that wrecker hoisting boom and hook are centered over the lifting shackles. Failure to do this may result in difficult turning during towing operations.

- Ensure safety chain is connected correctly to vehicles equipped with a front winch. Failure to do so may result in damage to front winch.

(2) Install chain assembly (8) through lifting shackles (7) and attach chain assembly (8) to the wrecker's hoisting hook (9).

(3) Secure safety chain (5) to towed vehicle's frame and to wrecker's rear tiedown bracket. Let safety chain (5) dip to about 1 ft (30.5 cm) from the ground.

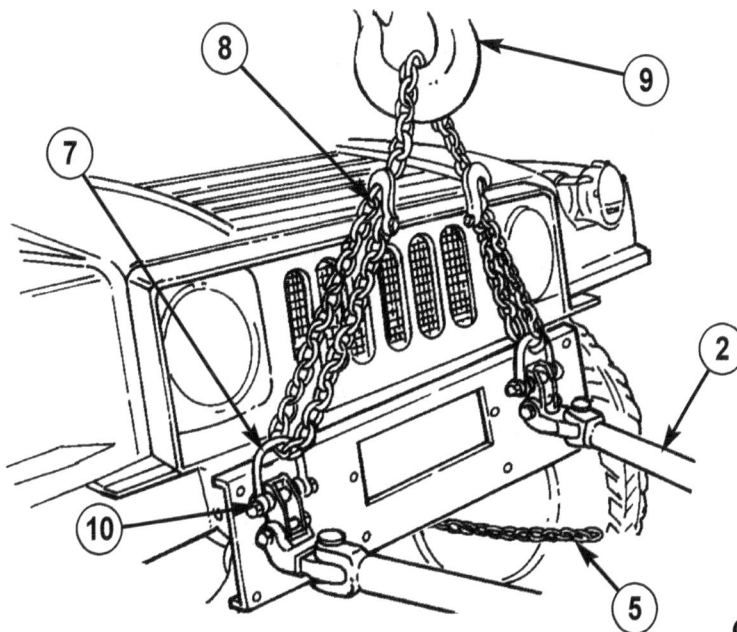

2-26. TOWING OPERATION (Cont'd)

(4) Hoist vehicle to be towed.

(5) Place transmission and transfer case shift levers in N (neutral) (para. 2-13).

(6) Turn on hazard warning lights on both towing and disabled vehicles (para. 2-15).

(7) Depress parking brake release button and release parking brake lever (para. 2-13).

(8) Lift front wheels from ground.

(9) Proceed with towing operation.

c. **Towing Vehicle from Rear (Rear Wheels Up).**

NOTE

It will be necessary to remove the wrecker lifting eyes (shackle) prior to attaching the towbar arms.

(1) Attach the eye of the towbar (3) to the pintle (4) of the vehicle requiring towing and the towbar arms to the wrecker rear lifting eye attaching bracket.

CAUTION

• Ensure safety chain is connected correctly to vehicles equipped with a rear winch. Failure to do so may result in damage to rear winch.

• Ensure that wrecker hoisting boom and hook are centered over the lifting shackles. Failure to do this may result in difficult turning during towing operations.

(2) Install chain assembly (1) through the rear lifting shackles (6) and attach the chain assembly (1) to the wrecker's hoisting hook (2).

(3) Secure safety chain (5) to towed vehicle's frame and to wrecker's rear tiedown bracket. Let safety chain (5) dip to about 1 ft (30.5 cm) from the ground.

WARNING

Prior to towing vehicle with rear wheels up, secure steering wheel to prevent front wheels from turning. Failure to do this may cause damage to vehicle and injury or death to personnel.

(4) Secure steering wheel.

(5) Place transmission and transfer case shift levers in N (neutral).

(6) Turn on hazard warning lights on both towing and disabled vehicles (para. 2-15).

(7) Depress parking brake release button and release parking brake lever.

(8) Lift rear wheels from ground.

(9) Proceed with towing operations.

d. Towing Shelter Vehicle from Rear (Rear Wheels Up).

NOTE

It will be necessary to remove the wrecker lifting eyes (shackle) prior to attaching the towbar arms.

(1) Attach the eye of the towbar (3) to the pintle (4) of the vehicle requiring towing and the towbar arms to the wrecker rear lifting eye attaching bracket.

CAUTION

Ensure that wrecker hoisting boom and hook are centered over the lifting shackles. Failure to do this may result in difficult turning during towing operations.

NOTE

If necessary, relocate the rear lifting shackles from the ends of the bumper to the location indicated.

(2) Install the two hooks of chain assembly (1) through the rear lifting shackles (6) and attach the chain assembly (1) to the wrecker's hoisting hook (2).

(3) Secure safety chain (5) to towed vehicle's frame and to wrecker's pintle. Let safety chain (5) dip to about 1 ft (30.5 cm) from the ground.

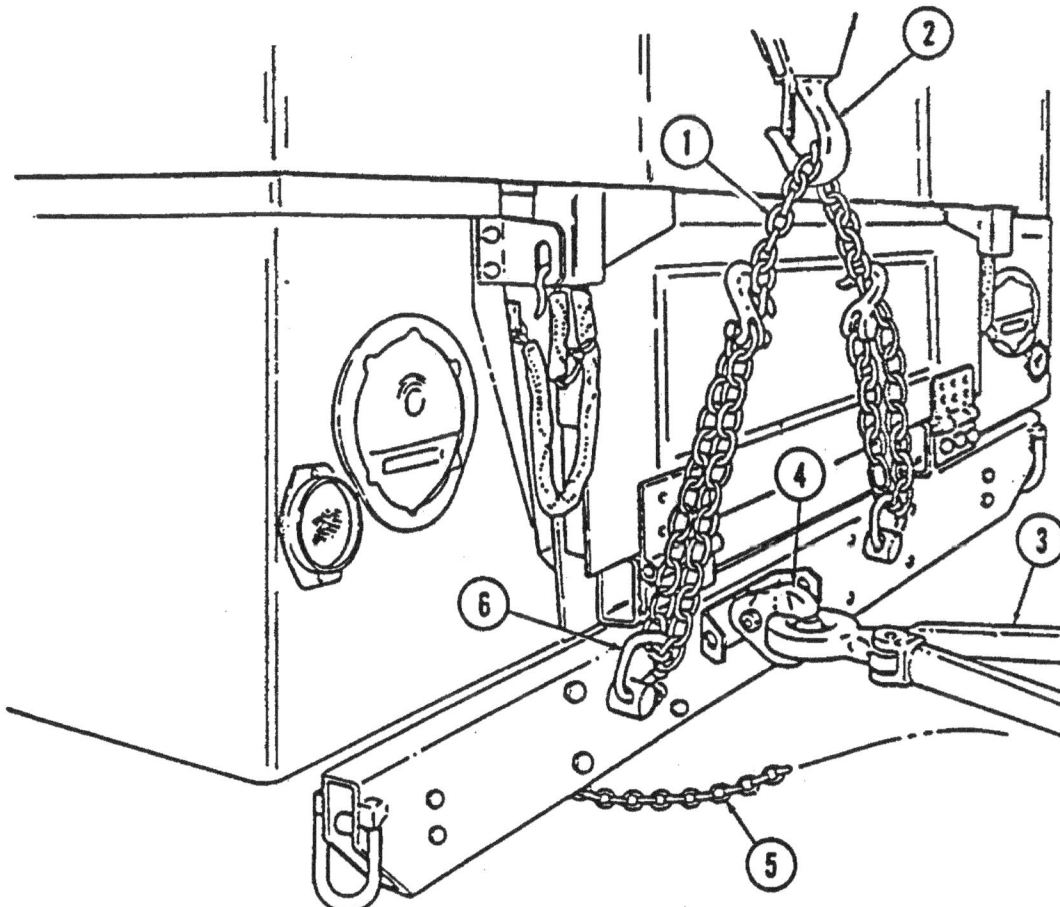

2-26. TOWING OPERATION (Cont'd)

WARNING

Prior to towing vehicle with rear wheels up, secure steering wheel to prevent front wheels from turning. Failure to do this may cause damage to vehicle and injury or death to personnel.

(4) Secure steering wheel.

(5) Place transmission and transfer case shift levers in N (neutral) (para. 2-13).

(6) Turn on hazard warning lights on both towing and disabled vehicles (para. 2-15).

(7) Depress parking brake release button and release parking brake lever (para. 2-13).

(8) Lift rear wheels from ground.

CAUTION

Do not exceed 15 mph (24 kph) towing speed when towing a shelter vehicle. Failure to comply may cause damage to equipment.

(9) Proceed with towing operations.

2-27. TRAILER TOWING

WARNING

- Towing trailers too large or too small for the vehicle capacity is dangerous.These trailers do not track vehicle properly, cargo shifting occurs, and the likelihood of trailer capsizing during movement is increased.This could result in damage to equipment and injury or death to personnel.

- M1113 and M1114 vehicles are not authorized to tow M1101 or M1102 trailers unless they have been retrofitted with MWO 9-2320-280-20-6;Airlift Bumper Reinforcement.Towing these trailers without the airlift bumper reinforcement MWO applied could result in damage to equipment or injury or death to personnel.

CAUTION

Be sure to close tailgate before towing trailer. Failure to do so may cause damage to the master cylinder.

NOTE

- Towing any trailer with a HMMWV other than the authorized M116 Series, M101 Series, M1101, and M1102 trailers or a M102 howitzer is not authorized.

- When towing a trailer, the maximum safe slope is reduced from 40% to 30%.

ECV vehicles are capable of towing a M116 Series, M101 Series, M1101, or M1102 trailers and the M102 howitzer. ECV vehicles have a maximum towed load capacity of 4,200 lb (1,907 kg) as specified in table 2-3.1.

Trailer payloads should be evenly distributed to prevent excessive tongue loads.

Table 2-3.1 Vehicle Trailer Towing Requirements.

HMMWV Model Number	Vehicle Description	MWO Requirements				Remarks
		M1101	M1102	<3400 lb (1544 kg)	<4200 lb (1907 kg)	
M1113	S250 Shelter Carrier	A	A	A	A	See Note
M1114	Up-Armored	A	A	A	A	—
M1151/M1151A1	Armament Carrier	B	B	B	B	—
M1152/M1152A1	Expanded Capacity	B	B	B	B	—
M1165/M1165A1	Command and Control	B	B	B	B	—
M1167	TOW ITAS Carrier	B	B	B	B	—

A = MWO 9-2320-280-20-6
   <3400 lb (1544 kg) = Any system mounted on LTT Chassis with GVW of 3400 lb (1544 kg) or less.
   <4200 lb (1907 kg) = Any system mounted on LTT Chassis with GVW of 4200 lb (1907 kg) or less.
B = Comes equipped. No MWO required.

Note: Pintle extensions are required on vehicles with Shelter Standardized Integrated Command Post System (SICPS) type II. If vehicle is equipped with SICPS (M788) mounted, MWO 9-2320-280-20-6 is not required.

2-28. ELECTRIC WINCH OPERATION (M1113, M1114)

NOTE

The electric winch is an optional kit which may or may not be installed on the vehicle.

a. General. The vehicle electrical system is used to power the winch. It is recommended to have engine running while operating the winch so the alternator recharges the battery. Increased engine rpm can be maintained by use of the hand throttle. When engaging or disengaging the clutch, it may be necessary to rotate the drum by hand to align gears.

b. Preparation for Use.

(1) Park vehicle directly facing object to be winched.

(2) Apply parking brake (para. 2-13).

(3) Start engine (para. 2-12).

(4) Chock wheels.

c. Unwinding Winch Cable.

CAUTION

Do not power out winch cable for more than 10 ft (3 m). Use FREE SPOOL for paying out long lengths of winch cable. Failure to FREE SPOOL long lengths of winch cable may cause damage to winch.

(1) Pull lever (4) out to FREE SPOOL.

WARNING

• Wear leather gloves when handling winch cable. Do not handle cable with bare hands. Failure to wear leather gloves may cause injury to hands.

• When fully extending winch cable, ensure that four wraps of winch cable remain on drum at all times. Failure to do this may cause injury, death to personnel, or damage to winch.

NOTE

Allow 1 ft (30.5 cm) of slack in winch cable prior to start of winching operations. This allows time for winch motor start up for maximum pulling power.

(2) Pull cable (5) out by hand to desired length. Connect to load leaving 1 ft (30.5 cm) of slack in cable (5).

d. Pulling Load.

WARNING

Direct all personnel to stand clear of winch cable during winch operation. A snapped winch cable may cause injury or death.

NOTE

Refer to table 1-11, Winch Data, for pulling load capacity.

(1) Install remote control switch (1) by connecting plug (2) to connector (3).

(2) Push lever (4) in to ENGAGED.

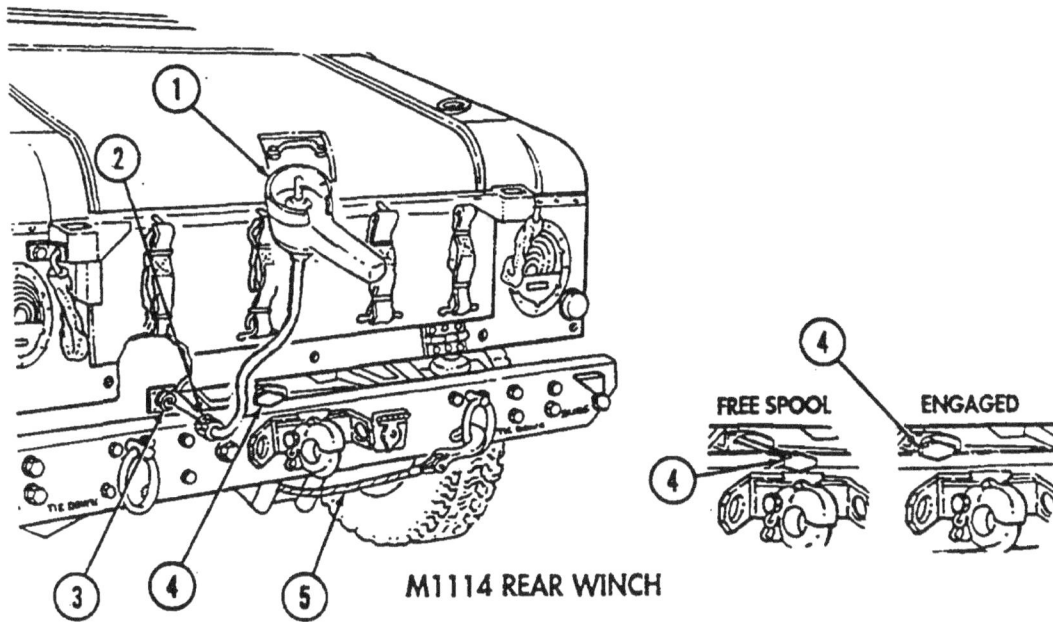

FREE SPOOL   ENGAGED

M1114 REAR WINCH

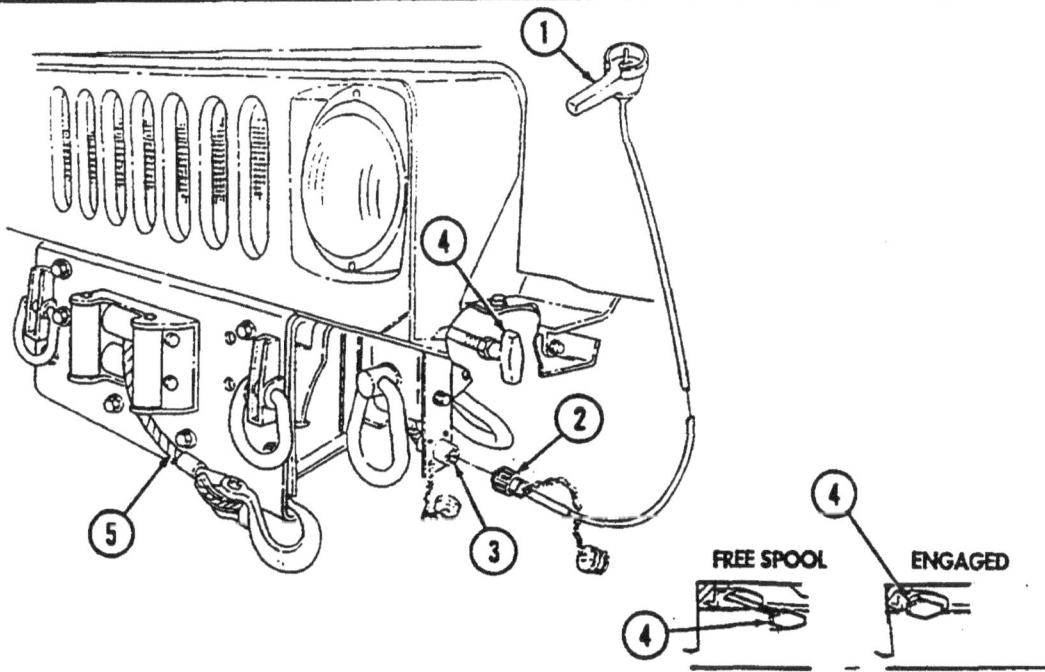

FREE SPOOL   ENGAGED

M1113 FRONT WINCH

2-28. ELECTRIC WINCH OPERATION (Cont'd)

d. Pulling Load (Cont'd).

CAUTION

Do not fully apply hand throttle during engine NO LOAD condition. Damage to engine may result.

NOTE

- The electric winch is equipped with an electronic current limiter switch to prevent winch overload. If winch stops repeatedly during operation and restarts in approximately five seconds, the electronic current limiter is being activated, indicating an overload condition.
- The electric winch is equipped with a thermal cutoff switch to prevent winch from overheating. If winch stops during operation, and does not restart within five seconds, wait approximately two minutes to let winch cool off and allow thermal switch to reset. If after five minutes winch is still inoperative, notify unit maintenance.
- Engine speed is maintained for battery charging only and will not change winch operating speed.

(3) Pull hand throttle (1) out until desired engine speed is obtained.

(4) Operate remote control switch (2) to IN or OUT until load has been retrieved.

e. Securing Winch After Operation.

CAUTION

Winch cable must be wound onto drum under a load of at least 500 lb (227 kg) or outer wraps will draw into the inner wraps and damage winch cable.

(1) Wind winch cable (6) until hook (7) is 4 ft (1.2 m) from cable guide (8).

(2) Pull lever (5) out to FREE SPOOL and rotate drum by hand to retrieve the remaining cable (6).

(3) Remove remote control switch (2) by disconnecting plug (3) from connector (4).

(4) Push lever (5) in to ENGAGED.

(5) Release hand throttle (1).

FREE SPOOL   ENGAGED

M1114 REAR WINCH

FREE SPOOL   ENGAGED

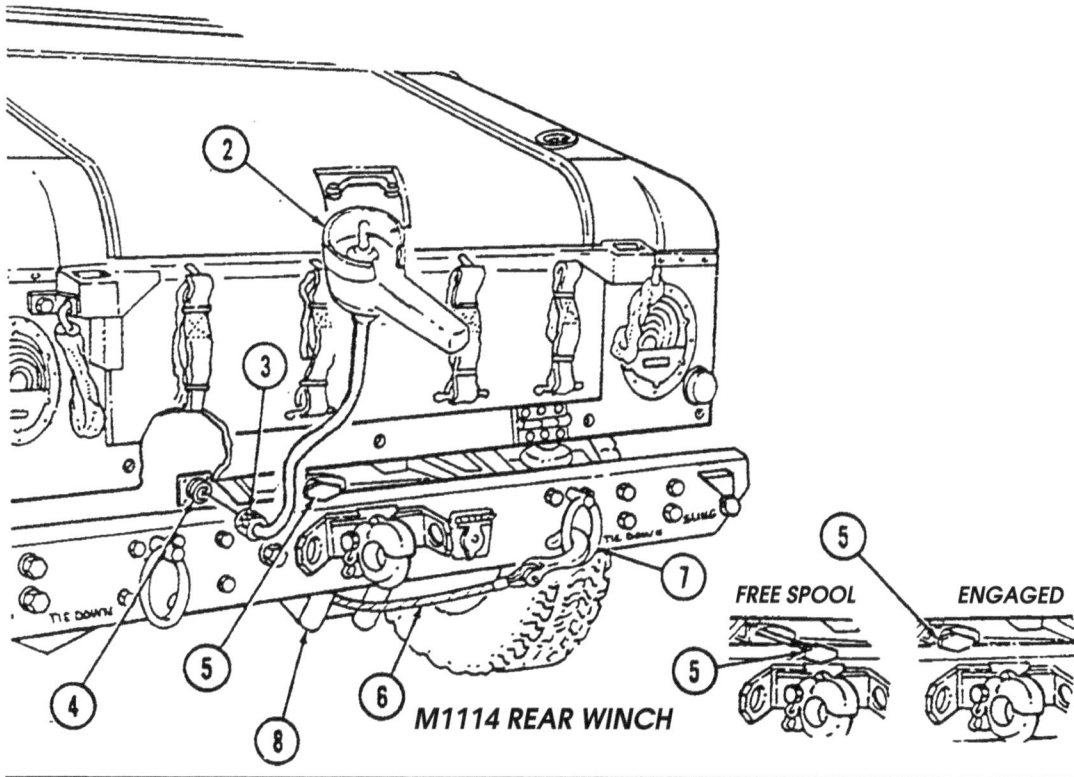

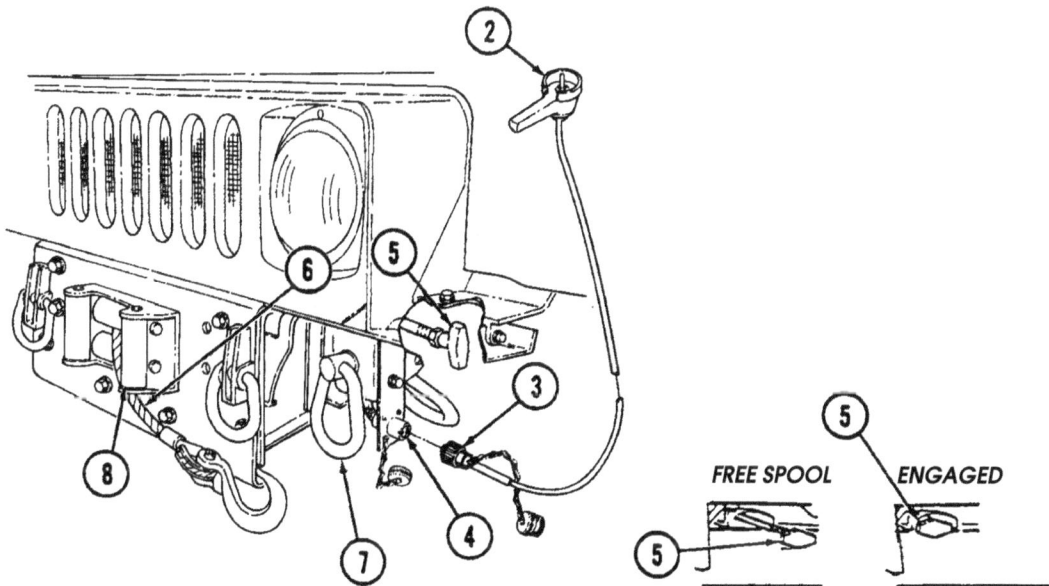

M1113 FRONT WINCH

2-28.1. HYDRAULIC WINCH OPERATION

a. General. Vehicle must be running for power steering system to power winch. Applying vehicle brakes or turning steering wheel can be done while operating winch, but this may cause winch to stop. To achieve 100% pulling power, do not brake or steer while operating winch. It is possible to pay out vehicle's winch cable without vehicle running. This is accomplished with the freespool method. Keep no less than five cable wraps on drum during all winch operations.

b. Preparation for Use.

(1) Park vehicle directly facing object to be winched and place transmission in P (park).

(2) Place transfer selector lever in low lock.

(3) Apply parking brake.

(4) Chock wheels.

(5) Start engine (para. 2-12).

c. Unwinding Winch Cable.

(1) Move selector levers (2) and (3) to FREE. This setting on winch is FREESPOOL.

WARNING

- When fully extending winch cable, ensure that five wraps of winch cable remain on drum at all times. Failure to do this may cause damage to equipment or injury or death to personnel.
- Wear leather gloves when handling winch cable. Do not handle cable with bare hands.
- Do not even slide gloved hands across winch cable. Injury can result.

NOTE

Allow 1 ft. (30.5 cm) of slack in winch cable prior to start of winching operations. This allows time for winch motor start up for maximum pulling power.

(2) Pull out winch cable (1) to desired length. Connect to load, leaving 1 ft. (30.5 cm) of slack in cable (1).

(3) If retrieving load by moving vehicle and not winding winch cable, move winch selector levers to WINCH LOCKED UP position.

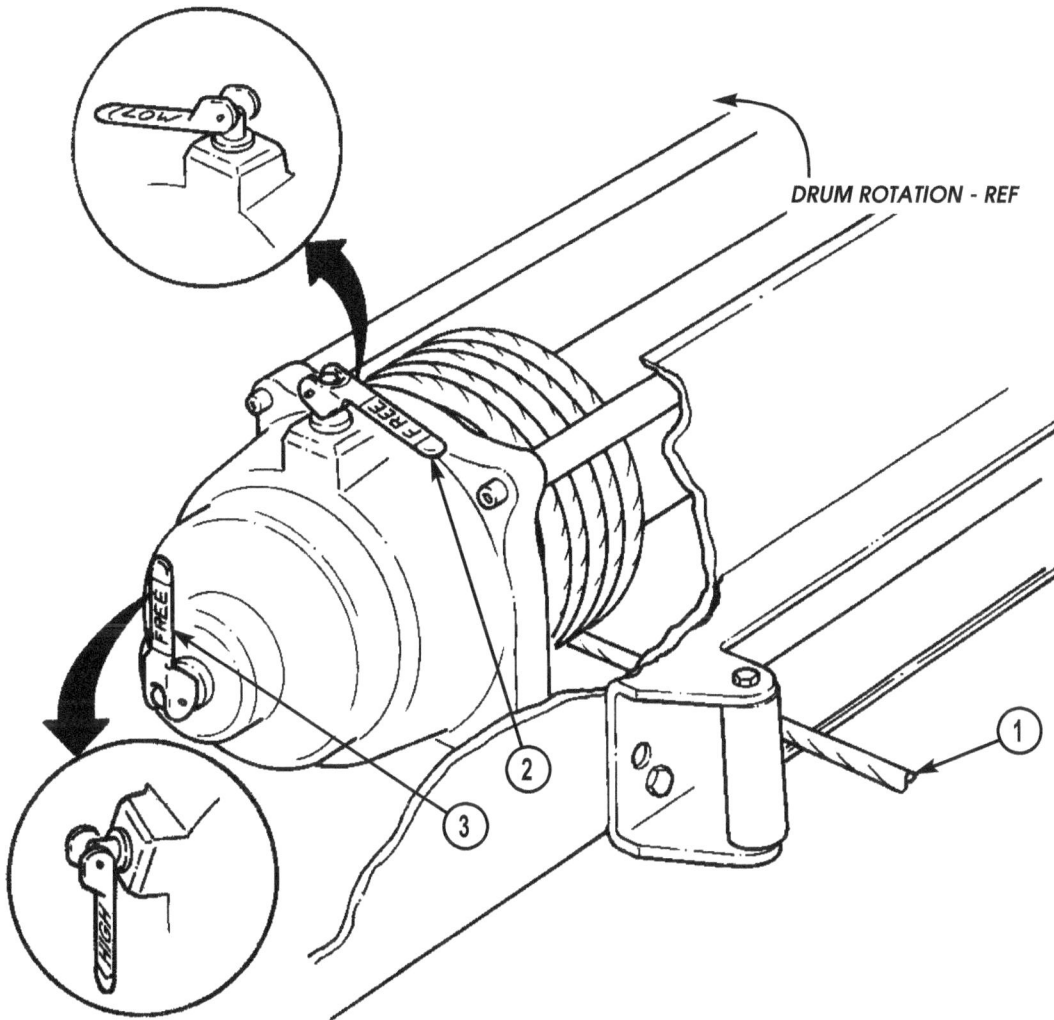

DRUM ROTATION - REF

d. Powering Winch In and Out.

CAUTION

• Do not power out winch cable for more than 10 ft. (3.5 m). Use free spool for paying out long lengths of winch cable. Failure to free spool long lengths of winch cable will cause damage to winch.
• Wind winch cable onto drum under a load of at least 500 lb (227 kg) or outer wraps will draw into inner wraps and damage winch cable.
• Do not move winch selector levers with load on winch or when powering winch in or out. Moving winch selector levers with load on winch or when powering winch may cause damage to winch.

NOTE

• Refer to table 1-11.1 (optional 10,500 lb.), Winch Data for Pulling Capacity.
• Wrapping winch cable is essential and should be done as soon as winch is applied to vehicle. This procedure ensures winch cable is tight on drum and should be accomplished during monthly PMCS, or if winch cable has been powered or freespooled out more than 50% of cable length.

(1) Align drum gears.

(a) Remove remote control (6) from storage compartment (7) under passenger seat.

(b) Remove cap (10) from controller plug (8) and connect hand controller connector (9) to controller plug (8).

NOTE

LOCK LOW GEAR is the preferred winch speed for all recovery and wrapping operations.

(c) Move winch selector lever (1) to LOW. Move winch selector lever (5) to FREE.

WARNING

Direct all personnel to stand clear of winch cable during winch operation. A snapped winch cable may cause injury or death.

CAUTION

Do not move winch levers with load on winch or when powering winch in or out.

(d) Power winch in and out with no load on cable for 1/2 second at a time until full engagement is accomplished.

(2) Attach winch cable (3) to load. Press OUT remote control button (11) or IN remote control button (12) until load has been retrieved.

NOTE

This winch uses right-lay bottom feed method of winch cable on winch drum. Ensure cable wrap begins from right side of winch drum.

(3) Power in winch cable (3), ensuring cable wraps (2) are tight.

(4) Continue powering in winch cable (3) under load until approximately four feet (1.2 m) from hook. Disconnect load and power in, keeping cable taut, until hook clevis touches fairlead roller (13). Clevis should hang freely with less than one foot (.31 m) of cable exposed.

e. Securing Winch After Operation.

(1) Move levers (1) and (5) to FREE positions and rotate drum by hand to retrieve remaining cable.

(2) Remove hand controller connector (9) from controller plug (8) and place remote control (6) in stowage compartment (7).

(3) Move lever (1) to LOW position and lever (5) to HIGH position to lockup winch (4)

(4) Install cap (10) on controller plug (8).

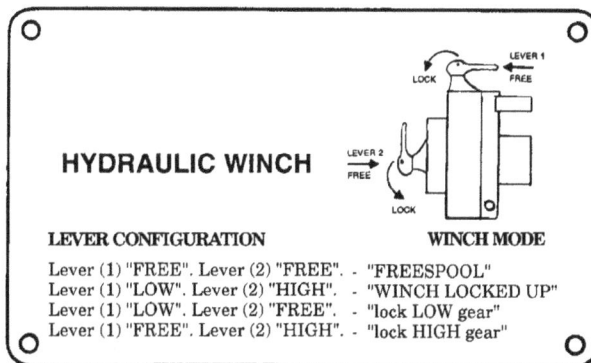

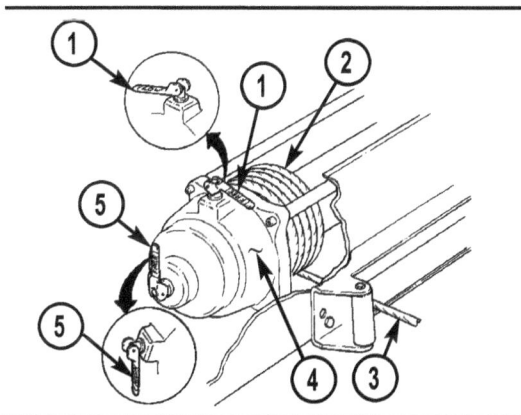 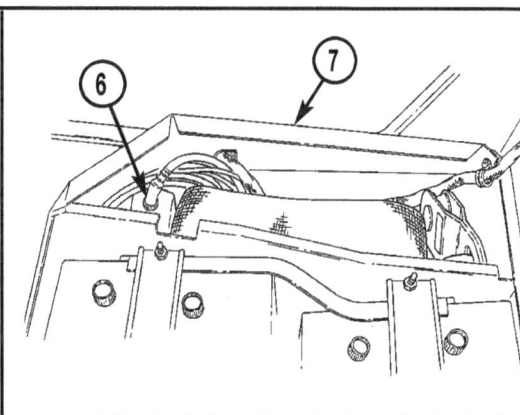

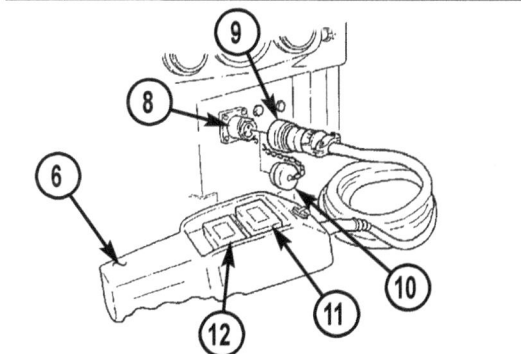 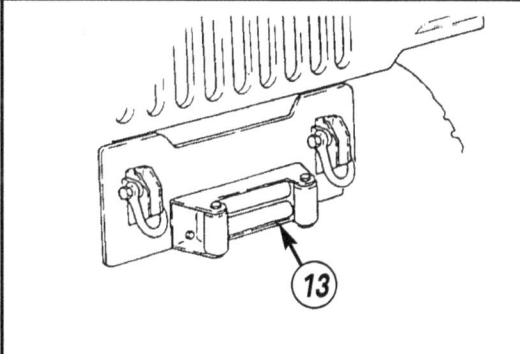

2-29. FIRE EXTINGUISHER OPERATION

a. Before Operation.

(1) Ensure operator PMCS have been accomplished and needle of indicator gauge (1) is in the green zone. Refer to table 2-2.

(2) Refer to para. 2-2a for fire extinguisher stowage location.

WARNING

- Avoid using fire extinguisher in unventilated areas. Prolonged inhalation exposure to extinguishing agent or fumes from burning materials may cause injury to personnel.

- Using fire extinguisher in windy areas will cause rapid dispersal of extinguishing agent and reduce effectiveness in fire control.

b. During Operation.

(1) To operate fire extinguisher (4), remove from stowage bracket. Remove locking pin (2).

(2) Squeeze handles (3) together and direct extinguishing agent to base of flames. To extinguish burning liquid in a container, direct extinguishing agent against inside of container, just above burning liquid.

c. After Operation.

(1) Stow fire extinguisher (4) in stowage bracket.

(2) Notify unit maintenance to replace or recharge fire extinguisher (4) after use.

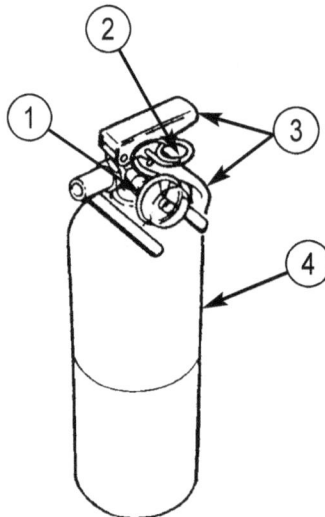

2-30. CREW DOOR OPERATION (M1114)

WARNING

• Use care when opening and closing doors. Do not rest fingers in door opening. Personnel injury may result.

• Ensure crew doors are locked during vehicle operation. Ballistic integrity cannot be maintained if crew doors are unlocked.

NOTE

Crew doors have a double-catch latching mechanism. Ensure door is completely closed and there is no gap between vehicle body and door.

a. Open crew door (5) by turning handle (1) on outside of crew door (5). Pull door closed using strap handle (8) on inside of crew door (5).

b. Lock crew door (5) by pushing sliding door bar (6) into door latch cover (9) while turning sliding door bar (6) downward, toward crew door (5). Unlock crew door (5) by pushing sliding door bar (6) into door latch cover (9) while turning sliding door bar (6) upward.

c. Lift up on window locking bar (10) until it is clear of locking hole (3) in window frame (4). Slowly raise or lower window (2) to desired position. Engage window locking bar (10) into locking hole (3) in window frame (4).

d. To exit vehicle, open crew door (5) by pulling up on crew door release handle (7) located on inside of crew door (5). Close crew door (5).

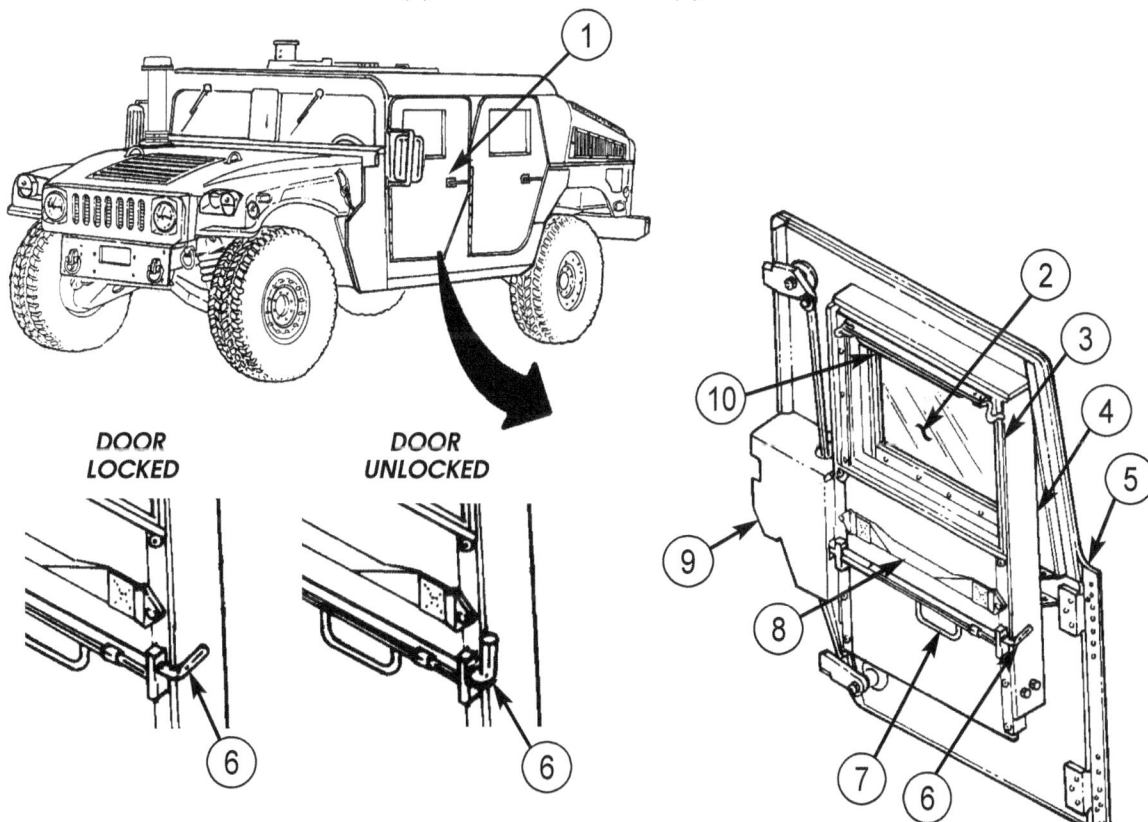

DOOR
LOCKED

DOOR
UNLOCKED

2-30.1. CREW DOOR OPERATION (M1151A1, M1152A1, M1165A1, M1167) (WITH PERIMETER ARMOR)

WARNING

- Use care when opening and closing doors. Do not rest fingers in door opening. Injury to personnel may result.
- Ensure crew doors are locked during vehicle operation. Ballistic integrity cannot be maintained if crew doors are unlocked. Failure to do so may result in injury to personnel.

NOTE

Crew doors have a double-catch latching mechanism. Ensure door is completely closed and there is no gap between vehicle body and door.

a. Open crew door (5) by turning handle (1) on outside of crew door (5). Pull door closed using strap handle (6) on inside of crew door (5).

b. Lock crew door (5) by pushing down combat lock handle (7) until engaging vehicle body by pulling and turning release pin (8). Unlock crew door (5) by pulling release pin (8) out and pulling up combat lock handle (7) until disengaging vehicle body.

c. Lift up on window locking bar (4) until it is clear of locking hole (3) in window frame (10). Slowly slide window (2) to desired position. Engage window locking bar (4) into locking hole (3) in window frame (10).

d. To exit vehicle, open crew door (5) by pushing in crew door release handle (9) located on inside of crew door (5). Close crew door (5).

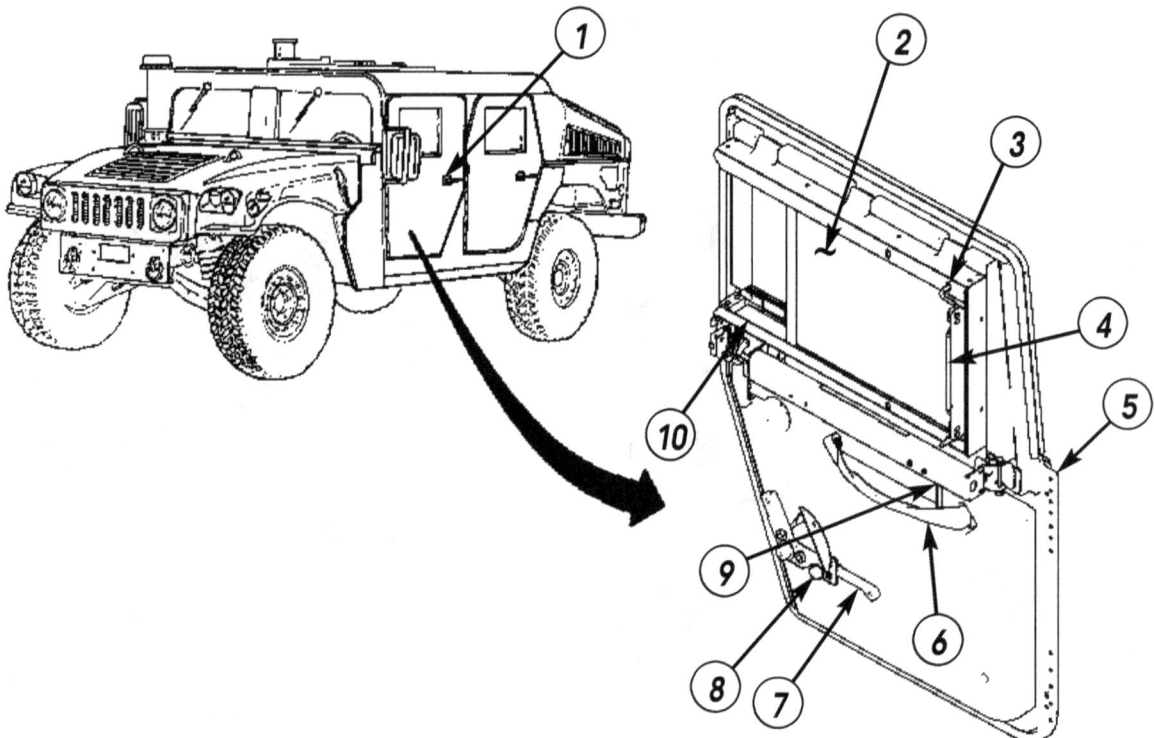

2-30.2. FRAG 5 CREW DOOR OPERATION (M1151A1, M1152A1, M1165A1, M1167)

WARNING

- Use care when opening and closing Frag 5 doors. Do not rest fingers in door opening. Injury to personnel may result.

- Ensure crew doors are locked during vehicle operation. Ballistic integrity cannot be maintained if crew doors are unlocked. Failure to do so may result in injury to personnel.

a. Opening Door from Exterior.

Push exterior handle (1) down toward bottom of vehicle.

b. Opening Door from Interior.

Lift interior handle (2) up toward top of vehicle.

NOTE

Engaging combat lock prevents access to interior of vehicle unless emergency rescue wrench is used.

c. Engaging Combat Lock.

Push interior handle (2) down toward bottom of vehicle.

d. Disengaging Combat Lock.

Lift interior handle (2) up toward top of vehicle.

2-30.2. FRAG 5 CREW DOOR OPERATION (M1151A1, M1151A1, M1165A1, M1167) (Cont'd)

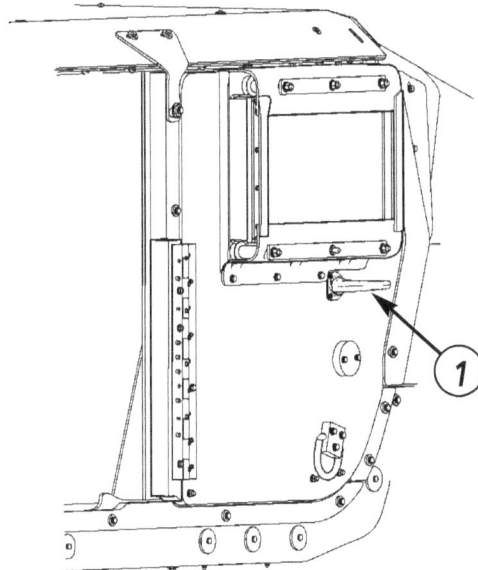

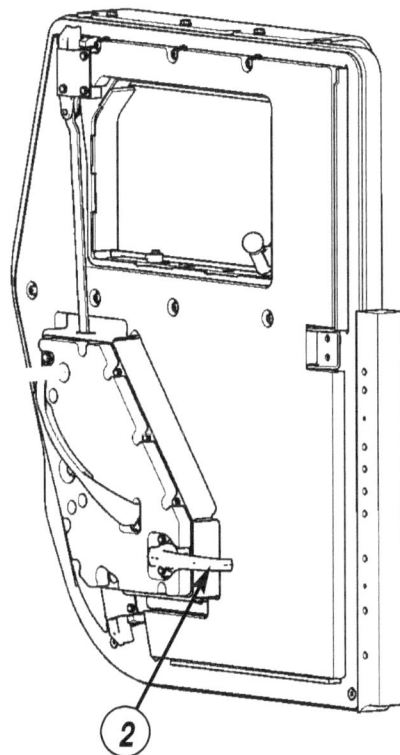

2-30.3. FRAG 5 CREW DOOR OPERATION (M1114)

WARNING

- Use care when opening and closing Frag 5 doors. Do not rest fingers in door opening. Injury to personnel may result.

- Ensure crew doors are locked during vehicle operation. Ballistic integrity cannot be maintained if crew doors are unlocked. Failure to do so may result in injury to personnel.

a. Opening Door from Exterior.
Push exterior handle (1) down toward rear of vehicle.

b. Opening Door from Interior.
Lift interior handle lock (2) and pull interior handle (3) toward rear of vehicle.

NOTE

Engaging combat lock prevents access to interior of vehicle unless emergency rescue wrench is used.

c. Engaging Combat Lock.
Push interior handle lock (2) and push interior handle (3) toward front of vehicle.

2-30.3. FRAG 5 CREW DOOR OPERATION (M1114) (Cont'd)

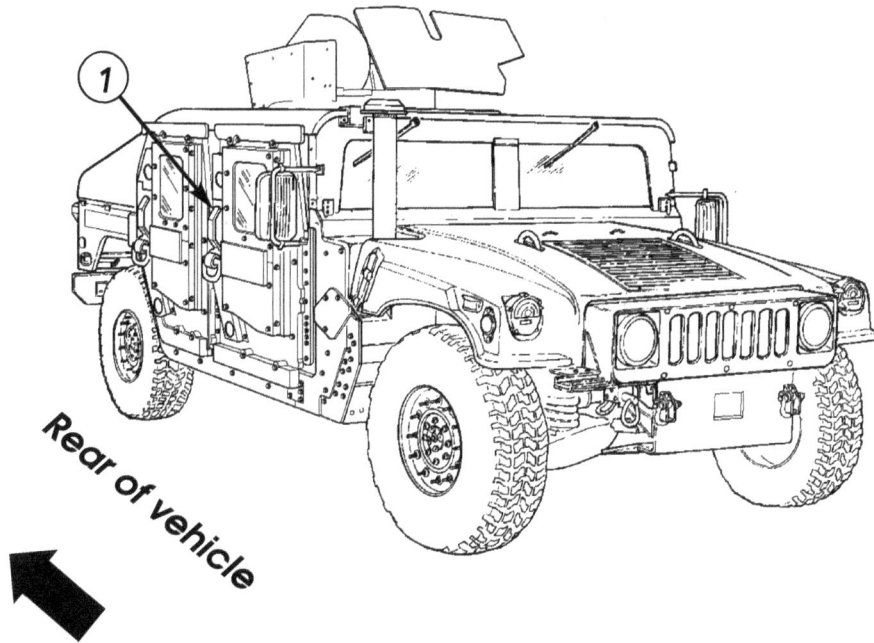

Rear of vehicle

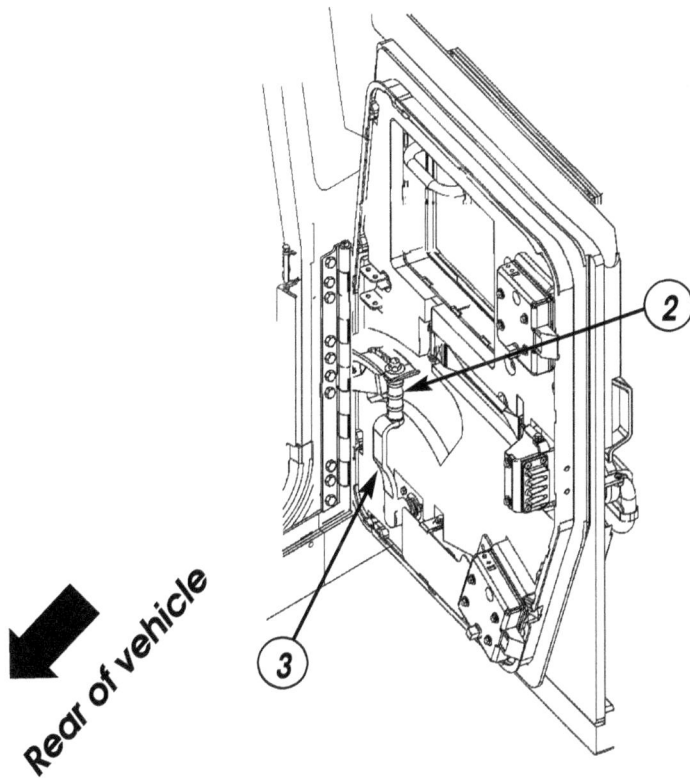

Rear of vehicle

2-31. C-PILLAR DOOR OPERATION (M1114)

NOTE

Second locking point is a safety feature which prohibits door from accidentally closing upon objects protruding through door opening if the door is not in a locked position.

a. To fully open, unlock C-pillar door (3) by depressing locking mechanism (1).

b. Using handle (2), move door (3) to right, keeping locking mechanism (1) depressed while sliding past safety locking point (4).

c. Release and engage locking mechanism (1) at far right. Door (3) is now secured.

d. To partially close, release locking mechanism (1) and engage at the second safety locking point (4).

e. Release and engage locking mechanism (1) at far left safety locking point (4). Door (3) is now secured.

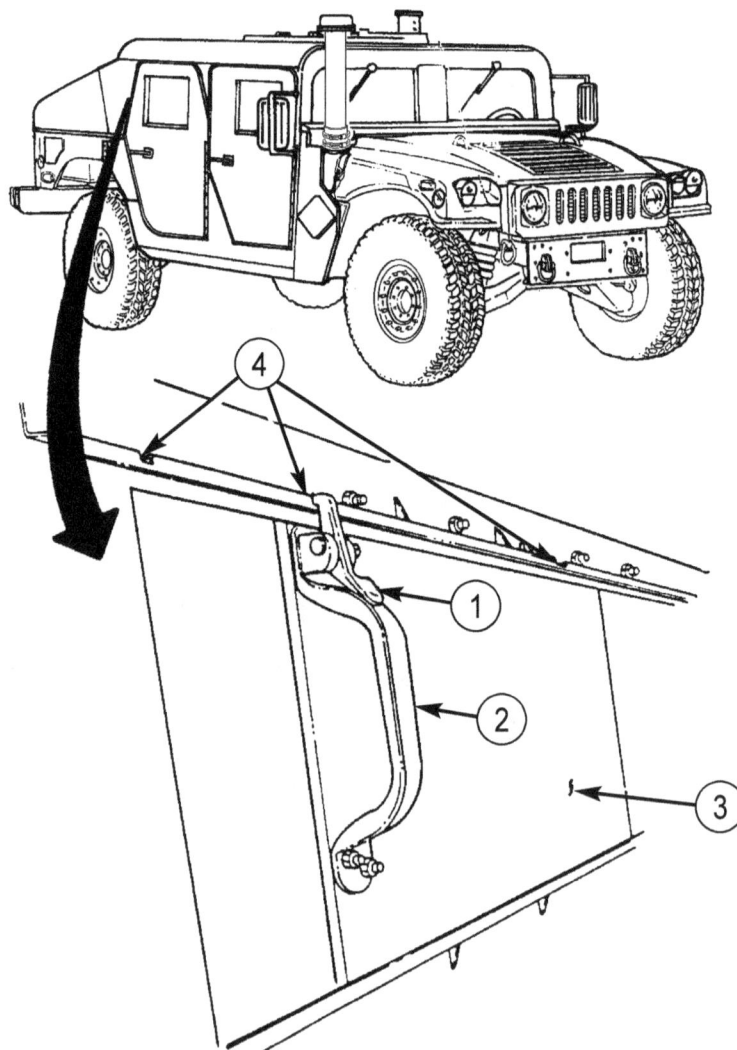

2-31.1. C-PILLAR DOOR OPERATION (M1151A1, M1167) (WITH PERIMETER ARMOR)

NOTE

- Second locking point is a safety feature which prohibits door from accidentally closing upon objects protruding through door opening if the door is not in a locked position.

- Operation of left and right C-pillar doors is basically the same. This procedure covers the right C-pillar door.

a. To fully open, unlock C-pillar door (3) by depressing locking mechanism (1).

b. Using handle (2), move door (3) to right, keeping locking mechanism (1) depressed while sliding past safety locking point (4).

c. Release and engage locking mechanism (1) at far right. Door (3) is now secured.

d. To partially close, release locking mechanism (1) and engage at the second safety locking point (4).

e. Release and engage locking mechanism (1) at far left safety locking point (4). Door (3) is now secured.

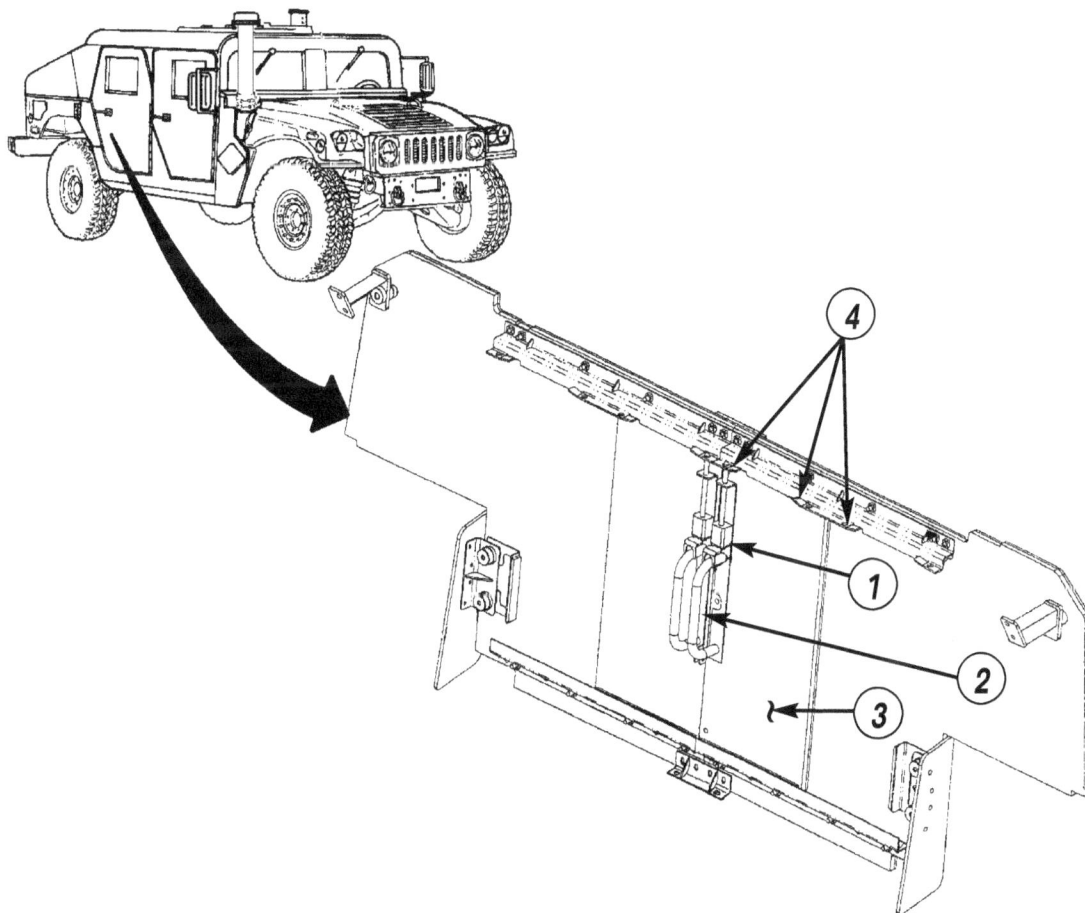

2-31.2. EMERGENCY RESCUE WRENCH OPERATION (OLD CONFIGURATION) (M1151A1, M1152A1, M1165A1)

WARNING

Emergency rescue wrench is for removal of capscrews in an emergency situation. Do not use wrench to install capscrews or for any other purpose. Failure to comply may result in injury to personnel or damage to equipment.

NOTE

Before utilizing this tool, attempt to open door using door handle, as combat locks may not be employed.

(1) Remove emergency rescue wrench (1) from stowage location (2).

(2) Identify door (3) that allows for most immediate access.

NOTE

One capscrew in each group protrudes out from door overlay.

NOTE

There is no order in which the upper and lower set of capscrews are to be removed. They are to be removed as groups, either upper or lower. Always begin with the two countersunk screws and ending with the protruding capscrew.

(3) Remove two countersunk capscrews (5) and capscrew (4).

(4) Open door (3) with handle (6) after capscrews (5) and (4) are removed.

(5) Return emergency rescue wrench (1) to vehicle stowage location (2) and secure.

2-31.2. EMERGENCY RESCUE WRENCH OPERATION
(OLD CONFIGURATION) (M1151A1, M1152A1, M1165A1) (Cont'd)

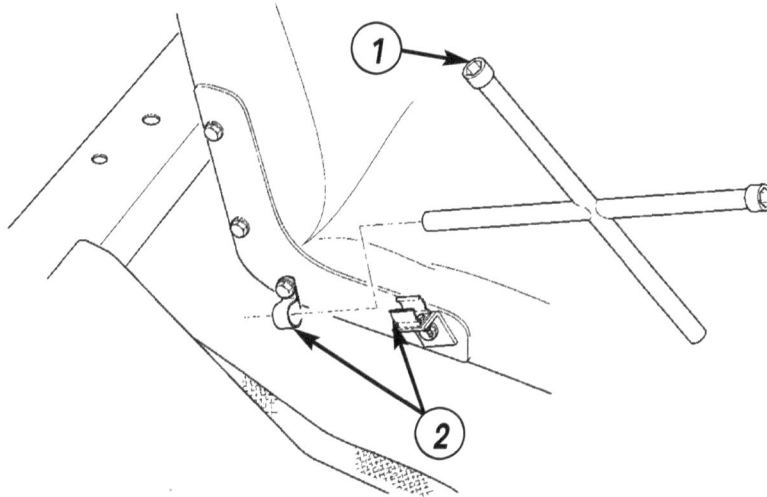

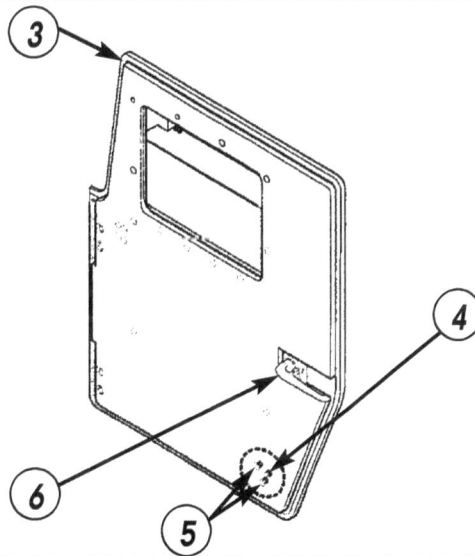

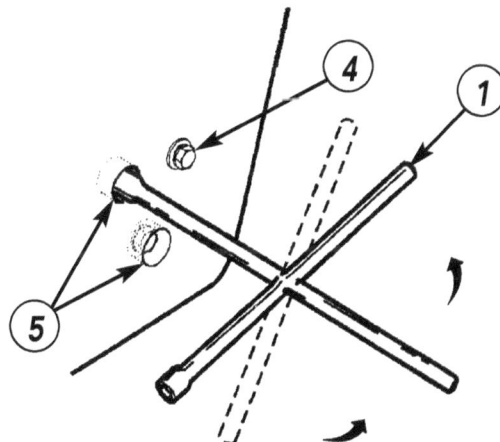

2-31.3. EMERGENCY RESCUE WRENCH OPERATION (NEW CONFIGURATION) (M1151A1, M1152A1, M1165A1, M1167)

WARNING

Door assemblies are extremely heavy.All personnel must take care during removal of door assembly.All personnel not participating in emergency door removal must stand clear of door assembly during removal. Failure to comply may result in injury or death to personnel or damage to equipment.

(1) Attempt to open door (2) using door handle (3) as combat locks may not be employed. If door (2) fails to open, proceed to step 2.

(2) Using emergency rescue wrench (1), remove two capscrews (8), lockwashers (9), washers (10), and combat lock access plate (5) from door (2).

NOTE

Perform step 3 for left side/driver's side of vehicle and step 4 for right side/passenger side of vehicle.

(3) **Left Side/Driver's Side of Vehicle.** Using emergency rescue wrench (1), turn capscrew (4) **counter-clockwise** approximately 1/4 turn until an audible click is heard (combat lock is disengaged).Attempt to open door (2) using door handle (3). If door (2) fails to open, proceed to step 5.

(4) **Right Side/Passenger Side of Vehicle.** Using emergency rescue wrench (1), turn capscrew (4) **clockwise** approximately 1/4 turn until an audible click is heard (combat lock is disengaged).Attempt to open door (2) using door handle (3). If door (2) fails to open, proceed to step 5.

(5) Attach chain (7) to D-ring (6) located on vehicle door (2).Attach chain (7) to rear bumper shackles of pulling vehicle and pull. If door (2) fails to open, proceed to step 6.

(6) Using prybar, attempt to pry door (2) open.

2-31.3. EMERGENCY RESCUE WRENCH OPERATION (NEW CONFIGURATION) (M1151A1, M1152A1, M1165A1, M1167) (Cont'd)

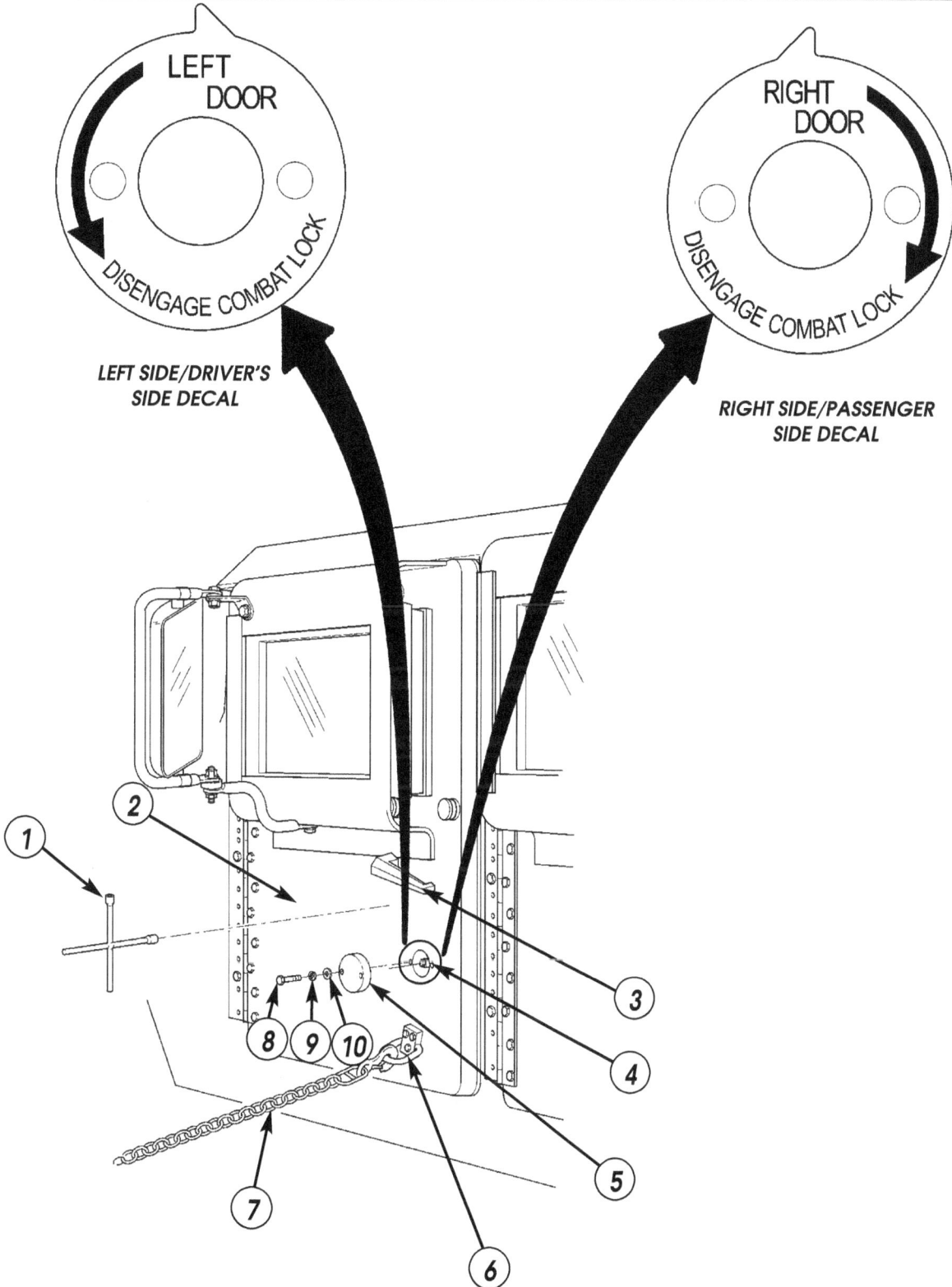

LEFT DOOR

DISENGAGE COMBAT LOCK

LEFT SIDE/DRIVER'S SIDE DECAL

RIGHT DOOR

DISENGAGE COMBAT LOCK

RIGHT SIDE/PASSENGER SIDE DECAL

2-31.4. EMERGENCY FRAG 5 DOOR REMOVAL (M1114)

NOTE

Emergency rescue wrench can be used in place of combat lock lever for the following procedures.

a. **Disengaging Combat Lock from Exterior.**

(1) Remove lanyard pin (5) and pull combat lock lever (1) off door assembly from first responder vehicle.

(2) Slide combat lock lever (1) onto square lug (8) in bottom exterior linkage cover (9) and turn combat lock lever (1) toward front of vehicle.

(3) Slide combat lock lever (1) onto square lug (6) in top exterior linkage cover (7) and turn combat lock lever (1) toward front of vehicle.

b. **Disengaging Combat Lock from Interior.**

(1) Remove lanyard pin (5) and pull combat lock lever (1) off door assembly (2).

(2) Slide combat lock lever (1) onto square lugs (3) and (4) and turn combat lock lever (1) toward rear of vehicle.

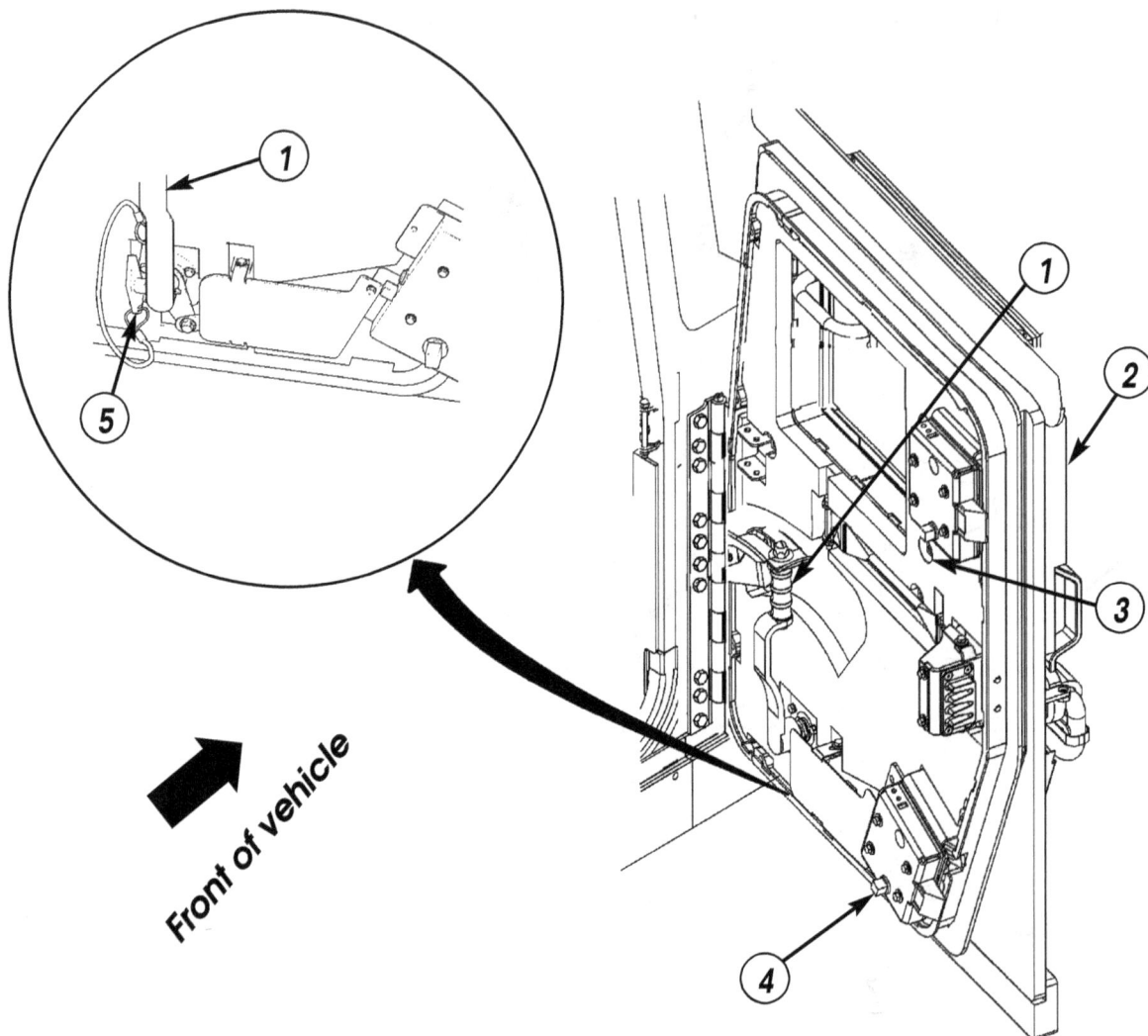

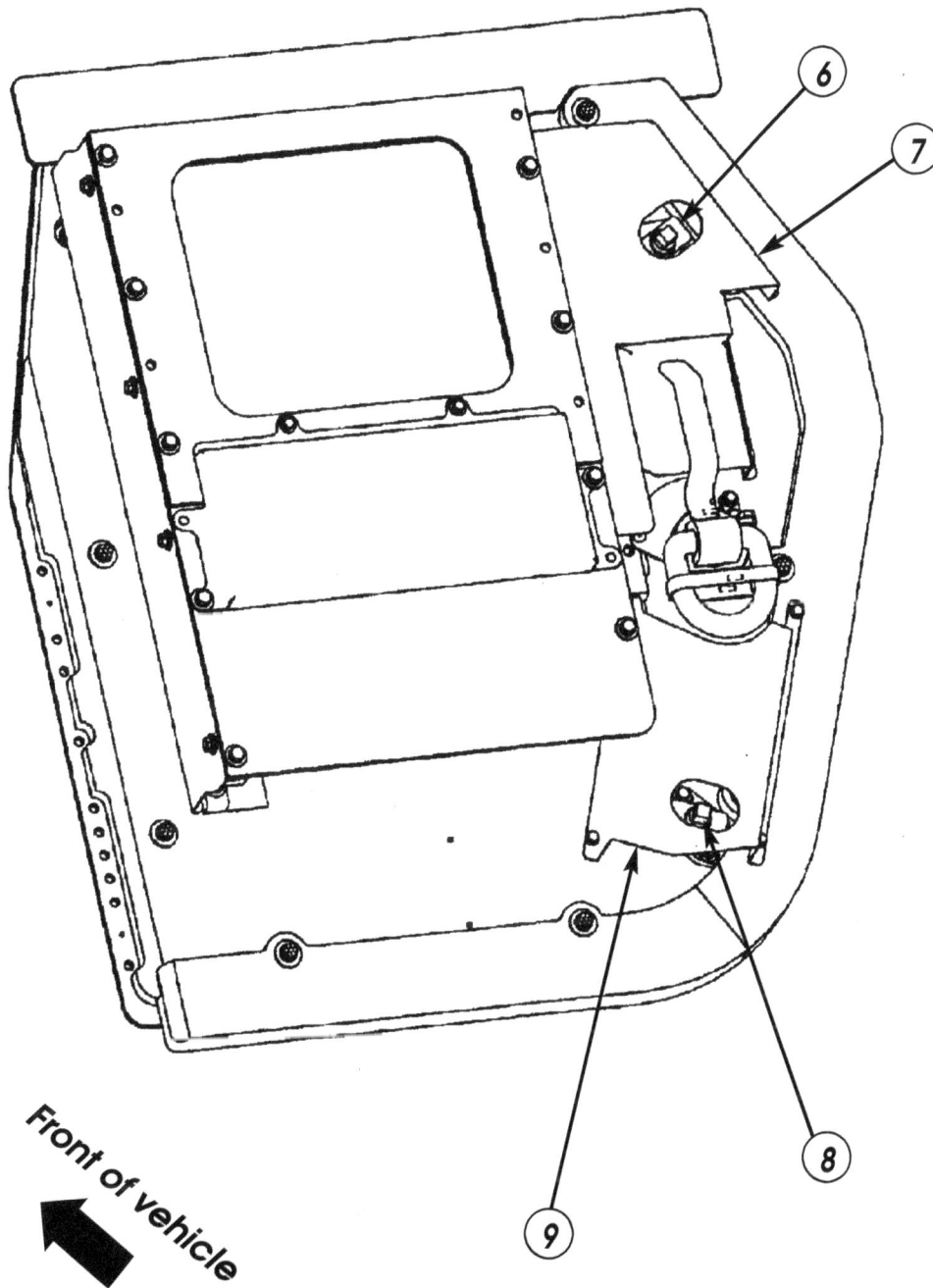

Front of vehicle

2-31.5. FRAG 5 DOOR CREW EXTRACTION BRACKET (CEB) OPERATION (M1114)

WARNING

- The Crew Extraction Bracket (CEB) is considered safe for use by properly trained soldiers, providing they are made aware of the following safety hazards, and appropriate caution is exercised. Failure to comply may result in injury or death to personnel.

- Door opening will result in hardware from the blast locks becoming secondary projectiles in the crew compartment.This method can be used, but flying debris may injure the crew. First Responders will have to assess the crew's condition inside, and assess likelihood and severity of casualties against possible loss of life or limb.

- The area should be cleared of all bystanders to prevent injuries from any potential hardware that may come from the vehicle during the door extraction. Failure to comply may result in injury or death to personnel.

- All rescue personnel will remain secured inside the rescue vehicle during actual removal of the door. Failure to comply may result in injury or death to personnel.

- The gunner should be inside the vehicle, and the hatch closed and secured. Failure to comply may result in injury or death to personnel.

NOTE

- The Emergency Crew Extraction Device is utilized in emergency situations to forcefully open/remove the door from an M1114 UAH in order to reach injured or incapacitated crew inside the vehicle.

- The removal operations of the door should be done with the vehicles positioned perpendicular to each other.

- The vehicle used to perform the door extraction should be an armored vehicle with acceptable tow rating for the load being extracted (Minimum working rating of 18,000 lbs).

- The vehicle that the selected door is being extracted from should be secured to prevent any unwanted movement during the extractions.

- Door extraction must be performed using a serviceable strap as specified and included in the M1114 HMMWV Winch BII. If a HMMWV BII strap is not available, a serviceable strap, chain, or cable with a working load rating of at least 18,000 pounds and a minimum length of 10 feet may be used.The strap/chain/cable should be inspected for defects prior to use in removing a door.A damaged or defective strap/chain/cable shall not be utilized to remove a door and should be disposed of.

2-31.5. FRAG 5 DOOR CREW EXTRACTION BRACKET (CEB) OPERATION (M1114) (Cont'd)

a. **Quick Pull Door Release.**

    (1) Secure tow strap or chain (2) to D-ring (6) with pinch mount or shackle to damaged vehicle (5).

NOTE

    The tow strap should have some slack in it before attempting door release.

    (2) Accelerate tow vehicle (1) and pull until door (4) is parallel with tow strap (2) and door handle (3) breaks off.

NOTE

    Combat lock lever can be used in place of emergency rescue wrench for the following step.

    (3) Slide emergency rescue wrench (9) onto square lug (8) in bottom exterior linkage cover (7) and turn combat lock lever (9) toward front of vehicle.

b. **Slow Pull Door Release.**

NOTE

    Slow pull door release may cause damaged vehicle to drag.

    (1) Secure tow strap (2) to D-ring (6) with pinch mount or shackle.

    (2) Slowly accelerate tow vehicle (1) until all slack is removed from tow strap (2).

    (3) Continue accelerating tow vehicle (1) until door is released (4).

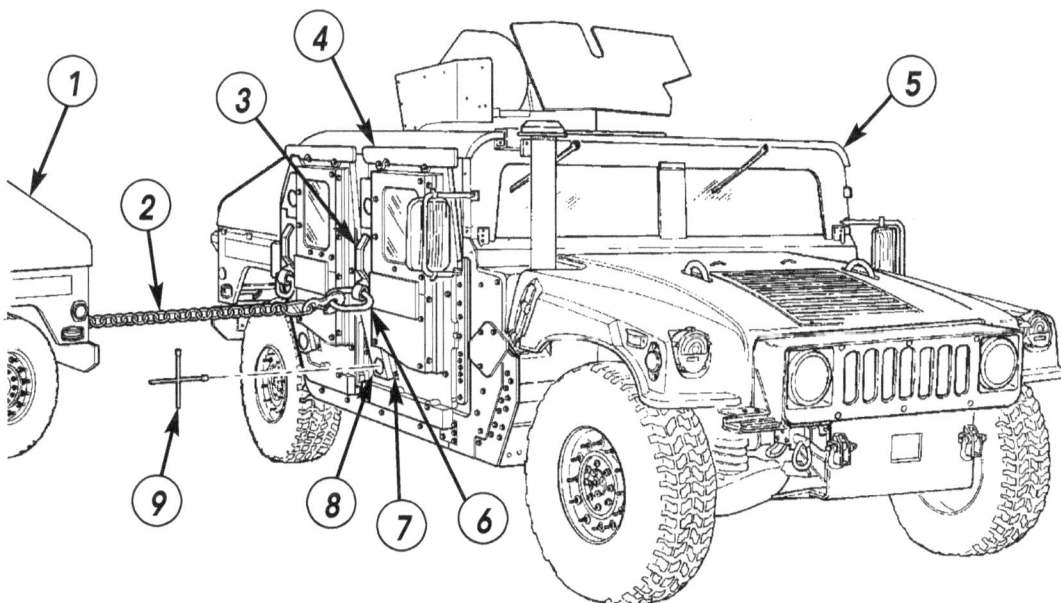

2-32. CARGO SHELL DOOR OPERATION (M1114, M1151, M1151A1, M1167)

WARNING

Do not attempt to operate cargo shell door forward latch. The cargo shell door is not to be opened from inside the vehicle. Opening cargo shell door from inside the vehicle may cause damage to equipment or injury to personnel.

NOTE

- For ease of operation, a grab-hold loop can be attached to the cargo door strap at the discretion of unit commander. Notify unit maintenance for installation of grab-hold loop.

- It may be necessary to lift cargo shell door past the FULL OPEN position for installation or removal of the rear hatch support.

a. Raising Cargo Shell Door.

Pull door handle (1) upward and release. Push door (2) open, until door (2) will open automatically to full open position. Insert rear hatch support (5) into stop plate (6).

b. Lowering Cargo Shell Door.

Remove rear hatch support (5) from stop plate (6). Pull on strap (3) to lower door (2) then slam shut. Ensure door (2) is locked by observing alignment of door surface with cargo shell surface side and bottom edges (4).

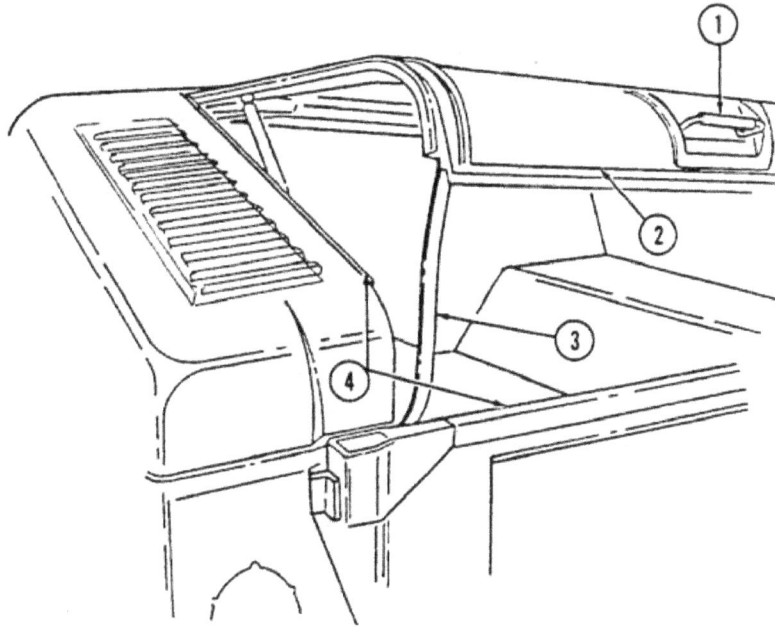

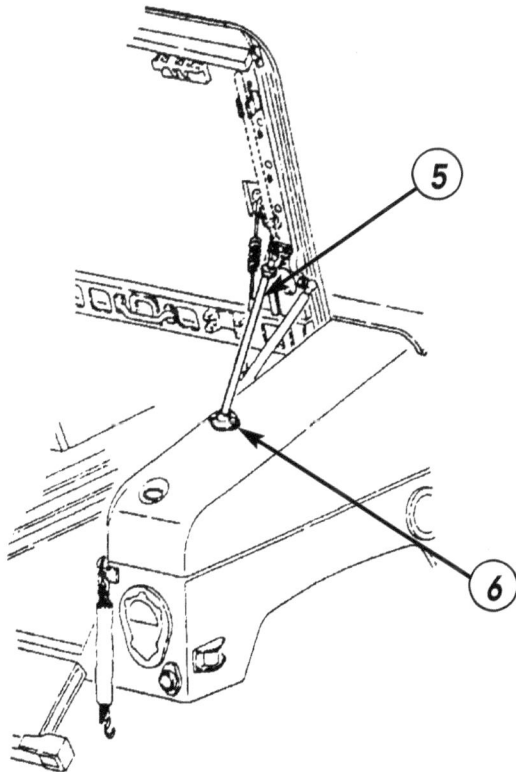

c. **Cargo Shell Door Emergency Service.**

NOTE

Should both ends of the cargo shell door open at the same time, a safety catch built into the rear latch mechanism will normally engage the rear strikers and prevent the door from slinging forward. Should this occur, procedures to reinstall the door are provided in steps 1 through 8. If door has separated from all four strikers, notify unit maintenance.

(1) Inspect rear latch mechanism and ensure rear strikers (2) have engaged the safety catch (3).

(2) Turn locking device (9), pull cargo shell door forward latch (6) from inside vehicle, and completely raise door (5).

(3) Release rear door latch locks (1), if locked.

(4) Grasp raised door (5) from rear of vehicle with both hands and pull right side of door (5) straight down. A distinctive latching sound should be heard.

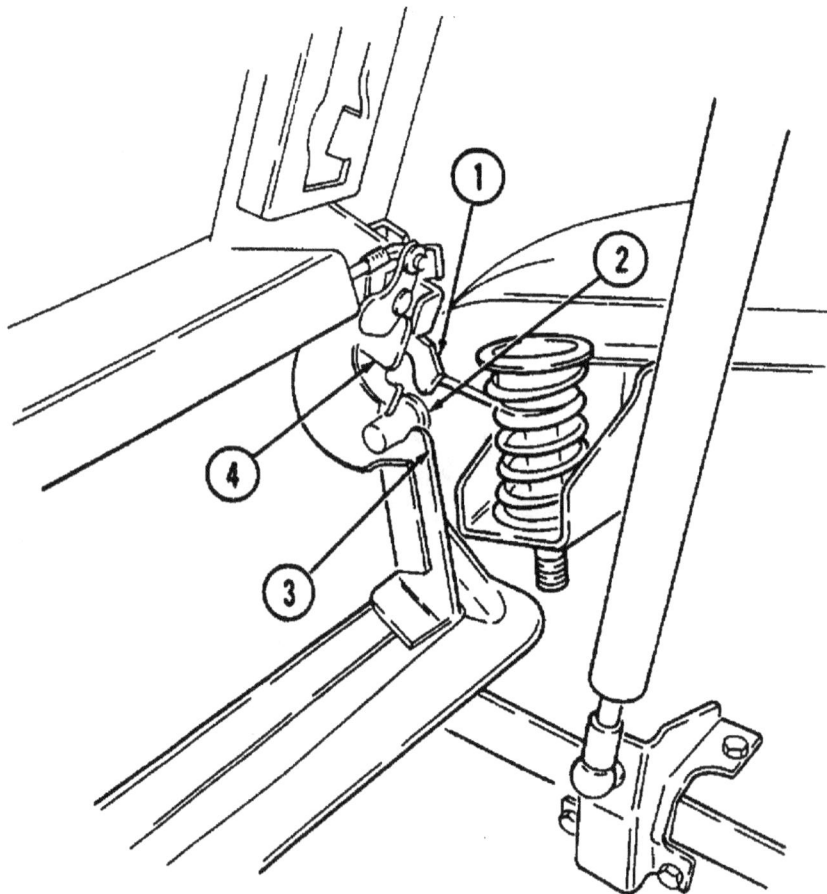

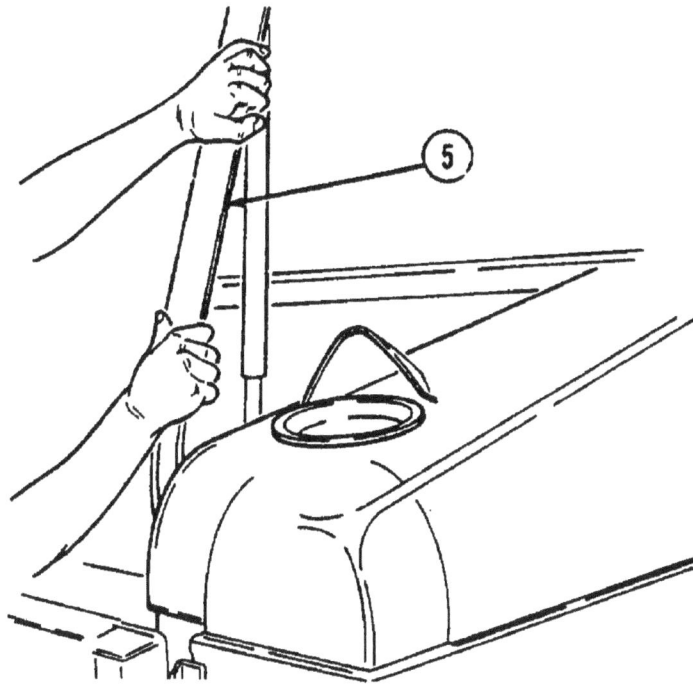

(5) Inspect right rear latch mechanism again to ensure striker (2) engagement with latch (4).

(6) Perform steps 3, 4, and 5 for left side of door.

(7) Enter vehicle, pull forward end of door (5) shut, and turn locking device (9). Inspect forward door latches (7) to ensure proper striker (8) engagement.

(8) Exit vehicle and raise door (5) from the rear. Operation of door (5) should be smooth. If operation is not smooth, or door components appear damaged, notify unit maintenance.

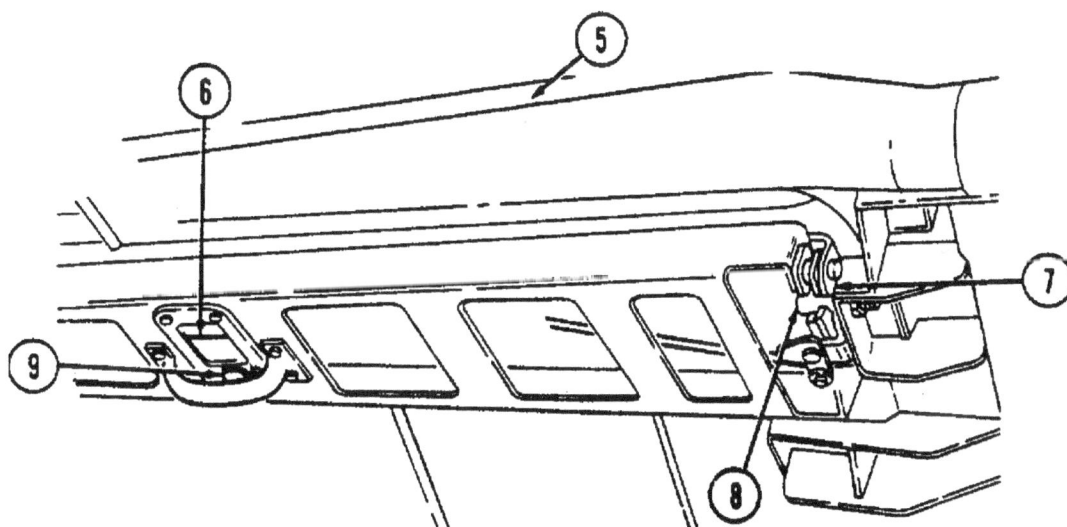

2-33. UP-ARMORED WEAPON STATION OPERATION (M1114)

a. Operation.

WARNING

Use care when opening and closing covers. Do not rest fingers in cover opening. Personnel injury may result.

(1) Open turret weapon station cover (1) by releasing three latches (5) on inside of vehicle (4).

(2) Push small section of cover (2) up and back until resting on large section of cover (1).

(3) Push up large section of cover (1). While holding cover (1), release two rods (3) from cover (1) and insert rods (3) into catch blocks (11).

(4) Slip pins at end of rods (3) into holes in catch blocks (11), then depress and rotate levers (10) until engaged.

NOTE

It may be necessary to rotate turret to set brake.

(5) Disengage turret brake (6) by lifting turret brake handle (7).

(6) Rotate turret (9) to desired position.

(7) Set turret brake (6) by applying downward pressure to turret brake handle (7) until locked.

NOTE

For new turret brake configuration, perform steps 7.1 through 7.3.

(7.1) Disengage turret brake (13) from turret stop ring (15) by pulling turret brake handle (12) outward.

(7.2) Rotate turret (14) to desired position.

(7.3) Engage turret brake (13) by pushing turret brake handle (12) inward.

CAUTION

Do not sit, stand, or place heavy objects on weapon station, tray, or cover. Components may bend and damage to equipment will occur.

(8) Use gunner's sling (8) as seat rest or restraint if gunner is positioned in the weapon station during travel or weapon operation.

b. After Operation.

(1) Depress and rotate levers (10) and remove two rods (3) from catch blocks (11).

(2) Insert rods (3) in cover (1).

(3) Close weapon station cover (1) and secure with three latches (5).

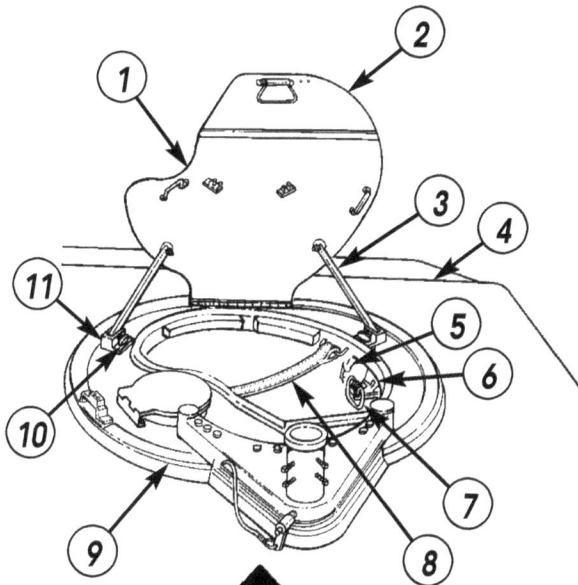

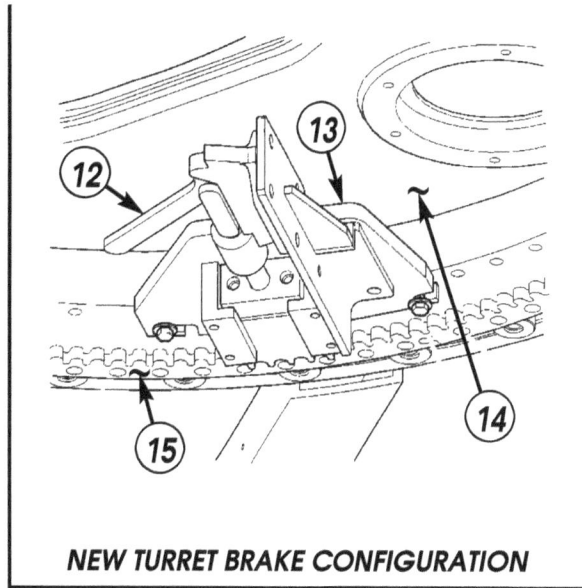

NEW TURRET BRAKE CONFIGURATION

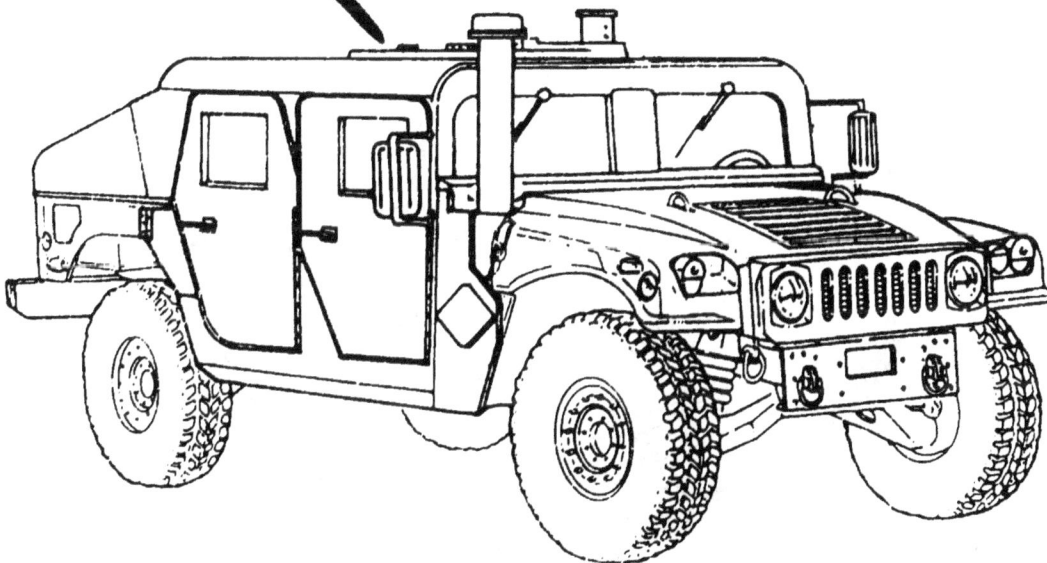

2-33.1. ARMAMENT WEAPON STATION OPERATION (M1151, M1151A1)

a. Before Operation.

(1) Release three weapon station cover securing latches (5) and push weapon station cover (3) open using weapon station cover handle (2).

(2) Secure weapon station cover (3) in the open position with retaining strap (1) and catch (9).

(3) Raise weapon station brake handle (4) to unlock weapon station (7).

(4) Use weapon station brake handle (4) and turret positioning handle (6) to rotate weapon station (7) to the desired azimuth.

(5) Lock weapon station (7) with brake handle (4).

CAUTION

Do not sit, stand, or place heavy objects on weapon station, tray, or cover. Components may bend and damage to equipment will occur.

(6) Use gunner's sling (8) as seat rest or restraint if gunner is positioned in the weapon station during travel or weapon operation.

b. After Operation.

(1) Remove two retaining strap (1) from catch (9).

(2) Close weapon station cover (3) and secure with three weapon station securing latches (5).

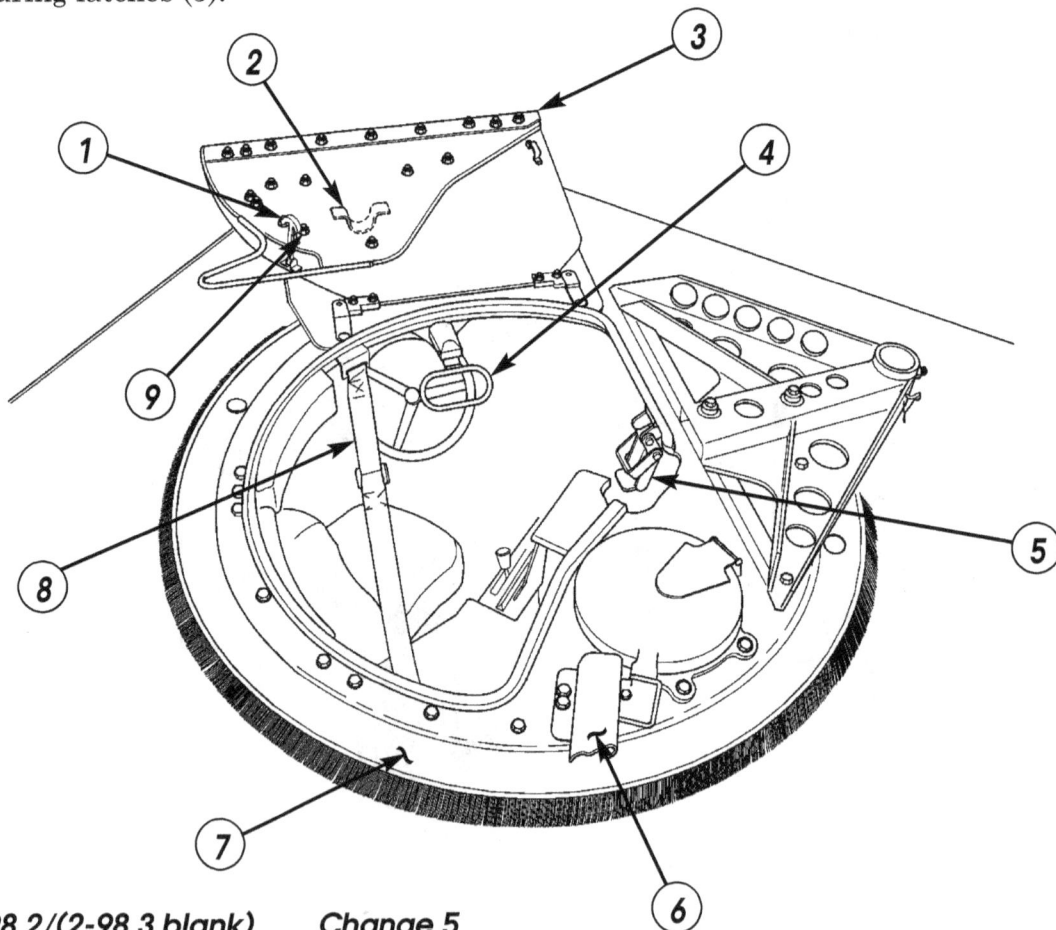

2-33.2. BATTERY POWERED MOTORIZED TRAVERSING UNIT (BPMTU)

a. Motorized Turret Rotation.

(1) Ensure motor engagement control (1) is in the horizontal position.

WARNING

Failure to disengage the manual traversing gear while operating the battery powered motorized traversing unit may result in injury to personnel or damage to equipment.

(2) Disengage the traversing gear.

(3) Magnetically attach joystick assembly (4) in desired position.

(4) To move joystick assembly (4), push down on joystick assembly (4) to break magnetic hold. Replace joystick assembly (4) in desired position.

(5) Pull out emergency stop switch (2) to "ON" position.

(6) Push joystick (5) to direction indicated on joystick assembly (4) to rotate turret left or right.

(7) Break-away connector (3) should stay connected when operating vehicle to allow constant charging of system.

NOTE

Turret rotation causes break-away connection to separate. The system is designed to continue functioning on its own power. As soon as mission allows, the break-away connectors should be re-connected to continue charging.

(8) When turret rotation is no longer necessary, depress emergency stop switch (2) to "OFF" position.

(9) Re-charge system after every mission.

(10) Push in emergency stop switch (2) to "OFF" position when BPMTU is not charging or in use.

b. Manual Turret Rotation.

CAUTION

Turret has no brake when motor engagement control is in the "NEUTRAL" vertical position.

(1) Push in and turn motor engagement control (1) to the vertical position to place motor and drive assembly in "NEUTRAL".

(2) Manually rotate turret as desired.

(3) When manual turret rotation is no longer needed, lock turret in place by turning the motor engagement control (1) in the horizontal "DRIVE" position.

2-33.2. BATTERY POWERED MOTORIZED TRAVERSING UNIT (BPMTU) (Cont'd)

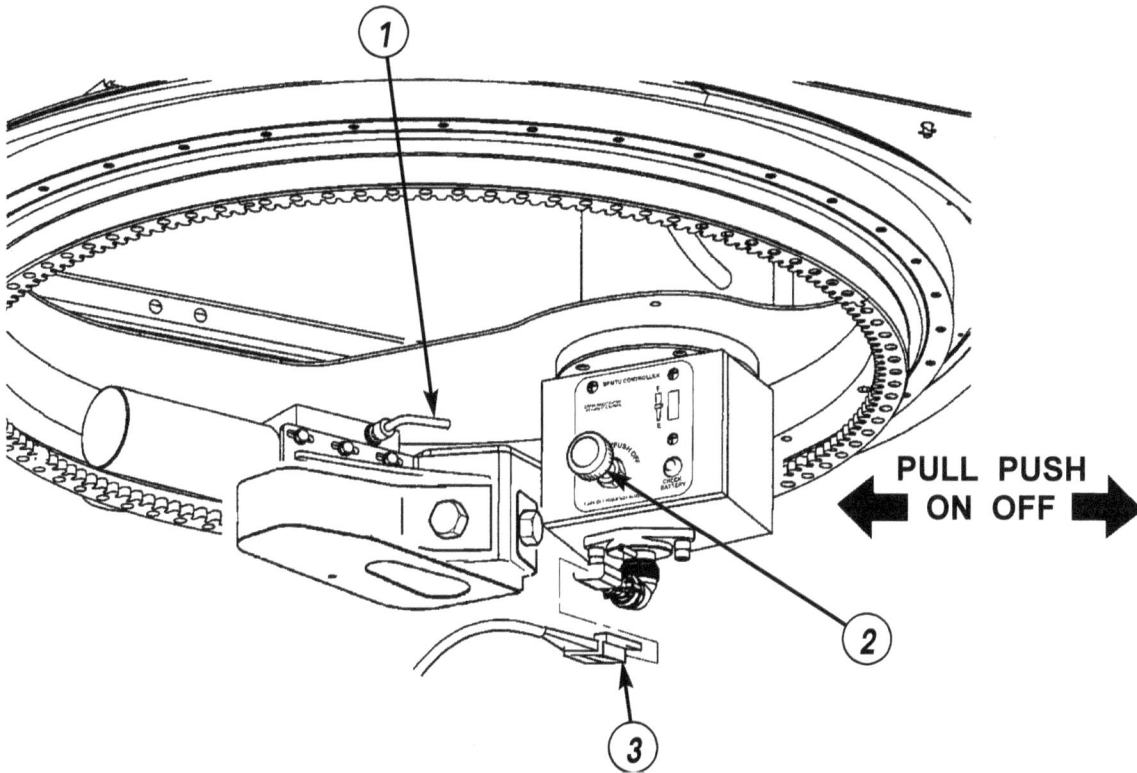

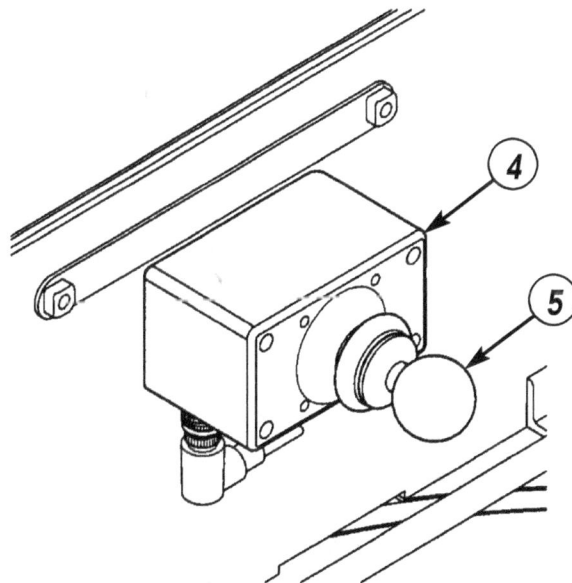

PULL PUSH
ON   OFF

2-33.2. BATTERY POWERED MOTORIZED TRAVERSING UNIT (BPMTU) (Cont'd)

---

c. Checking Charge Level Of Turret Batteries.

NOTE

Emergency stop switch must be in the "ON" position to charge batteries.

(1) Battery status indicator (1) is normally off – "lights out". Pushing battery status indicator button (2) will turn LED display of battery status indicator (1) "ON" for ten seconds. Holding battery status indicator button (2) down will keep LED display on, until ten seconds after button (2) is released.

(2) Seven LED segments are provided for battery capacity indication. When all seven segments are illuminated, battery is fully charged. As battery discharges, fewer LED segments will be illuminated. When battery is low, only one LED segment will be illuminated and will be flashing. There may still be significant battery life remaining when flashing low, but running below this point will degrade battery life.

(3) The battery status indicator button (2) may be used at any time to indicate battery capacity, but to get the most accurate indication, perform the following test:

    a.) Disconnect break-away connector (5).

    b.) In order to reset system, push in the emergency stop switch (4) to the "OFF" position, pull out emergency stop switch (4) to the "ON" position.

    c.) Activate the turret motor by moving the joystick rotating turret for approximately five seconds.

    d.) Press battery status indicator button (2) and read the capacity from the LED display.

d. Charging the system using vehicle power system.

NOTE

- Charge batteries whenever possible to maintain capacity. Batteries not in a fully charged state will have decreased capacity.
- Turret batteries have more capacity when warm than cold. Turret batteries have more capacity when new than old.

(1) Rotate turret so that the break-away connector (5) is aligned and able to fully engage when connected.

(2) Depress emergency stop switch (4) to "OFF" position.

(3) Connect break-away connector (5) connecting the vehicle buss bar cable to the system control assembly (3).

(4) Pull out emergency stop switch (4) to the "ON" position.

(5) Start vehicle (para. 2-12).

NOTE

Vehicle system must be above 25 amps to allow BPMTU charging to external batteries.

(6) Ensure all unnecessary electrical equipment in vehicle is turned off.

2-33.2. BATTERY POWERED MOTORIZED TRAVERSING UNIT (BPMTU) (Cont'd)

(7) Allow vehicle to idle at high RPM by pulling throttle lock until LED display (1) shows batteries are fully charged.

(8) Full charge allows approximately ten continuous hours of motorized traversing unit usage.

e. Charging the system using external charging unit.

(1) Rotate turret so that connector (5) is aligned and able to be fully engaged when connected.

(2) Depress emergency stop switch (4) to "OFF" position.

(3) Connect break-away connector (5) connecting the battery switching system to the system control assembly (3).

(4) Ensure all unnecessary electrical equipment in vehicle is turned off.

(5) Remove cover from slave charging receptacle.

(6) Connect external battery charging unit cable to the vehicle slave receptacle.

(7) Pull out emergency stop switch (4) to the "ON" position.

(8) Charge the system until battery status indicator shows batteries are fully charged.

(9) Remove slave cable and replace cover.

(10) Full charge allows approximately ten continuous hours of motorized traversing unit usage.

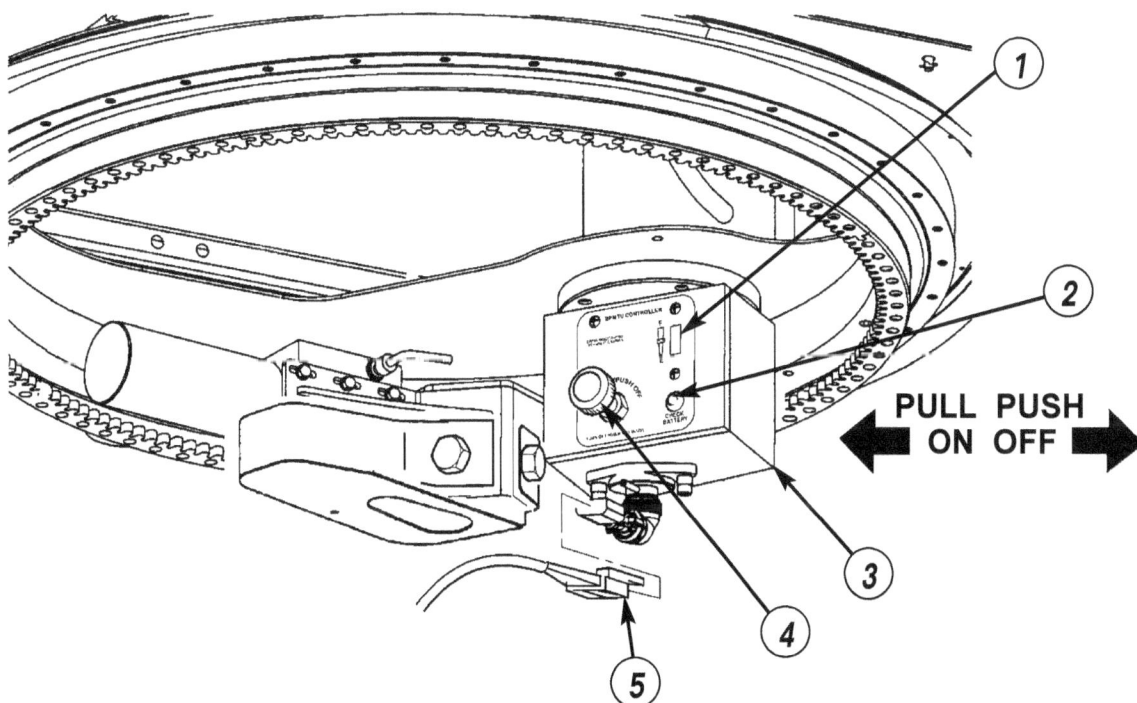

2-33.3. MANUAL TRAVERSING UNIT OPERATION

Operation.

WARNING

- Ensure roof is clear of tools and personnel prior to operating weapon station. Failure to comply may result in injury to personnel or damage to equipment.
- For TOW ITAS vehicles, always utilize the manual crank handle to rotate the turret. Failure to do so may result in injury to personnel.
- For TOW ITAS vehicles, always secure the locking mechanism while on an incline to prevent the T-GPK from unexpectedly rotating, which may result in injury to the crew or damage to the ITAS.

(1) Use the turret brake lever (3) to unlock the turret brake (2) (para. 2-33.4).

(2) To engage manual traversing gear on manual gear unit (5), pull back and lift up on the engage/disengage lever (6) and let gear swing in to engage ring gear (1).

(3) If not installed, install manual crank handle (4) on manual traversing gear unit (5) with detent pin (7) and swing hand lever down on manual crank handle (4).

(4) To rotate weapon station to the right, turn the manual crank handle (4) clockwise.

(5) To rotate weapon station to the left, turn the manual crank handle (4) clockwise.

(6) Use the turret brake lever (3) to lock the turret brake (2) to desired position. (para. 2-33.4).

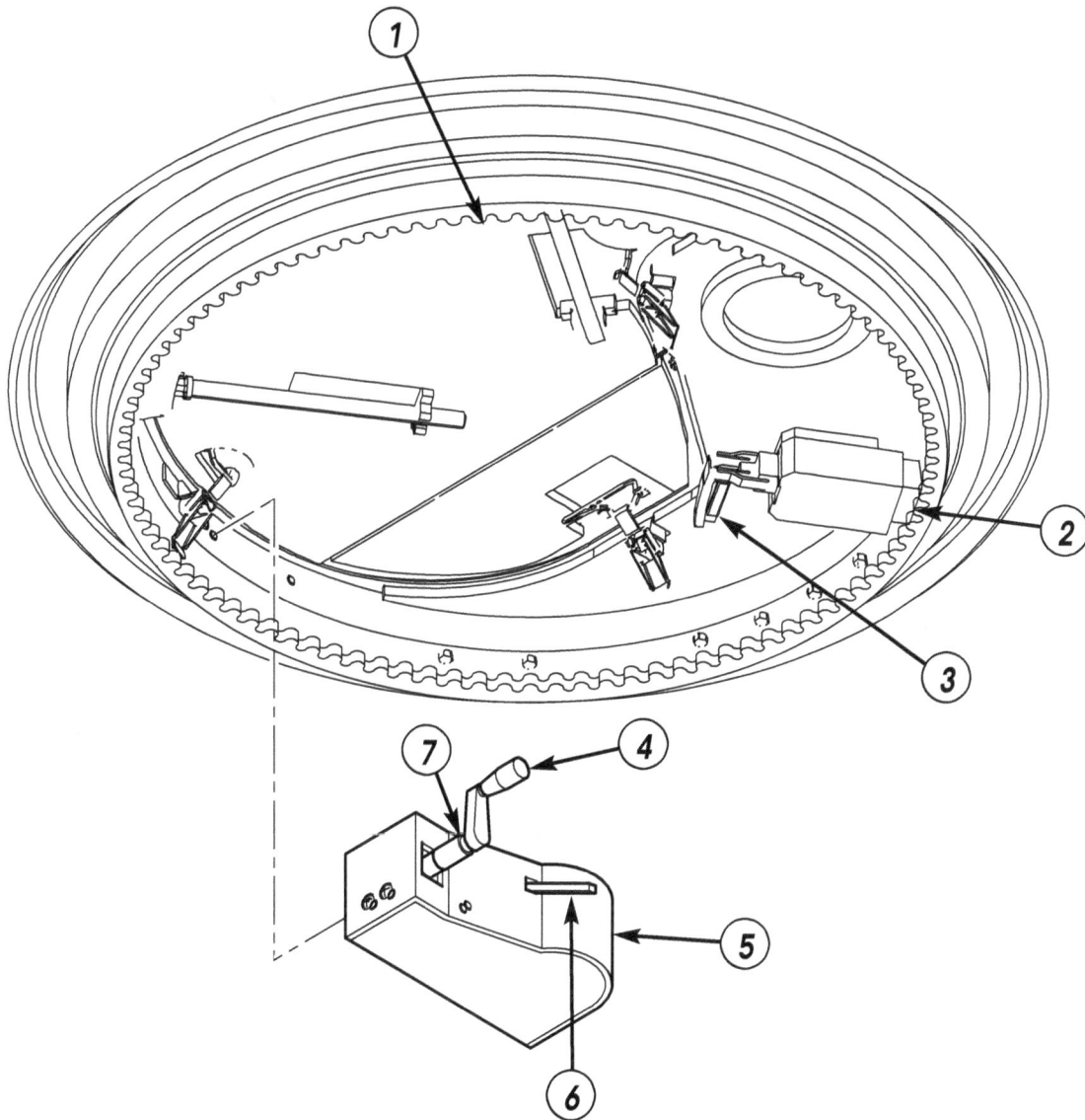

2-33.4. TURRET BRAKE OPERATION

Operation.

WARNING

- Ensure roof is clear of tools and personnel prior to operating weapon station. Failure to comply may result in injury to personnel or damage to equipment.
- For TOW ITAS vehicles, always utilize the manual crank handle to rotate the turret. Failure to do so may result in injury to personnel.
- For TOW ITAS vehicles, always secure the locking mechanism while on an incline to prevent the T-GPK from unexpectedly rotating, which may result in injury to the crew or damage to the ITAS.

NOTE

A hand brake lever is used to lock and unlock the turret brake.

(1) Squeeze hand brake lever (3) and rotate it to unlocked position.

(2) Release hand brake lever (3) to allow turret stop ring (1) to rotate freely.

NOTE

The turret stop ring should turn easily when force is applied but stop quickly when no force is exerted.

(3) With the turret brake (2) unlocked, use body motion to rotate turret stop ring (1) to desired position.

(4) Use the gunner's stand and sling seat to help rotate turret stop ring (1).

(5) Turn turret stop ring (1) slightly, if necessary, to align teeth (4) on turret stop ring (1) with locking notches (5) on turret brake (2).

(6) Apply hand brake lever (3) to lock turret stop ring (1) in place for travel and weapon maintenance. Rotate turret stop ring (1) to where weapon is facing forward before locking turret brake (2) for travel to keep roof hatch clear.

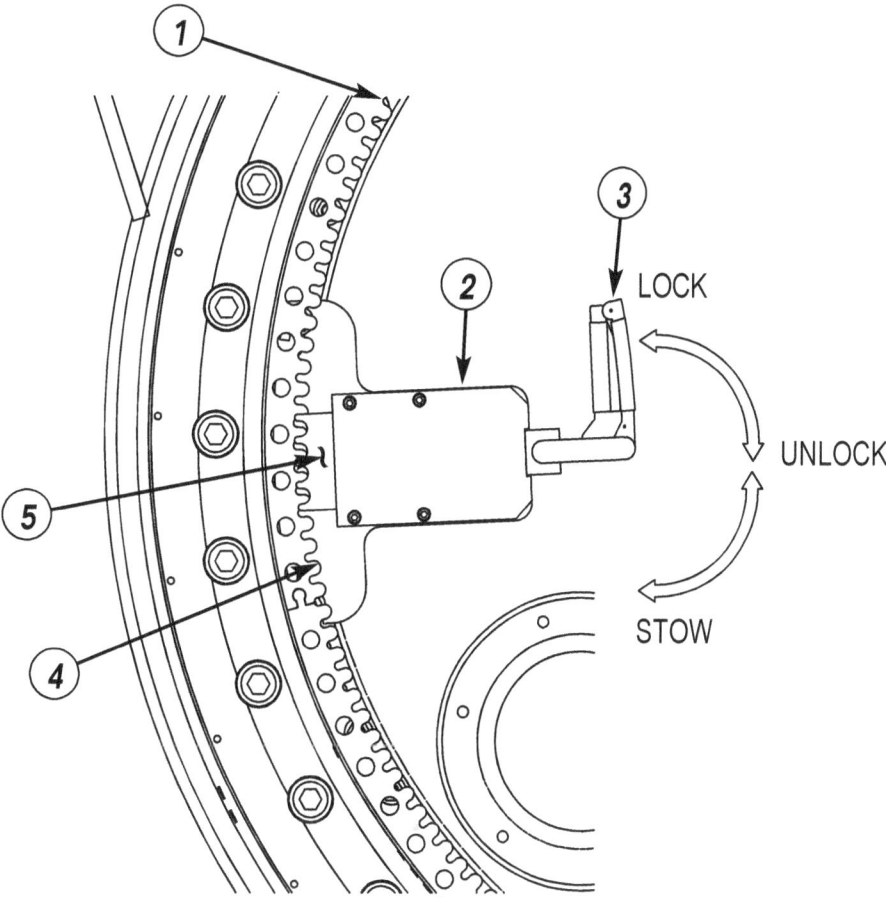

2-34. FUEL DOOR OPERATION (M1114)

a. Operation.

(1) Open right rear crew door (1).

(2) Release catch (2) holding fuel door (3) closed.

(3) Pull fuel door (3) open.

b. After Operation.

(1) Close fuel door (3).

(2) Engage catch (2) into fuel door (3) and lock catch (2).

(3) Close right rear crew door (1).

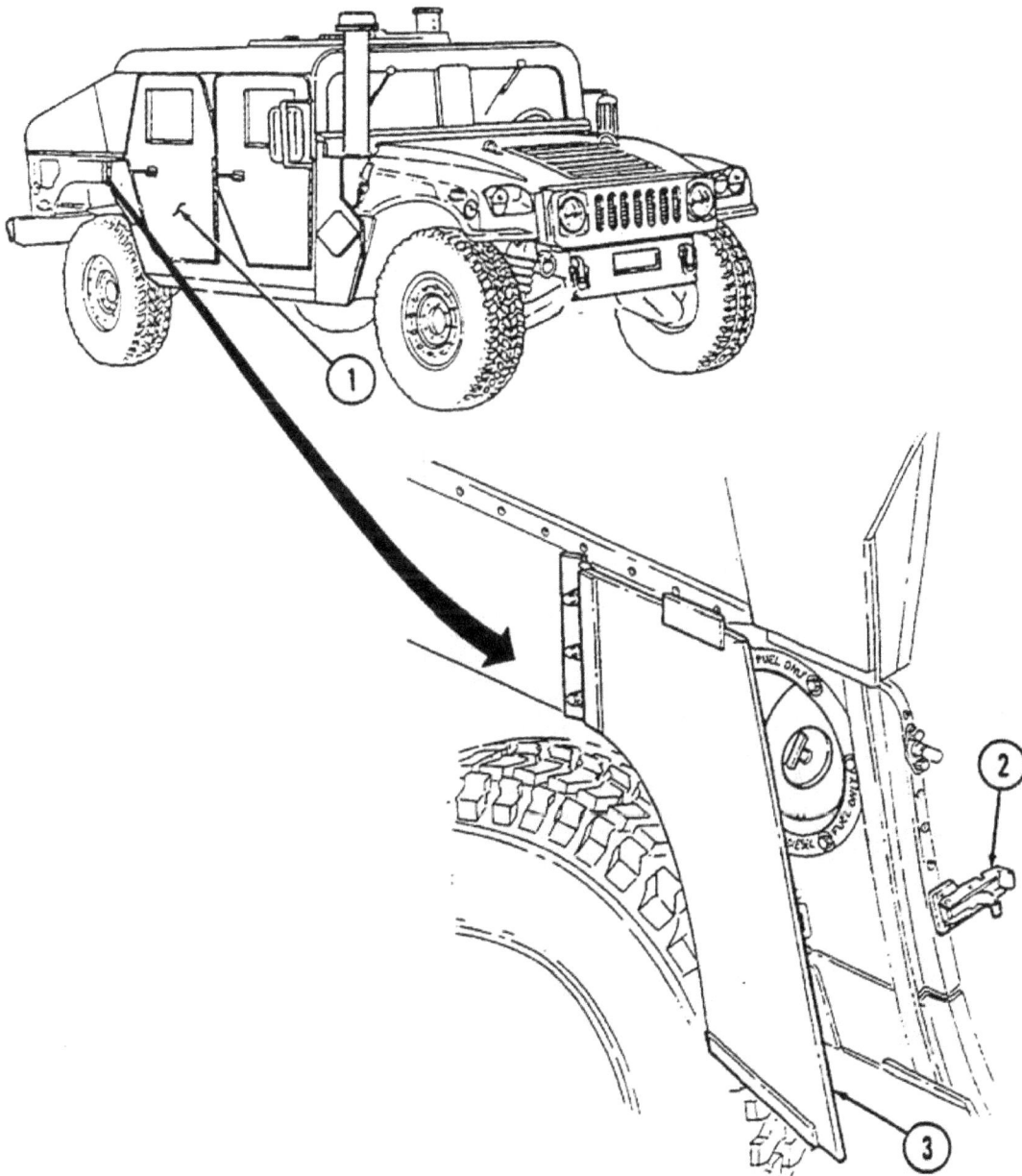

2-35. AIR CONDITIONER OPERATION (M1114)

a. Operation.

(1) Start engine (para. 2-12).

(2) Flip air conditioner ON/OFF switch (2) to the right to turn on air conditioner.

(3) Flip heater fan switch (6) up to turn on.

(4) Pull fresh air control knob (4) out and slide defroster register (7) to the right to open vents (1).

(5) Push up floor outlet lever (3) to open floor vents.

(6) Position fan control knob (5) to desired setting.

(7) Adjust vents (1) for air flow directions.

b. After Operation.

(1) Push down floor outlet lever (3) to close floor vents.

(2) Slide defroster register (7) left and push knob (4) in to close vents (1).

(3) Flip heater fan switch (6) down to turn off.

(4) Flip air conditioner ON/OFF switch (2) to the left to turn off air conditioner.

(5) Stop engine (para. 2-14).

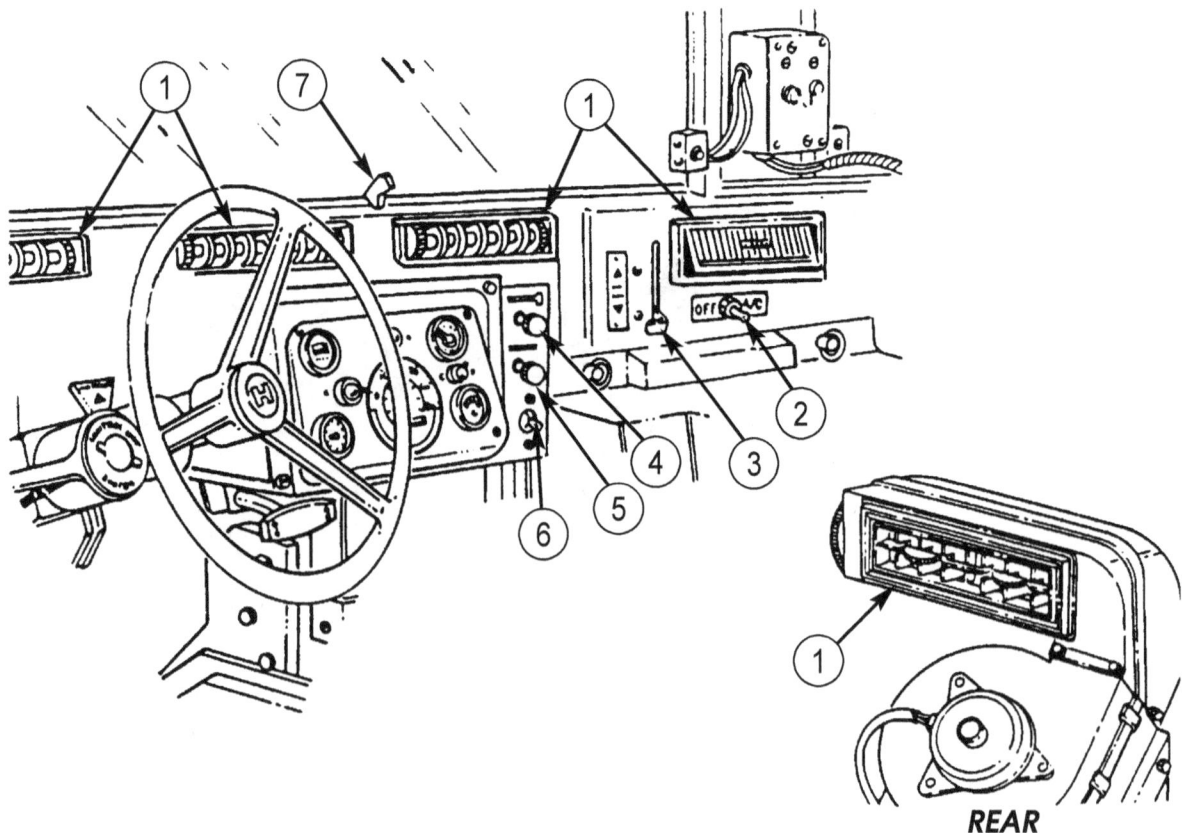

REAR

2-35.1. AIR CONDITIONER OPERATION (M1151, M1151A1, M1152, M1152A1, M1165, M1165A1, M1167)

a. Operation.

    (1) Start engine (para. 2-12).

    (2) Flip A/C / HEAT switch (2) up to turn on air conditioner.

    (3) Flip fan switch (3) to desired setting, HIGH/LOW, to activate A/C air output.

    (4) Adjust vents (1) for air flow direction.

b. After Operation.

    (1) Flip fan switch (3) down to OFF position.

    (2) Stop engine (para. 2-14).

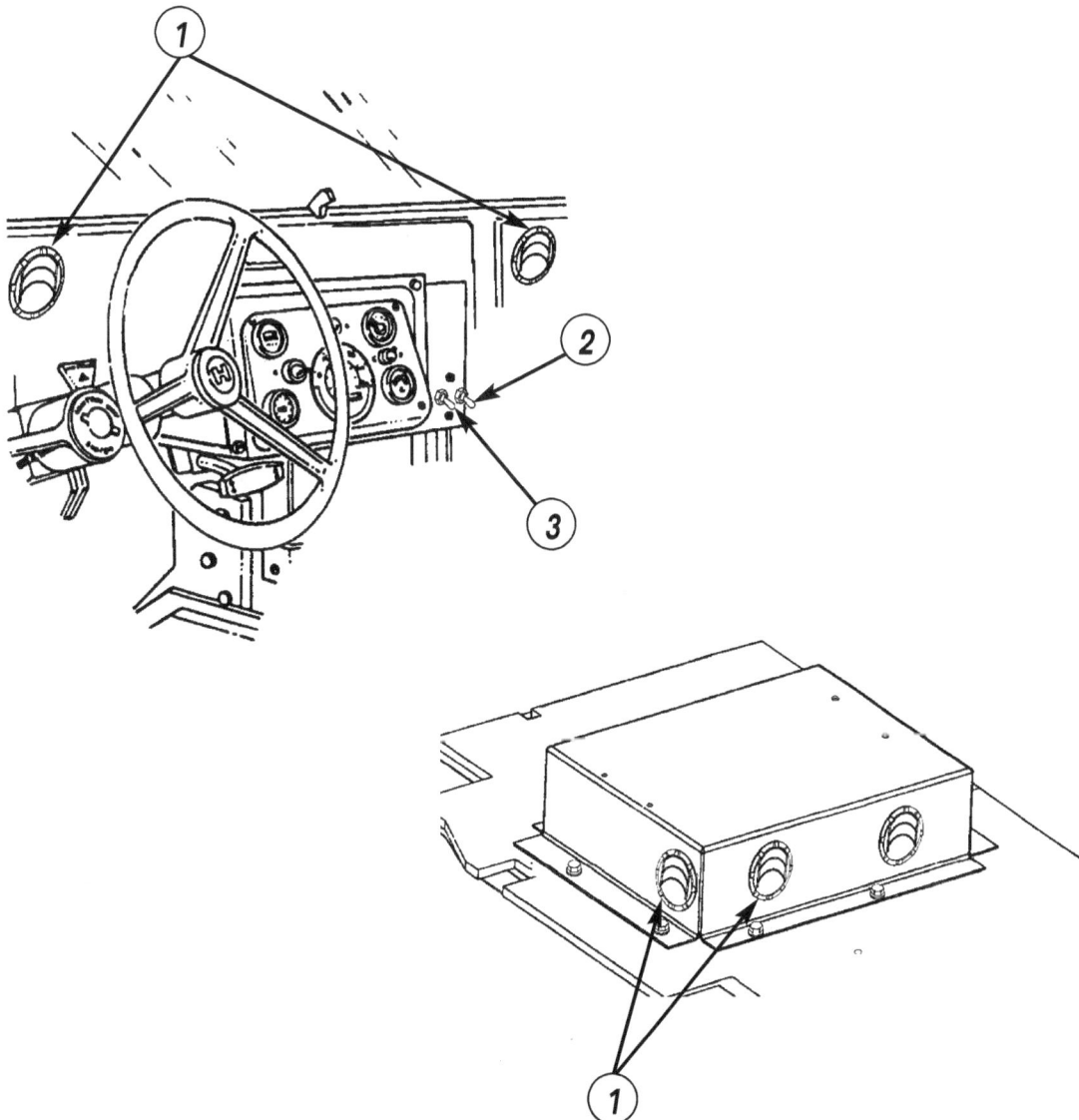

2-36. WINDSHIELD DE-ICER OPERATION (M1114)

NOTE

De-icer will not operate if vehicle is not running.

a. Start engine (para. 2-12).

b. Flip toggle switch (1) on de-icer unit (2) to engage position. Hold until indicator light (3) illuminates, then release.

NOTE

De-icer will continue to operate until toggle switch is moved to OFF position or engine is turned off.

c. Toggle switch (1) will revert to ON position and the de-icer unit (2) will be in operation.

d. To discontinue de-icer unit (2), flip toggle switch (1) to OFF position, heating element will disengage and indicator light will go out.

e. Stop engine (para. 2-14).

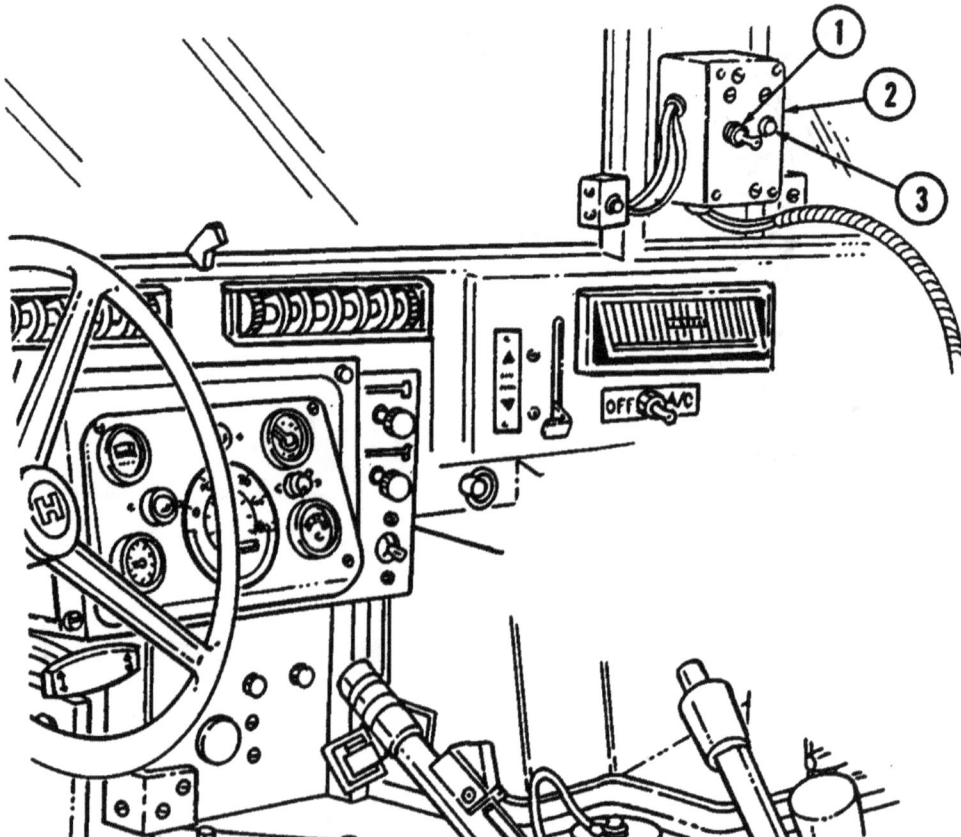

2-36.1. ADJUSTABLE GUNNER PLATFORM OPERATION (M1114, M1167)

a. Operation.

(1) Depress two lockbuttons (6) on locking pins (5) and remove locking pins (5) from locking lugs (9) and holes (4) securing platform (1) in stowed position (13).

NOTE

Ensure footman loops are laid out flat behind platform when securing platform in stowed position.

(2) Grasp hand hold (2) on platform (1) to lift platform (1) to either half-height position (12) or full-height position (11).

(a) If half-height position (12) is required, lift platform (1) all the way up and turn two latches (7) so ledges of latches (7) are facing up and parallel to bottom edge of platform (1). Lower platform (1) and allow to rest on ledges of latch (7) at half-height position (12). Insert locking pins (5) through locking holes (4) in platform (1) and holes (10) in platform risers (8).

(b) If full-height position (11) is required, lift platform (1) to full-height position (11), and ensure locking holes (3) in left and right sides of platform (1) are aligned with locking holes (10) in platform risers (8). Insert locking pins (5) through locking holes (3) in platform (1) and holes (10) in platform risers (8).

b. After operation.

(1) Remove locking pins (5) securing platform (1) in either half-height position (12) or full-height position (11) by pressing lockbuttons (6) and removing locking pins (5). Lift up on platform (1), and turn two latches (7) so ledges of latches (7) are facing downward, to floor.

(2) Lower platform (1) to stowed position (13) and align holes (4) with locking lugs (9). Insert locking pins (5) through holes (4) and lugs (9).

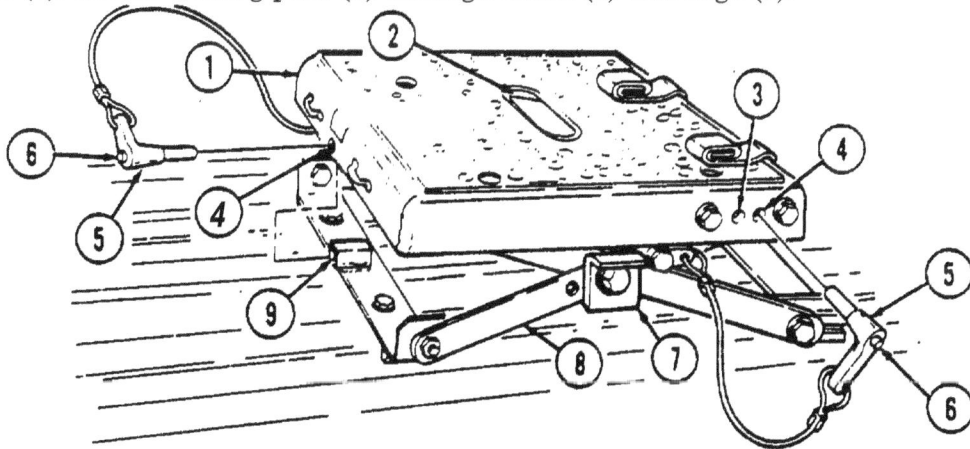

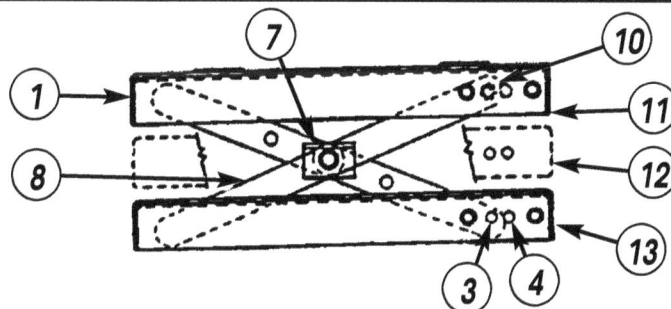

2-36.2. GUNNER'S RESTRAINT SYSTEM OPERATION (M1114, M1151, M1151A1, M1167)

NOTE

- The gunner's restraint system is only designed to prevent the gunner from being ejected from the vehicle; it will not pull the gunner back into the vehicle.

- All straps can be adjusted to provide a secure fit.

- Straps are color-coded to aid in installation.

a. Position red shoulder strap (1) over right shoulder and attach to rotary buckle (5).

b. Position green shoulder strap (3) over left shoulder and attach to rotary buckle (5).

c. Position rotary buckle (5) to front of body, waist high.

d. Position red lap strap (6) around right side of waist and attach to rotary buckle (5).

e. Position green lap strap (4) around left side of waist and attach to rotary buckle (5).

f. Adjust all straps (1), (3), (4), and (6) to provide a secure fit.

WARNING

- Ensure anchor strap is positioned in front of gunner's sling seat. Failure to do so may result in injury or death to personnel.

- Ensure anchor strap is adjusted so that no slack remains in strap. Failure to do so may result in injury or death to personnel.

g. Attach anchor strap (9) to lower retractor belt (8).

h. To release gunner's restraint harness (2), push and rotate rotary release button (7) on rotary buckle (5) in either direction.

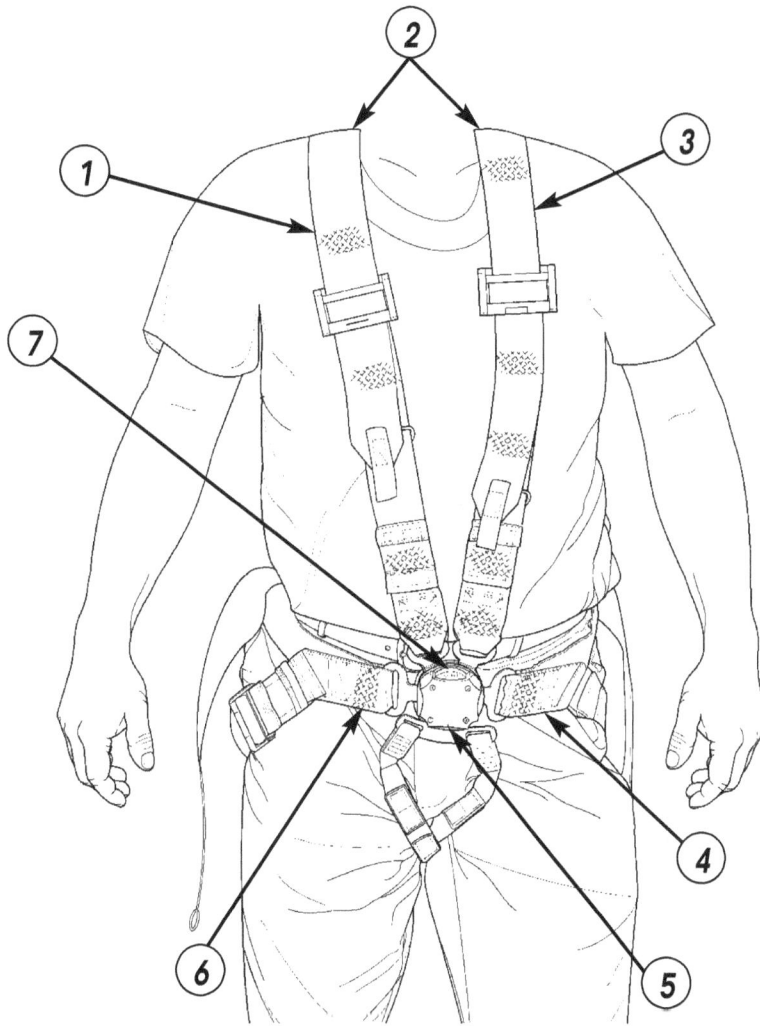

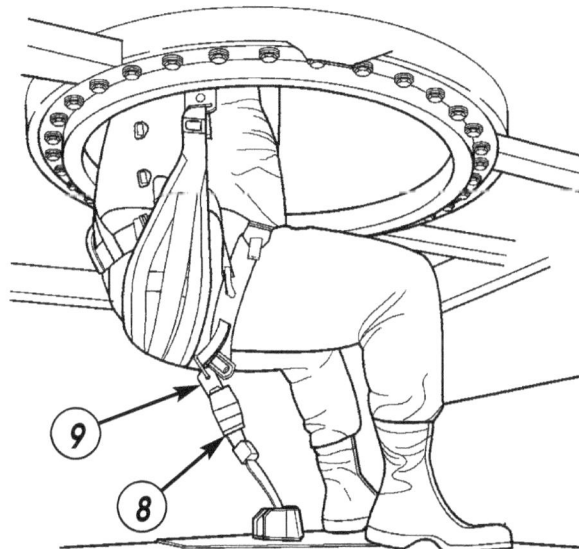

Section IV. OPERATION UNDER UNUSUAL CONDITIONS

2-37. GENERAL

Special instructions for operating and maintaining vehicles under unusual conditions are included in this section. Unusual conditions are extreme temperatures, humidity, and/or terrain. Special care in cleaning and lubrication must be taken in order to keep vehicles operational when operating under unusual conditions.

2-38. OPERATION UNDER UNUSUAL CONDITIONS REFERENCE INDEX

2-39. SPECIAL INSTRUCTIONS

WARNING

This vehicle has been designed to operate safely and efficiently within the limits specified in this TM. Operation beyond these limits is prohibited by IAW AR 750-1 without written approval from the Commander, U.S. Army Tank-automotive and Armaments Command, ATTN: AMSTA-CM-S, Warren, MI 48397-5000.

NOTE

Except where noted, all normal operating procedures will apply in addition to special instructions for unusual operating conditions.

a. Cleaning. Refer to para. 2-5 for cleaning instructions and precautions.

b. Lubrication. Refer to appendix-G for proper lubricating instructions.

c. Driving Instructions.

(1) FM 21-305, Manual for the Wheeled Vehicle Driver, contains special driving instructions for operating wheeled vehicles.

(2) FM 55-30, Army Motor Transport Units and Operations, contains instructions on driver selection, training, and supervision.

(3) FM 9-207, Operation and Maintenance of Ordnance Materiel in Cold Weather, contains instructions on vehicle operation in extreme cold of 0° to -65°F (-18° to -54°C).

(4) Other documents with information on vehicle operation under unusual conditions are:

(a) FM 31-70 Basic Cold Weather Manual   (d) FM 90-5 Jungle Operations

(b) FM 31-71 Northern Operations   (e) FM 90-6 Mountain Operations

(c) FM 90-3 Desert Operations

d. Special Purpose Kits. Paragraphs describing special purpose kits for operation under unusual conditions are:

(1) Deep water fording operation, para. 2-48.

(2) Vehicular heater operation, para. 2-50.

e. Transmission Range Selection. For proper transmission range selection, refer to table 1-6. If transmission range selection is peculiar to an unusual operating condition, it will be specified in the applicable paragraph.

f. Transfer Case Range Selection.

CAUTION

• Vehicle must be stopped and transmission shift lever in N (neutral) before transfer case can be shifted. Failure to do this will result in damage to drivetrain.

• Damage to drivetrain will result if transfer case is operated in L (low range) or H/L (high/lock range) on high-traction surfaces with no wheel slippage.

CAUTION

* When necessary to temporarily operate transfer case in H/L (high/lock range) or L (low range) when additional traction is needed to prevent wheel slippage, avoid sharp, continuous turns. Failure to avoid sharp, continuous turns while operating transfer case in locked ranges may cause damage to drivetrain.

* Immediately after operation in H/L (high/lock range) or L (low range), ensure transfer case is shifted to H (high range) to avoid damage to drivetrain. If any noises from drivetrain components are heard, ensure that transfer case range is properly selected.

For proper transfer case range selection, refer to table 1-7. If transfer case range selection is peculiar to an unusual operating condition, it will be specified in the applicable paragraph.

2-40. OPERATING ON UNUSUAL TERRAIN

a. General Rules. Driving off-road over rough or unusual terrain basically requires using good driving sense. Experience is the best teacher, but there are a few good rules to keep in mind when you are in that kind of driving situation.

WARNING

* Use extreme caution when transporting personnel. Although certain design characteristics of the vehicle, such as vehicle width, ground clearance, independent suspension, etc., provide improved capabilities, accidents can still happen.

* Operators are reminded to observe basic safe driving techniques/skills when operating the vehicle, especially when transporting personnel. Vehicle speed must be reduced consistent with weather and road/terrain conditions. Obstacles such as stumps and boulders must be avoided. Failure to use basic safe driving techniques/skills may result in loss of control and an accident or rollover resulting in injury or death to personnel and damage to equipment. Since the troop/cargo area has minimal overhead protections and does not have seatbelts, personnel seated here are at greater risk of serious injury.

    (1) Use H/L (high/lock range) or L (low range) only when absolutely required by the conditions identified in table 1-7. After operations on unusual terrain, be sure to shift the transfer case from H/L (high/lock range) or L (low range) to H (high range) to avoid damaging drivetrain components.

    (2) Select proper transmission and transfer case driving ranges. Refer to table 1-6 for transmission range selections. For transfer case range selections, refer to table 1-7.

    (3) Keep engine at a moderate speed. The engine works at its best pace in mid-range revolutions per minute (rpm). You can slow down or speed up quickly without changing gears if you get into a tight spot. Use transmission shift lever and transfer case shift lever to control engine speed.

    (4) Attempt to keep wheels from spinning. If wheels start to spin, ease off the accelerator pedal until wheels regain traction.

    (5) Instructions for placing vehicle in motion (para. 2-13) also apply to operating on unusual terrain.

2-40. OPERATING ON UNUSUAL TERRAIN (Cont'd)

b. **Unusual Terrain Driving Techniques.**

CAUTION

- Do not shift into any lower gear than is necessary to maintain headway. Attempt to maintain a constant engine speed. Over-revving engine will cause the wheels to slip, and traction will be lost.
- Before ascending or descending steep hills, stop vehicle, place transmission to N (neutral), and shift transfer case to L (low range). Failure to shift transfer case to L (low range) before ascending or descending steep hills may result in damage to drivetrain.
- Operating vehicle in H/L (high/lock range) or L (low range) with uneven tire pressures on hard top surfaces where the wheels cannot slip to equalize rotation may cause driveline torque buildup, difficulty in shifting, or damage to vehicle components.

(1) Before climbing a steep hill, shift the transfer case into L (low range) and the automatic transmission into 1 (first). If wheels start to slip, walk the vehicle the last few remaining feet of a hill by swinging the front wheels sharply left and right if situation permits. This action will provide fresh "bite" into the surface and will usually result in enough traction to complete the climb.

CAUTION

When L (low range) is used for engine braking while descending steep grades, avoid sharp, continuous turns. Failure to avoid sharp, continuous turns while operating transfer case in range may cause damage to drivetrain.

(2) You can proceed safely down a steep grade by shifting the transfer case into L (low range) and the transmission into 2 (second) or 1 (first). Let the vehicle go slowly down the hill with all four wheels turning against engine compression.

WARNING

Do not travel diagonally across a hill unless it is absolutely necessary, or injury to personnel or damage to equipment may result.

(3) When moving across a slope, choose the least angle possible, keep moving, and avoid turning quickly.

(4) After shifting from locked ranges, always back up for a distance of approximately 5 ft. (1.5 m) before proceeding to prevent driveline torque buildup.

2-41. COLD WEATHER STARTING BELOW +32°F

WARNING

Starting aids will not be used on the engine. Use of starting aids may cause damage to vehicle and injury or death to personnel.

a. Start engine (para. 2-12).

WARNING

- Do not use the hand throttle as an automatic speed or cruise control. The hand throttle does not automatically disengage when brake is applied, resulting in increased stopping distances and possible hazardous and unsafe operation.
- Do not fully apply hand throttle when engine is not under load.

NOTE

If engine cranks slowly and voltmeter indicates low battery charge level, attempt to slave start vehicle (para. 2-23). If vehicle still will not start, perform troubleshooting procedures in table 3-1.

b. After engine starts, pull out hand throttle until desired engine speed is obtained. Twist handle to lock hand throttle.

c. Allow engine to warm up at an increased speed for approximately three minutes.

d. After warm-up period, unlock and push in hand throttle. Allow engine speed to decrease.

2-42. OPERATING IN EXTREME COLD, ON ICE OR SNOW

a. Before Operation.

   (1) Operate vehicular winterization equipment (para. 2-50).

   (2) Scrape off any ice accumulated on vehicle.

   (3) Remove ice and snow from area around air cleaner intake cap.

   (4) Refer to para. 2-41 for cold weather starting instructions.

   (5) Refer to para. 2-40 for techniques that can be used when operating on unusual terrain.

   (6) Refer to para. 3-22 for tire chain installation and operation.

   (7) Operate troop/cargo winterization heater, if applicable (para. 2-50.1).

b. During Operation.

WARNING

- Vehicle operation in snow is a hazardous condition. Operator must travel at reduced speeds and be prepared to meet sudden changes in road conditions. Failure to maintain safe stopping distances may cause damage to vehicle and injury or death to personnel.
- Pump brakes gradually when stopping vehicle on ice or snow. Sudden braking will cause wheels to lock and vehicle to slide out of control, causing damage to vehicle and injury or death to personnel.
- Chock blocks shall be used when parking a vehicle in extreme cold conditions. Failure to do so may result in injury to personnel or damage to equipment.

NOTE

Keep area around air cleaner intake cap clear of snow and ice. Snow and ice may melt, refreeze, and cause restriction in air intake system. If necessary, remove intake cap and clear ice and snow without damaging intake cap screen. Hold the cap near the vehicle exhaust to quickly melt ice without damaging screen.

   (1) Place transmission shift lever in Ⓓ(overdrive), and transfer case shift lever in H/L (high/lock range). Place vehicle in motion slowly to prevent wheels from spinning.

NOTE

If additional power is needed to extract vehicle when mired in snow, place transmission in 1 (first) and transfer case in L (low range). After vehicle is extracted from mired condition, immediately return transfer case to H/L (high/lock range) position and transmission to Ⓓ(overdrive).

   (2) If rear skidding occurs:

   (a) Let up on accelerator pedal.

   (b) Turn steering wheel in direction of skid until control has been regained.

   (c) Apply brake pedal in a gradual pumping manner.

c. After Operation.

   (1) Remove all ice and snow from underside of vehicle and fuel tank filler cap.

   (2) Drain fuel filter (para. 3-11).

2-43. OPERATING IN DUSTY, SANDY AREAS

a. General. Vehicles operating in dusty or sandy areas require frequent servicing of the air filter and cooling system.

NOTE

For dusty conditions, a air cleaner, NSN 2940-01-302-8028, can be inserted in the air intake shield assembly at the discretion of the unit commander.

b. Before Operation.

(1) When operating on loose sand or soft ground, place transfer case shift lever in H/L (high/lock range) position and transmission shift lever in (D)(overdrive).

(2) Refer to para. 2-40 for techniques that can be used when operating on unusual terrain.

c. During Operation.

NOTE

If additional power is needed to extract vehicle when mired in sand, place transmission in 1 (first) and transfer case in L (low range). After vehicle is extracted from mired condition, immediately return transfer case to H/L (high/lock range) position and transmission to (D)(overdrive).

(1) Frequently check air restriction gauge. If indicator shows red, park vehicle, stop engine, and refer to para. 3-15 for emergency air cleaner servicing.

(2) If engine overheating occurs:

(a) Park vehicle, and allow engine to idle.

(b) Observe coolant temperature gauge for steady cooling.

CAUTION

Stop engine if coolant temperature gauge suddenly increases beyond approximately 250°F (120°C) and/or overheat lamp illuminates. Failure to comply will result in damage to engine.

(c) If coolant temperature continues to increase or does not lower, stop engine. Perform applicable troubleshooting procedures in table 3-1.

(3) Accelerate slowly so wheels will not spin and dig into sand.

CAUTION

Use a wrecker or second vehicle equipped with winch to recover vehicles mired in deep sand. Do not attempt to rock vehicles out of deep sand with quick transmission shift changes. Damage to transmission will occur.

d. After Operation.

(1) At end of daily operation, clean all sand from accelerator linkage and brake components.

(2) Park vehicle in shade whenever possible to protect tires, soft tops, paint, wood, and seals from sun, dust, and sand.

(3) If shade is not available, cover vehicle with tarpaulin. When entire vehicle cannot be covered, protect windows and hood with tarpaulin to prevent entry of sand or dust.

(4) Vehicles completing operation in dusty, sandy areas must be lubricated and serviced by unit maintenance as soon as possible.

2-44. OPERATING IN MUD

a. **Before Operation.**

(1) Before operating in mud, place transfer case shift lever in H/L (high/lock range) and transmission shift lever in (D)(overdrive).

(2) Refer to para. 2-40 for techniques that can be used when operating on unusual terrain.

b. **During Operation.**

CAUTION

- Do not repeatedly shift transmission or overspeed the engine during operation in deep mud. Damage to drivetrain may result.
- Use wrecker or a second vehicle equipped with winch to recover vehicles mired in deep mud. Do not attempt to rock vehicles out of deep mud with quick transmission shift changes. Damage to transmission will occur.

NOTE

If additional power is needed to extract vehicle when mired in mud, place transmission in 1 (first) and transfer case in L (low range). After vehicle is extracted from mired condition, immediately return transfer case to H/L (high/lock range) position and transmission to (D)(overdrive).

Skidding and sudden loss of steering control are operating problems in mud. When rear end skidding occurs, immediately turn wheel in direction of skid until control has been regained.

c. **After Operation.**

WARNING

Do not rely on service brakes until they dry out. Keep applying brakes until uneven braking ceases. Failure to do this may cause damage to vehicle and injury or death to personnel.

CAUTION

Do not allow water to enter air intake cap or air cleaner assembly. Damage to engine will occur.

(1) Wash the following components as soon as possible with low-pressure water:

   (a) Radiator and oil cooler
   (b) Propeller shaft U-joint and halfshafts
   (c) Steering linkage and ball joints
   (d) Brake rotors and pads (service)
   (e) Parking brake linkage
   (f) Service lights
   (g) Transmission control linkage
   (h) Accelerator control linkage and "V" of engine block
   (i) Swaybar bushings
   (j) Towing pintle
   (k) Fuel filler cap
   (l) Vehicle exterior
   (m) Geared hubs
   (n) Advance solenoid rocker arm and fuel injection pump

2-44. OPERATING IN MUD (Cont'd)

(2) Remove mud from air cleaner dump valve (para. 3-16).

(3) Remove mud from drain hole (1) on converter housing cover (2).

(4) Remove mud from battery box drain holes.

NOTE

To prevent parking brake linkage from binding, lithium grease should be used after operating in mud.

(5) Clean mud, grit, and debris from linkage. Apply lithium grease (appendix D, item 14), and move linkage back and forth to work into joints.

(6) After operation in deep mud, vehicle must be lubricated and serviced by unit maintenance as soon as possible.

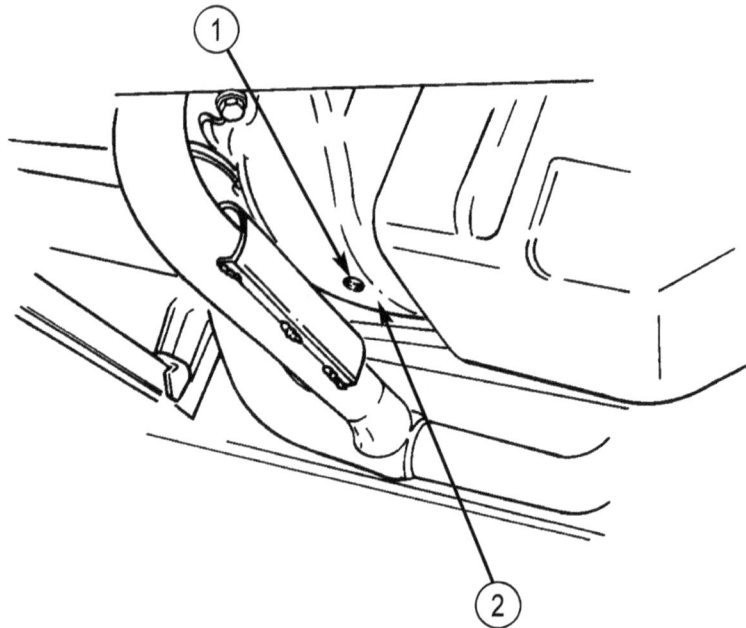

2-45. OPERATING IN EXTREME HEAT

a. General. Extreme heat exists when ambient temperatures reach 95°F (35°C) or more. The effect of extreme heat on vehicle engine is a decrease in engine efficiency.

b. Before Operation.

(1) Perform before operation checks and services in table 2-2.

(2) Check for foreign objects in front of radiator and clean as required.

(3) Check batteries more frequently. If electrolyte is low, add distilled water (appendix D, item 7).

c. **During Operation.**

CAUTION

Avoid continuous vehicle operation at high speeds. Avoid long, hard pulls on steep grades with transfer case shift lever in L (low range) position. Damage to transfer case will result.

(1) Frequently check coolant temperature gauge (2) and oil pressure gauge (1). Engine is overheating if one or more of the following conditions exist:

(a) Engine coolant temperature is more than approximately 250°F (120°C) as indicated by temperature gauge (2) and/or overheat lamp illuminates.

(b) Engine oil pressure drops below approximately 15 psi (103 kPa) with engine under a load.

(c) Engine oil pressure drops below approximately 6 psi (41 kPa) with engine at idle.

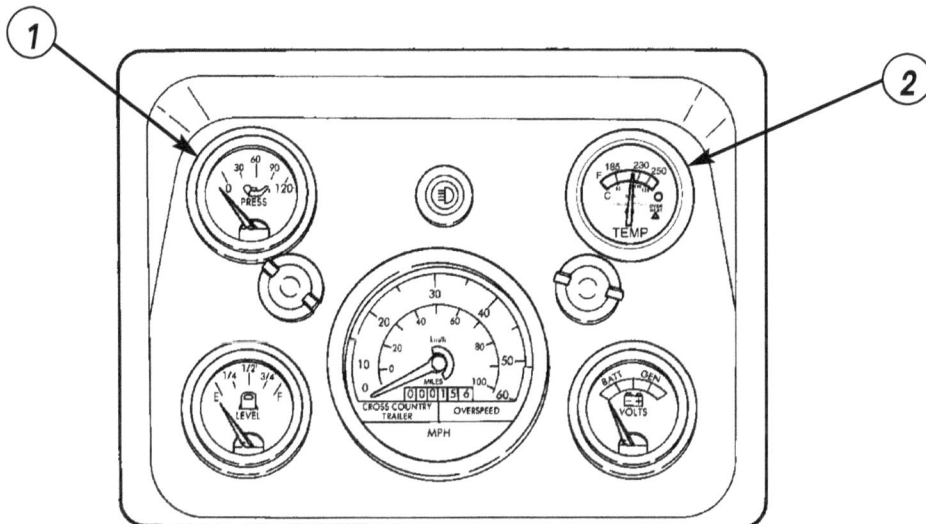

(2) If engine overheating occurs:

(a) Park vehicle, allowing engine to idle.

(b) Observe coolant temperature gauge (2) for steady cooling.

CAUTION

• Stop engine if coolant temperature gauge suddenly increases beyond approximately 250°F (120°C) and/or overheat lamp illuminates. Failure to comply will result in damage to engine.

• Notify unit maintenance to check differentials, transfer case, and transmission fluids for "oil breakdown" caused by overheating.

(c) If engine coolant temperature continues to increase or does not lower, stop engine. Perform troubleshooting procedures in table 3-1.

2-46. OPERATING IN RAINY OR HUMID CONDITIONS

a. General. Materiel inactive for long periods during rainy or humid conditions can rust rapidly. Fungus may develop in the fuel tanks as well as on soft tops, seats, and other components. Frequent inspections, cleaning, and lubrication are necessary to maintain the operational readiness of vehicles.

b. Before Operation. Fuel filter must be drained frequently because of high condensation in fuel system. To drain fuel filter, refer to para. 3-11.

c. During Operation.

(1) If necessary, place transfer case shift lever in H/L (high/lock range) to start without spinning wheels.

(2) Do not spin wheels when placing vehicle in motion in heavy rain.

(3) Refer to para. 2-40 for techniques that can be used when operating on unusual terrain.

2-47. SHALLOW WATER FORDING OPERATION

a. General. The vehicles have a 30 in. (76 cm) shallow water fording capability.

CAUTION

Never attempt shallow water fording unless water depth is known to be 30 in. (76 cm) or less, and bottom is known to be hard. Do not exceed 5 mph (8 kph) during fording operation. Damage to vehicle may result.

b. Before Operation.

(1) Ensure oil dipstick, transmission dipstick, oil filler cap, and fuel tank cap are secure.

(2) Secure all loose objects on vehicle.

(3) Ensure battery caps are all present and tight.

(4) Place transfer case shift lever in H (high range).

c. During Operation.

(1) Enter water slowly and maintain even vehicle speed while fording.

(2) Exit water in area with gentle slope.

NOTE

• Hydrostatic lock is caused by the entry of substantial amounts of water into the engine through the air intake system and subsequent contamination of the fuel system. Hydrostatic lock most frequently occurs during or just after fording. Water is forced into the air intake system, drawn into the engine, and locks up the engine.

• Notify unit maintenance if you suspect hydrostatic lock and they will test the engine.

d. **After Operation.**

WARNING

Do not rely on service brakes after fording until the brakes dry out. Keep applying brakes until uneven braking ceases. Failure to do this may cause damage to vehicle or injury or death to personnel.

NOTE

If accumulated water drains slowly through floor drain holes, refer to unit maintenance for drilling and improving drain holes.

(1) If fording operation was through salt water, wash and wipe off salt deposits as soon as possible.

NOTE

To prevent parking brake linkage from binding, lithium grease should be used after operating in water.

(2) Clean mud, grit, and debris from brake linkage. Apply lithium grease (appendix D, item 14), and move linkage back and forth to work into joints.

(3) After shallow water fording operation, vehicle must be lubricated and serviced by unit maintenance as soon as possible.

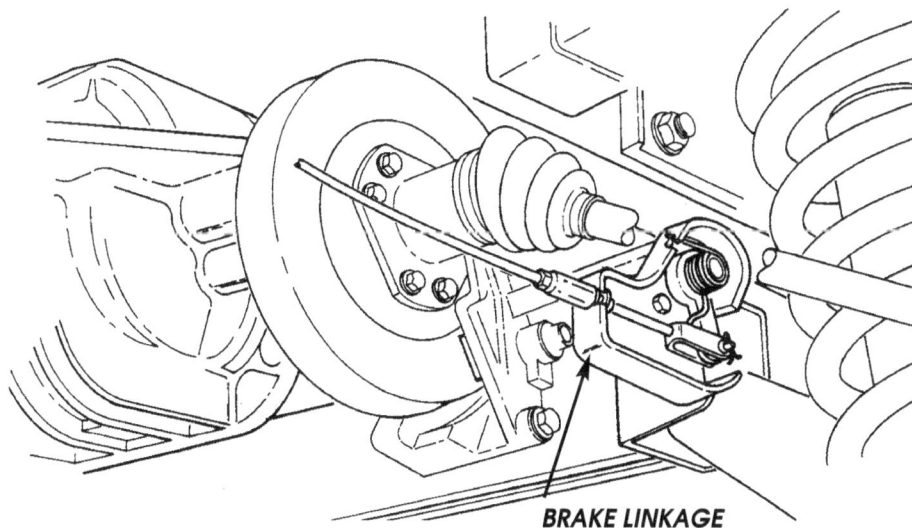

BRAKE LINKAGE

2-48. DEEP WATER FORDING (M1113, M1151, AND M1152) OPERATION

a. General. The deep water fording kit allows M1113, M1151, and M1152 vehicles to ford water up to 60 in. (152 cm) deep.

CAUTION

- Never attempt deep water fording unless water depth is known to be 60 in. (152 cm) or less, and bottom is known to be hard. Do not exceed 5 mph (8 kph) during fording operation. Damage to vehicle may result.

- Do not operate fan switch during deep water fording operation. Damage to vehicle may result.

b. Before Operation.

(1) Raise and secure hood (para. 3-9).

(2) Ensure rubber cap (2) on bottom of air cleaner body (1) is secure.

(3) Lower and secure hood (para. 3-9).

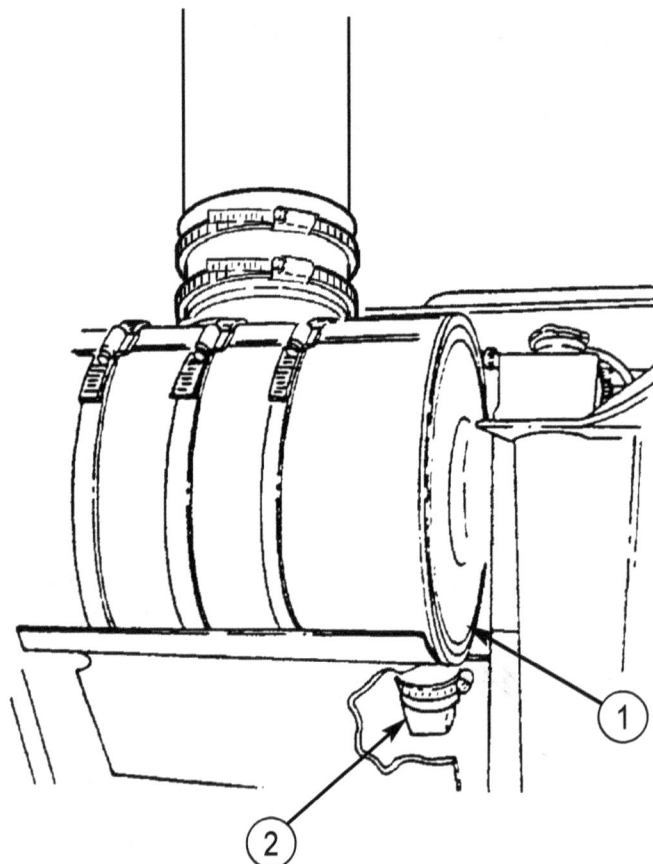

WARNING

Exhaust system components are hot after prolonged vehicle use. Ensure exhaust system components are cool before removing/installing exhaust assembly. Failure to do this may result in injury to personnel.

NOTE

Any items removed for fording must be stowed for reuse.

(4) Remove three locknuts (19), washers (11), capscrews (12), and washers (11) securing tailpipe (13) to muffler (18).

(5) Remove two nuts (17), lockwashers (16), and U-bolt (14) securing tail-pipe (13) to clamp (15).

(6) Remove tailpipe (13) and gasket (10) from muffler (18).

(7) Install two rubber isolators (7) into wheelhouse (8). To ease installation, wet rubber isolators (7) with water.

(8) If isolators (7) cannot be installed easily, check alignment of holes in wheelhouse (8) and reinforcement bracket (5). To align holes, loosen capscrews (6) securing reinforcement bracket (5) to wheelhouse (8). Align holes in wheelhouse (8) and reinforcement bracket (5) and tighten capscrews (6). Install isolators (7).

(9) Install exhaust assembly (9) and gasket (10) on muffler (18) with three washers (11), capscrews (12), washers (11), and locknuts (19).

(10) Install exhaust assembly (9) on wheelhouse (8) with two washers (4) and locknuts (3).

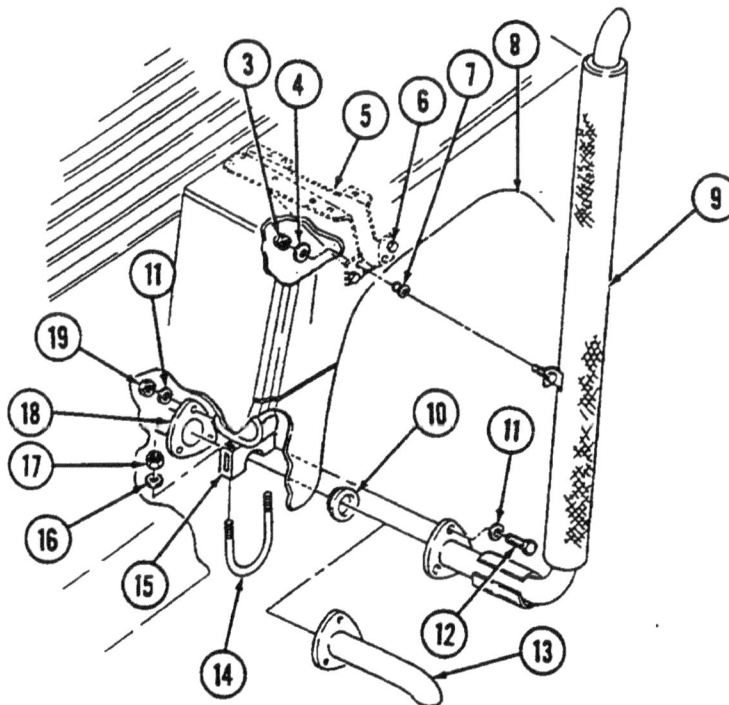

2-48. DEEP WATER FORDING (M1113, M1151, AND M1152) OPERATION (Cont'd)

(11) Ensure oil dipstick, transmission dipstick, oil filler cap, and fuel tank cap are secure.

(12) Secure all loose objects on vehicle.

(13) Ensure battery caps are all present and tight.

(14) Place transfer case shift lever in appropriate range (table 1-7).

(15) Turn off all non-essential electrical loads (lights, heater/defroster, A/C fan).

(16) Place fording selector switch (1) in DEEP FORD position prior to entering water.

(17) Pull out hand throttle (2) until desired engine speed is obtained. Twist hand throttle (2) to lock in position.

(18) Open driver and passenger windows.

c. **During Operation.**

(1) Enter water slowly and maintain even vehicle speed, 5 mph (8 kph) maximum.

(2) Exit water in area with gentle slope.

(3) Place fording selector switch (1) in VENT position after leaving water.

WARNING

After fording, do not use the hand throttle as an automatic speed or cruise control. The hand throttle does not automatically disengage when brake is applied, resulting in increased stopping distances and possible hazardous and unsafe operation. Injury to personnel or damage to equipment may result.

(4) Unlock and push in hand throttle (2).

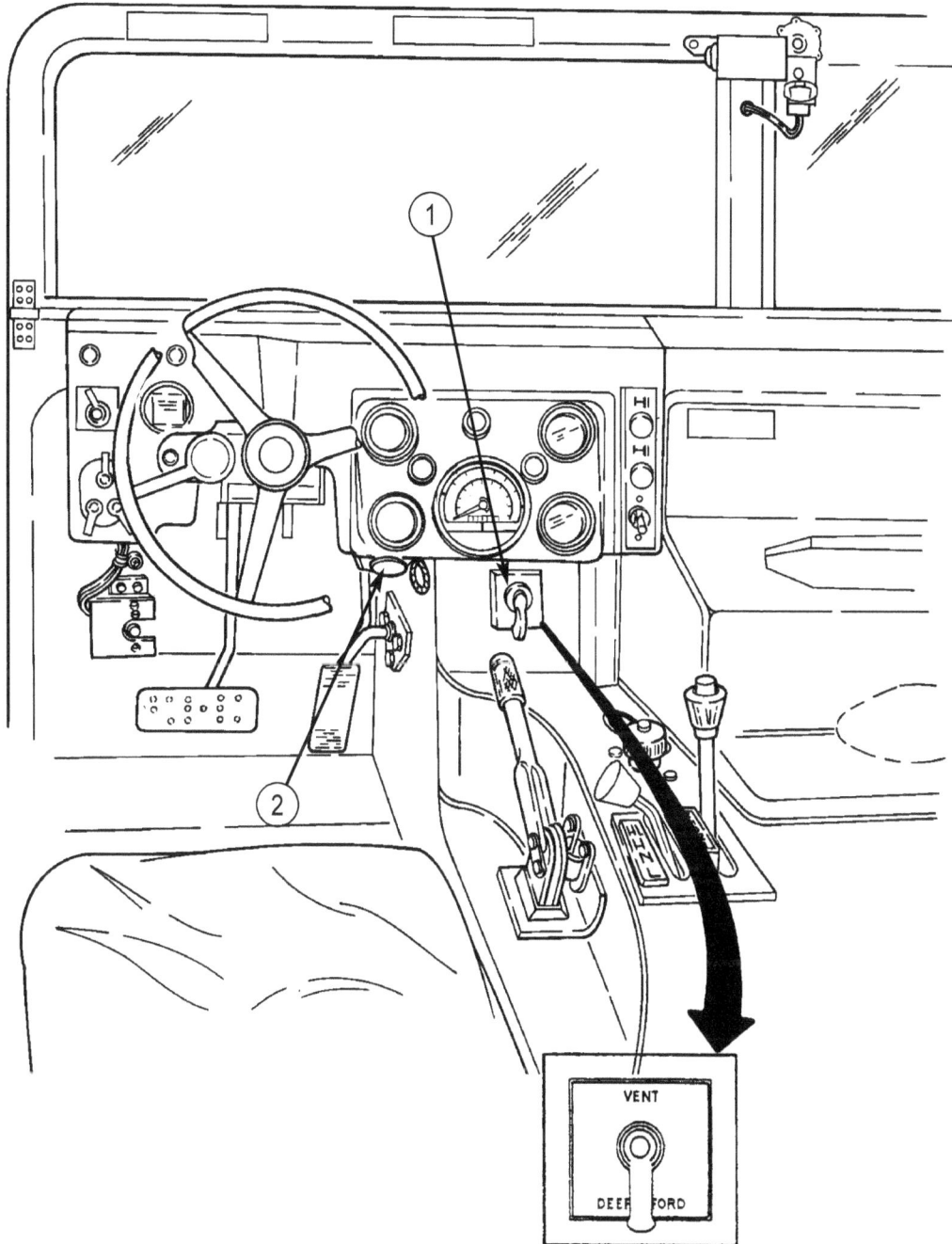

2-48. DEEP WATER FORDING (M1113, M1151, AND M1152) OPERATION (Cont'd)

WARNING

Do not rely on service brakes after fording until the brakes dry out. Keep applying brakes until uneven braking ceases. Failure to do this may cause damage to vehicle, and injury or death to personnel.

(5) Place transfer case shift lever in desired range (table 1-7).

NOTE

- Hydrostatic lock is caused by the entry of substantial amounts of water into the engine through the air intake system and subsequent contamination of the fuel system. Hydrostatic lock most frequently occurs during or just after fording. Water is forced into the air intake system, drawn into the engine, and locks up the engine.

- Notify unit maintenance if you suspect hydrostatic lock and they will test the engine.

d. After Operation.

(1) Stop engine.

WARNING

Exhaust system components are hot after prolonged vehicle use. Ensure exhaust system components are cool before removing/ installing exhaust assembly. Failure to do this may result in injury to personnel.

NOTE

- Steps 2 through 9 are performed only if required.

- If accumulated water drains slowly through the holes, refer to unit maintenance for drilling and improving drain holes.

(2) Remove three locknuts (14), washers (15), capscrews (7), washers (6), gasket (5), and exhaust assembly (4) from muffler (13).

(3) Remove two locknuts (1), washers (2), and exhaust assembly (4) from wheelhouse (3).

(4) Install gasket (5) and tailpipe (8) on muffler (13) with three washers (6), capscrews (7), washers (15), and locknuts (14).

(5) Install tailpipe (8) on clamp (10) with U-bolt (9), two lockwashers (11), and nuts (12).

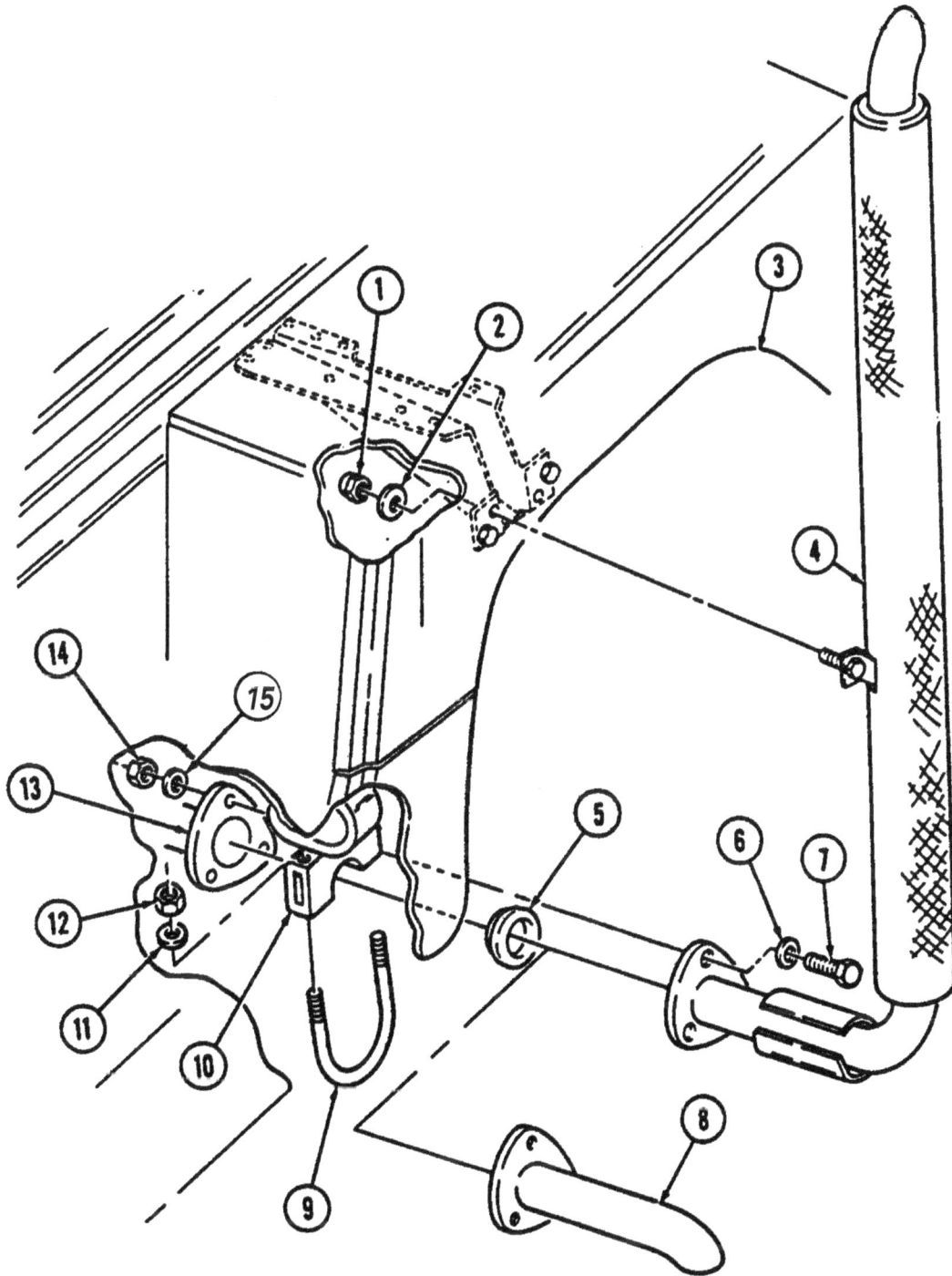

2-48. DEEP WATER FORDING (M1113, M1151, AND M1152) OPERATION (Cont'd)

(6) Clean and stow intake and exhaust assembly components.

(7) If fording operation was through salt water, wash and wipe off all salt deposits as soon as possible.

NOTE

To prevent parking brake linkage from binding, lithium grease should be used after operating in water.

(8) Clean mud, grit, and debris from brake linkage. Apply lithium grease (appendix D, item 14), and move linkage back and forth to work into joints.

(9) After deep water fording operation, vehicle must be lubricated and serviced by unit maintenance as soon as possible.

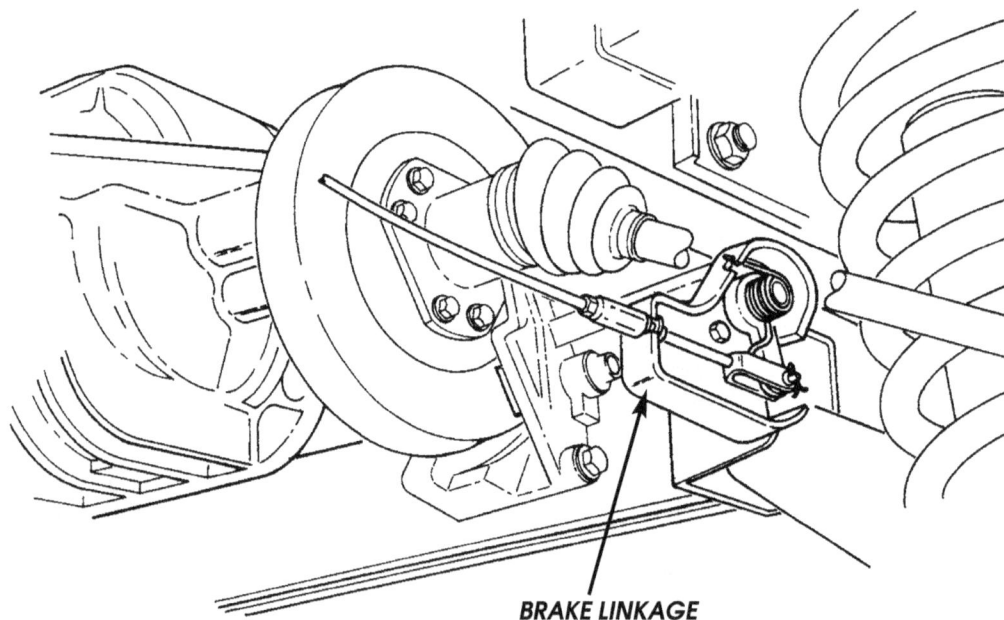

BRAKE LINKAGE

2-49. RUNFLAT OPERATION

The vehicles are equipped with runflat devices, allowing the vehicle to be driven with one or more tires flat. For runflat operations, refer to table 2-4.

WARNING

- Do not exceed 30 mph (48 kph) during any runflat operation. Do not exceed 20 mph (32 kph) with both rear tires flat. Loss of vehicle control may occur, causing damage to equipment and injury or death to personnel.
- Speeds indicated in table 2-4 are maximum and must be reduced when traveling on secondary roads, cross-country, or in traffic. Failure to reduce speeds could cause loss of control of vehicle, causing damage to equipment and injury or death to personnel.
- When driving vehicle, existing conditions are constantly changing. Never drive at a speed greater than is reasonable and prudent for these conditions. Loss of vehicle control may occur, causing damage to equipment and injury or death to personnel.

CAUTION

- A wheel that has been run flat must be replaced and inspected by unit maintenance as soon as possible before reuse, or damage to equipment may result.
- Runflat operation may cause the tread to separate from the tire and/or wheel. If abnormal handling is experienced, or noise such as flapping or pounding around the wheel well occurs, the tread needs to be cut away from the wheel before continuing operation. Failure to do so could result in damage to the vehicle.

Table 2-4. Runflat Operation.

NOTE

Runflat travel distance will improve with rubber runflat. If additional travel is required, tire tread may be cut away from tire.

Combination of Flat Tires	Recommended Vehicle Speed	Distance
Two tires flat – rear	20 mph max. (32 kph)	30 mi (48 km)
One tire flat – any location	30 mph (48 kph)	30 mi (48 km)
Two tires flat – same side	30 mph (48 kph)	30 mi (48 km)
Two tires flat – front only	30 mph (48 kph)	30 mi (48 km)

2-50. VEHICULAR HEATER OPERATION

NOTE
The vehicular heater is an optional kit which may or may not be installed on the vehicle.

a. General. The vehicular heater is used for pre-heating and boosting the heat of the water-cooled engine. The heater works in conjunction with the heating system of the vehicle to heat driver's cabin, passenger compartment, and to defrost windows and windshield of the vehicle.

The heater operates independently of the vehicle engine. It is connected to the coolant circuit or the heating circuit, fuel supply system, and the electrical system of the vehicle.

When the operating temperature has been reached, the temperature sensor will send a signal to the control unit. The heat output will then be reduced to 25 percent. Should the temperature still remain at the upper limit, the heater will turn off, but the coolant pump will remain running.

The heater will automatically restart once the system temperature has dropped to the lower switch point of the sensor.

WARNING
• Do not operate the vehicular heater in a closed area without proper exhaust evacuations. Damage to heater or injury to personnel may result.
• Do not operate the vehicular heater when refueling. Damage to heater or injury to personnel may result.

NOTE
• After shutting off the on/off switch there is a cool down period of 90 seconds. The blower and coolant pump works for 90 seconds to purge the heater.
• DFA fuel should be used at temperatures below -40°F (-4°C).

b. Operation.

(1) Activate heater by moving on/off switch (1) on the vehicle control panel (3) to ON.

(2) An amber light (2) indicates heater is in operation.

(3) Pull heater control knob (4) all the way out for maximum heat output.

(4) If, during heater operation, supplied voltage drops for more than 20 seconds, fault lock out will occur and the heater will switch off. If this happens, notify unit maintenance.

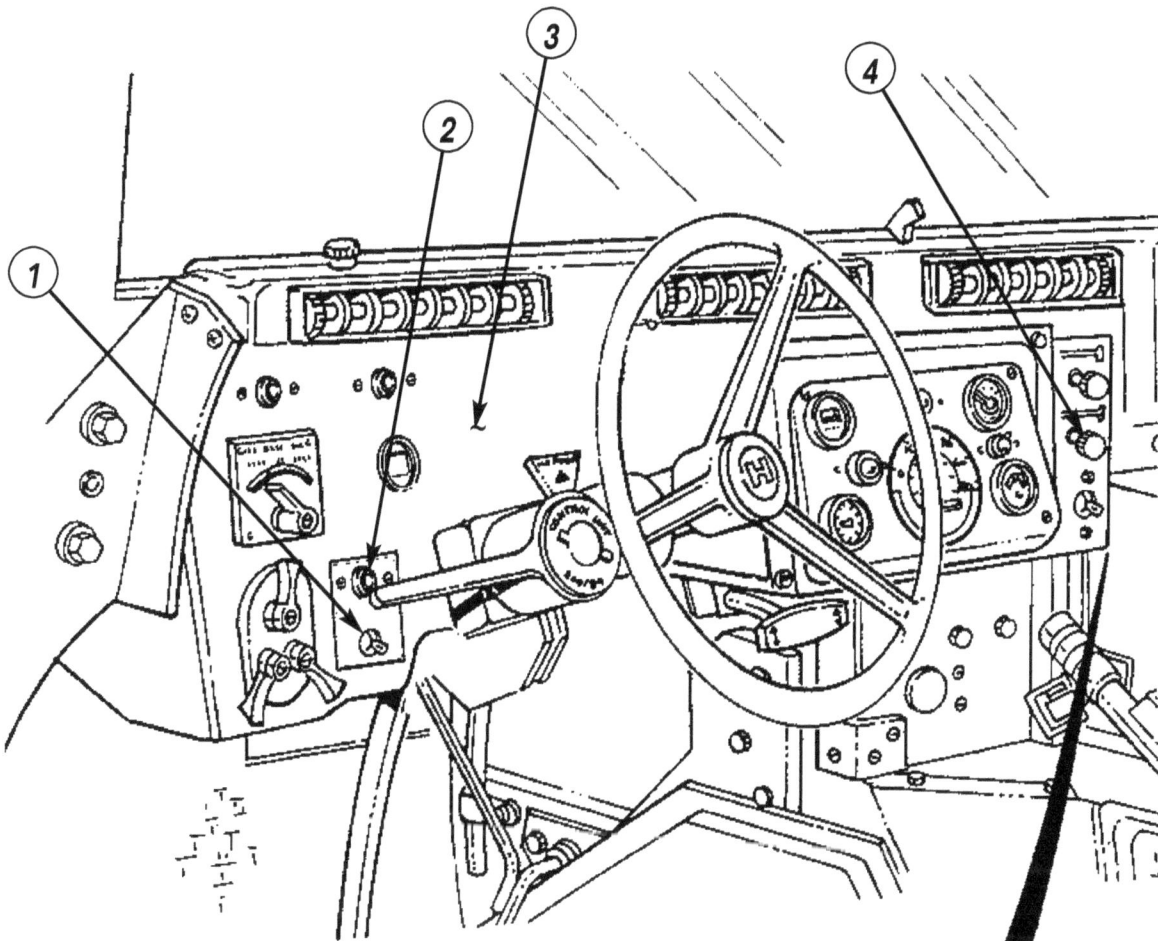

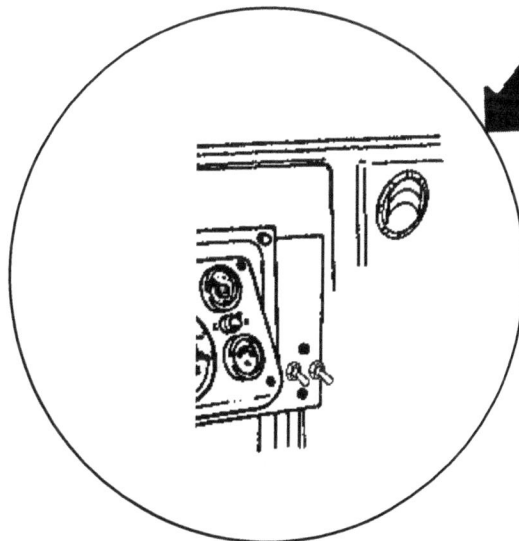

M1151, M1151A1, M1152,
M1152A1, M1165, M1165A1, M1167

2-50.1. TROOP/CARGO WINTERIZATION HEATER OPERATING INSTRUCTIONS

a. General.

Troop/cargo winterization heater is for use during extremely cold temperatures from 0°F (-18°C) to -50°F (-46°C).

CAUTION

Do not operate troop/cargo heater in ambient temperatures above 0°F (-18°C), or with fan switch in HI position, when temperature is above -25°F (-32°C). Damage to equipment will result.

NOTE

Heater may be operated with engine running or engine off.

b. Starting Heater.

(1) Set heater fan switch (3) to LO position.

(2) Move heater control switch (2) to START position and hold for approximately two minutes or until heater starts and heater indicator light (1) comes on.

(3) When heater starts, move heater control switch (2) to RUN position. The heater indicator light (1) will remain illuminated.

(4) Set heater fan switch (3) to desired position (HI or LO).

c. If Heater Fails to Start:

NOTE

If the start switch is moved from the START to RUN position too quickly, the heater flame will be extinguished and the heater will need restarting.

(1) Move heater control switch (2) to OFF position for 10 seconds, and then back to START for 60 seconds. Repeat procedure if heater doesn't start.

NOTE

If heater indicator light works and heater still fails to start in approximately three minutes, service is required

(2) Press heater indicator light (1) to test electrical circuit.

d. Shutting Heater Off.

Move heater control switch (2) to OFF.

NOTE

The heater indicator light will remain on and the fan will continue to run until the heater has purged itself of fuel.

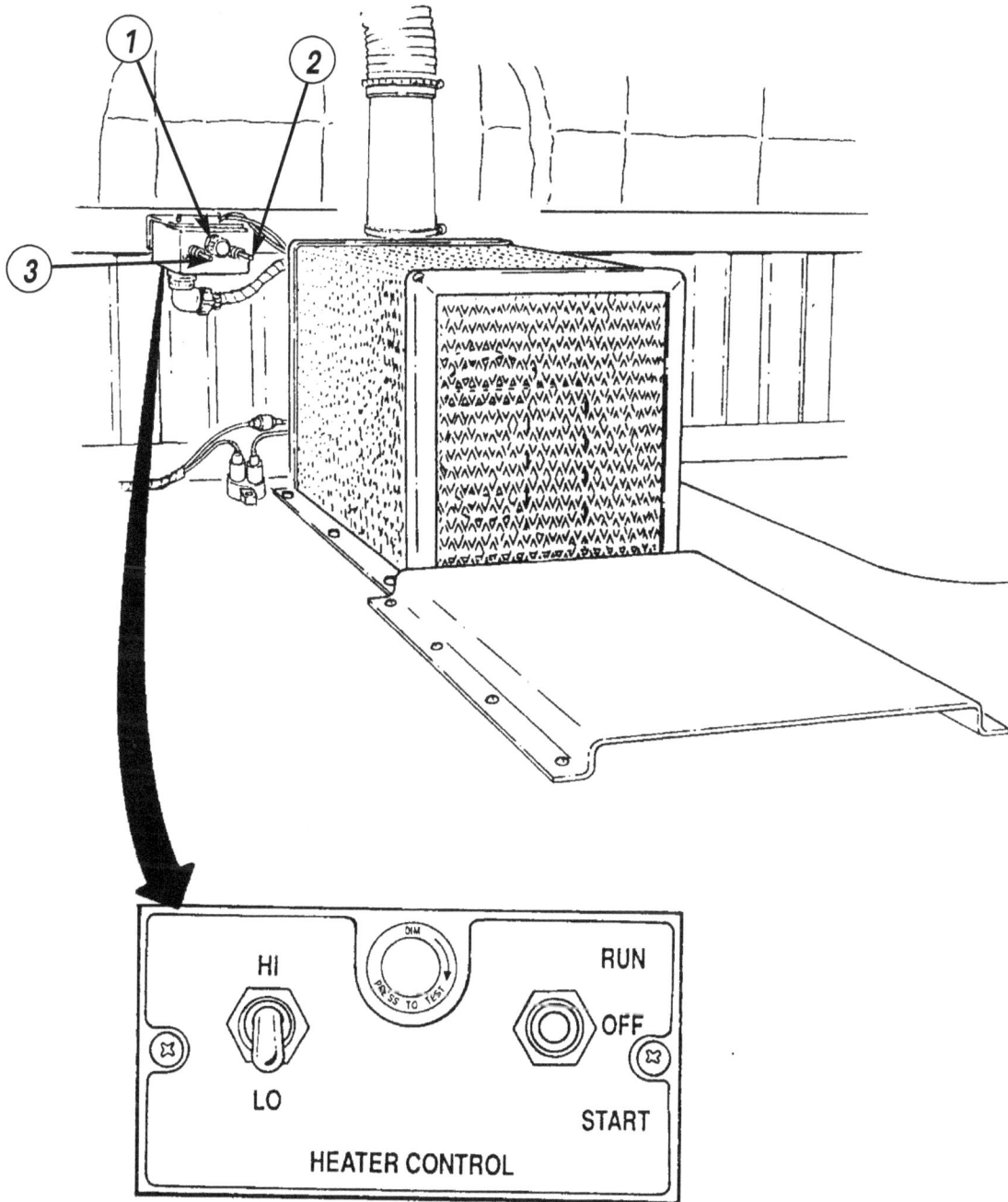

HI

RUN

OFF

LO

START

HEATER CONTROL

Section V. CARGO/TROOP CARRIER AND
S250 SHELTER CARRIER OPERATION

2-51. GENERAL

a. This section provides operating instructions for components found on M1113, M1152, and M1152A1 (if equipped) shelter carriers.

b. Refer to para. 2-2a for stowage location of fire extinguisher on S250 shelter carrier. Refer to para. 2-29 for operation of fire extinguisher.

2-52. CARGO/TROOP CARRIER AND S250 SHELTER CARRIER OPERATION REFERENCE INDEX

2-53. WINDSHIELD ASSEMBLY OPERATION

WARNING

Do not operate vehicle without windshield assembly positioned upright and the B-pillar securely attached.Operation of vehicle without these structures in place may result in injury to personnel and damage to equipment.

NOTE

Before lowering windshield on vehicles equipped with retention bracket, the windshield retention bracket must be removed. Notify unit maintenance.

a. Lowering Windshield Assembly.

(1) Remove two hitch pins (3) from inside hinge pins (4) and remove hinge pins (4).

(2) Lower windshield assembly (5) to hood and secure windshield assembly (5) to four footman loops (2) with two straps (1).

b. Raising Windshield Assembly.

(1) Remove two straps (1) from four footman loops (2) securing windshield assembly (5) to hood.

(2) Raise windshield assembly (5).

NOTE

Notify unit maintenance to install windshield retention bracket.

(3) Secure windshield assembly (5) with two inside hinge pins (4) and hitch pins (3).

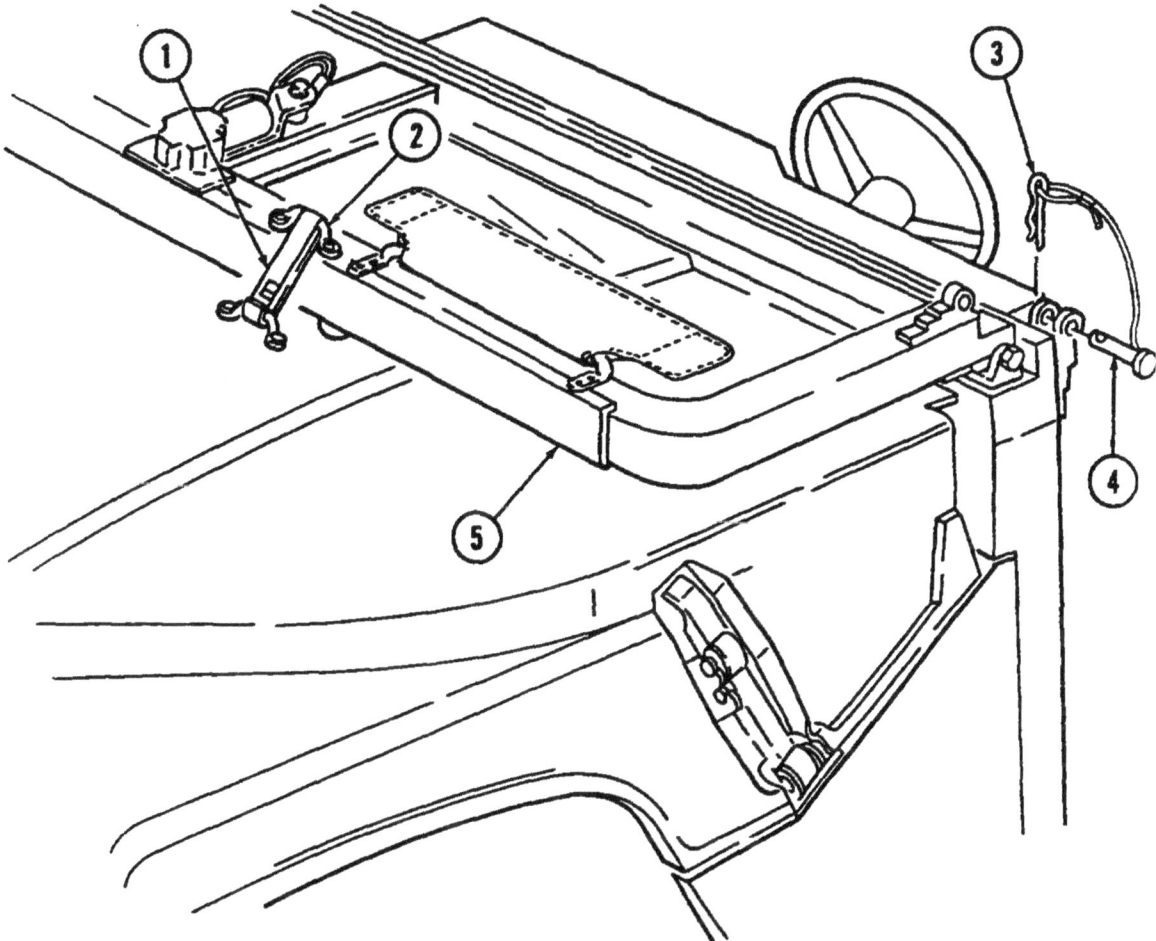

2-53.1. TROOP SEAT KIT OPERATION (M1152, M1152A1)

a. **General.** The troop seat kit is used to convert M1152 and M1152A1 vehicles into troop carriers.

b. **Lowering Troop Seat.**

(1) Remove two lockpins (7) from troop seat (6).

(2) Lower troop seat (6).

(3) Install two lockpins (7) to secure troop seat (6).

c. **Raising Troop Seat.**

(1) Remove two lockpins (7) from troop seat (6).

(2) Raise troop seat (6) and secure with two lockpins (7).

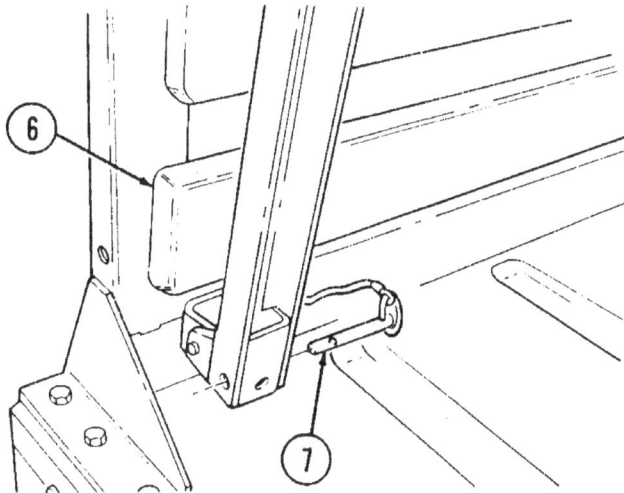

2-54. REMOVAL AND INSTALLATION OF TWO-MAN CREW AREA SOFT TOP ENCLOSURE

NOTE
- For ease of installation, soft top components should be installed when temperature is above 72°F (22°C).
- To keep hinge screws tight, lockwasher NSN 5310-00-527-3634 and locknut NSN 5310-00-241-6658 can be used for installation of soft top doors.
- To prevent seams from leaking, coat with adhesive as needed (appendix D, item 1).

a. General. The two-man soft top enclosure consists of two soft doors, rear curtain, cab roof cover, two rails, and bow assembly.

b. Removal of Soft Top Doors.

(1) Open door (1) and detach door holding check strap (3) from mounting plate (2).

(2) Raise door (1) to remove hinge pins (5) from hinge brackets (4).

(3) Remove door (1) from vehicle.

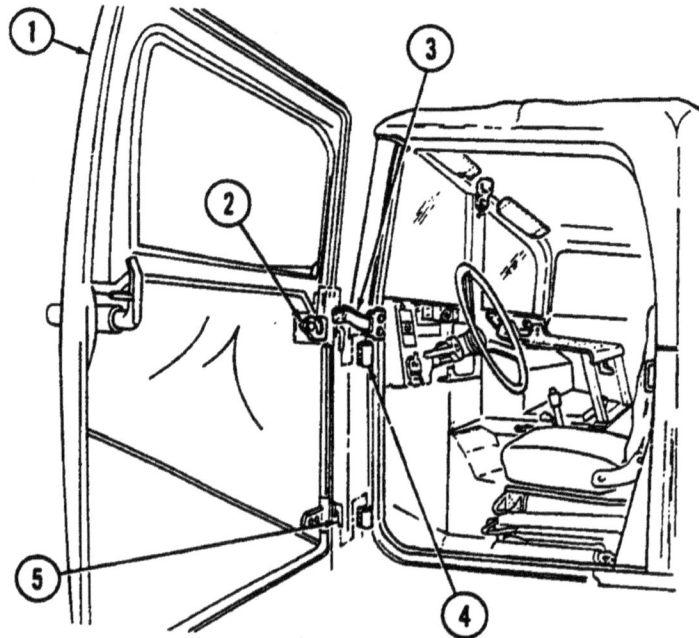

c. Removal of Cab Roof Cover, Rails, and Bow Assembly.

(1) Unfasten eyelets (7) from turnbuttons (15) on horizontal rails (14), and unfasten hook and loop attachments (9) securing cab roof cover (10) to bow assembly (8).

(2) Remove cab roof cover (10) from horizontal rails (14).

(3) Unfasten eyelets (11) from turnbuttons (12) on B-pillar (13) and A-pillar (18).

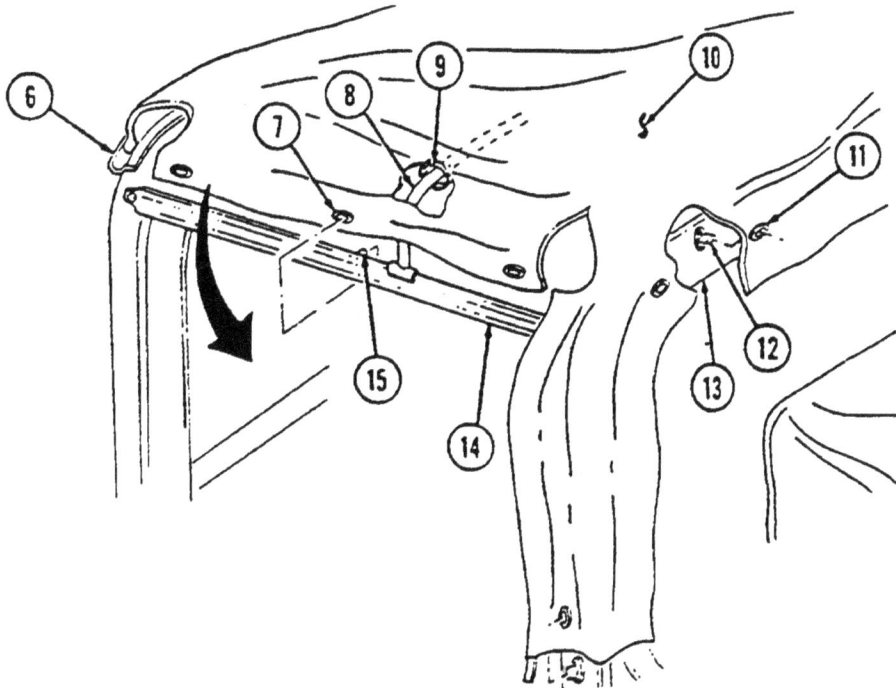

(4) Roll cab roof cover (10) over windshield and slide cab roof cover (10) from channel (6).

(5) Remove four screws (17) and bow assembly (8) from two horizontal rails (14).

(6) Remove four screws (16) securing two horizontal rails (14) to A-pillar (18) and B-pillar (13). Remove horizontal rails (14).

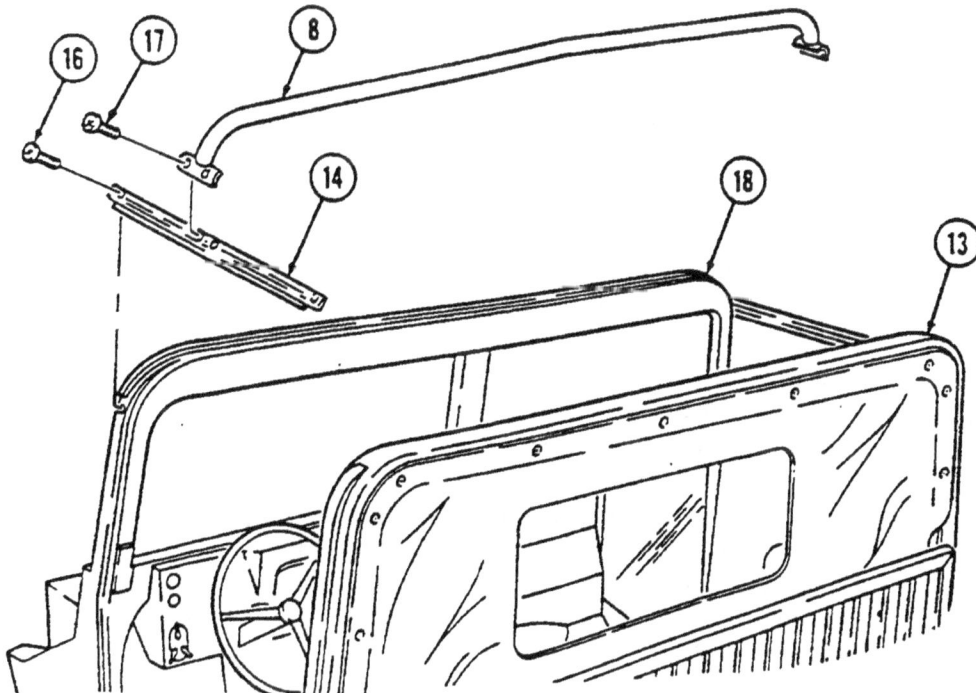

2-54. REMOVAL AND INSTALLATION OF TWO-MAN CREW AREA SOFT TOP ENCLOSURE (Cont'd)

d. Removal of Rear Curtain.

(1) Detach curtain straps (5) from footman loops (7) behind seats by depressing locking tabs (6).

NOTE

Perform step 2 if vehicle is equipped with two-man arctic cab.

(2) Peel curtain (4) back from fastener tape (8).

(3) Unfasten eyelets (3) from turnbuttons (2) on B-pillar (1).

(4) Remove rear curtain (4) from B-pillar (1).

e. Installation of Rear Curtain.

(1) Install rear curtain (4) on B-pillar (1) by fastening eyelets (3) to turn-buttons (2).

NOTE

Perform step 2 if vehicle is equipped with two-man arctic cab.

(2) Attach curtain (4) to fastener tape (8).

(3) Attach curtain straps (5) to footman loops (7) located behind seats. Tighten straps (5) equally, but do not overtighten.

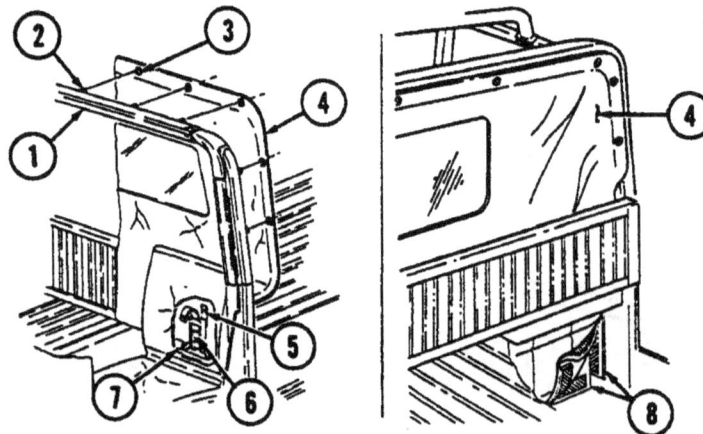

f. Installation of Cab Roof Cover, Rails, and Bow Assembly.

(1) Install two horizontal rails (12) on A-pillar (13) and B-pillar (1), and secure with four screws (9). The short end of horizontal rails (12) goes toward front of vehicle.

(2) Loosen end bracket screws (14) and install bow assembly (11) on horizontal rails (12). Secure with four screws (10) and tighten bracket screws (14).

(3) Slide cab roof cover (18) into channel (15) on A-pillar (13) and roll cab roof cover (18) over cab.

(4) Install cab roof cover (18) over rear curtain (4) at B-pillar (1) and fasten eyelets (3) to turnbuttons (2) on B-pillar (1) and A-pillar (13).

(5) Position cab roof cover (18) around two horizontal rails (12) and fasten eyelets (16) to turnbuttons (19). Secure cab roof cover (18) to bow assembly (11) with hook and loop attachments (17).

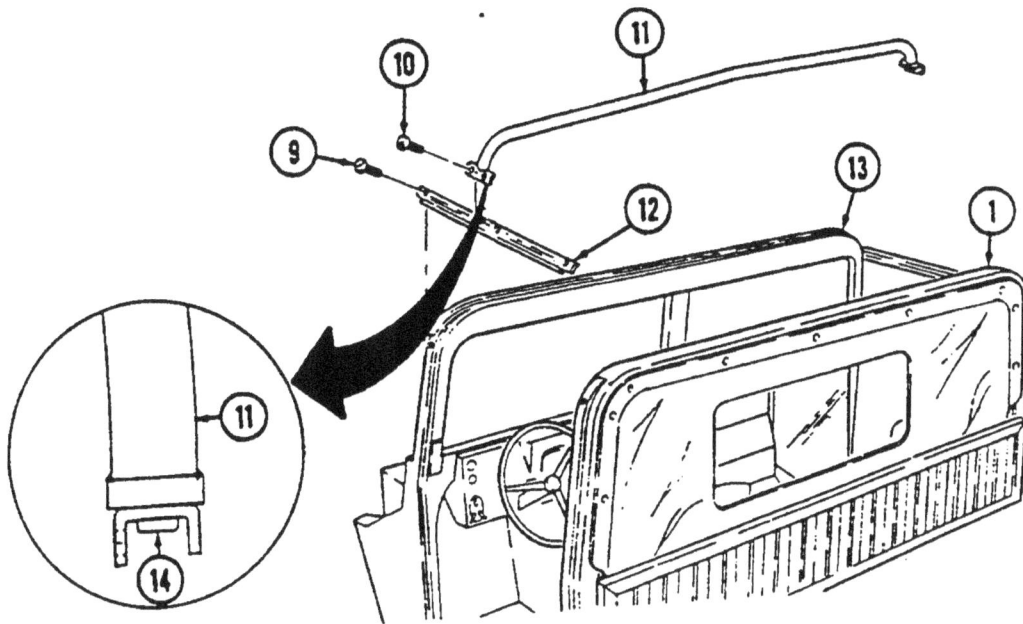

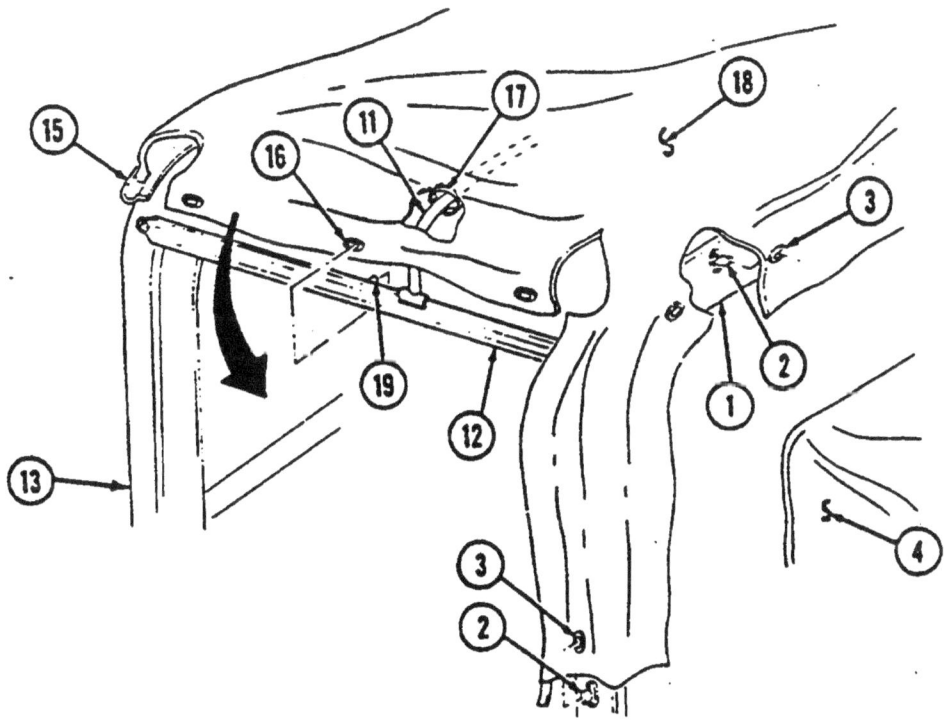

2-54. REMOVAL AND INSTALLATION OF TWO-MAN CREW AREA SOFT TOP ENCLOSURE (Cont'd)

g. Installation of Soft Top Doors.

(1) Apply a small amount of seasonal grade OE/HDO oil to hinge pins (6), and install door (1) by inserting hinge pins (6) into hinge brackets (5).

(2) Install door holding check strap (3) on mounting plate (2).

CAUTION

If door is jammed, do not force shut. Excessive force may damage door.

(3) Close door (1). It may be necessary to adjust door hinges (5) or door latch striker (4) to achieve a tight seal. If adjustment is necessary, notify unit maintenance.

NOTE

When lowering soft top door window, always fold window to inside of vehicle and secure between door and door frame crossmember.

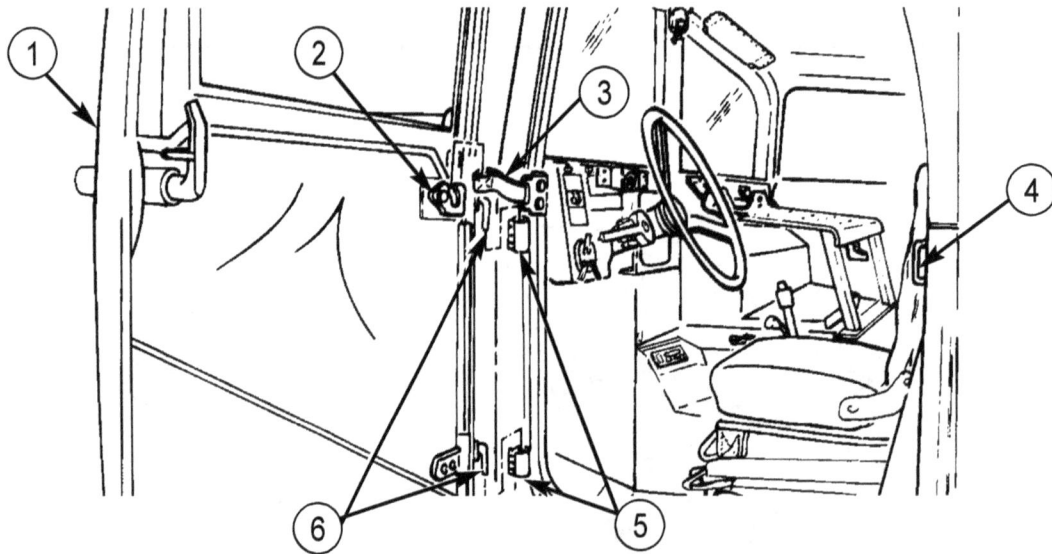

2-54.1. REMOVAL AND INSTALLATION OF TROOP AREA SOFT TOP ENCLOSURES (M1152, M1152A1)

CAUTION

Remove any accumulation of rain, snow, and ice from the cargo cover as soon as possible. Failure to do so could result in damage to the cargo cover assembly. If the vehicle is to be parked for a long period of time, the cargo cover can be removed.The decision whether or not to remove the cover should be based on the length of time the vehicle is to be parked and the expected weather conditions.

NOTE

- For ease of installation, soft top components should be installed when temperatures are above 72°F (22°C).
- To prevent seams from leaking, coat with adhesive as needed. (Refer to appendix D, item 1.)

a. **Removal of Troop Area Enclosure.**

   (1) Unfasten eyelets (9) from turnbuttons (8) on B-pillar (16), front bow (10), and rear bow (13).

   (2) Press locking tab (20) and unhook straps (14) from footman loops (19).

   (3) Detach hook and loop attachment securing front flap (7) of troop area soft top (12) from two-man crew area soft top (17). Remove grommets (18) from footman loops (19).

   (4) Remove troop area soft top (12).

b. **Removal of Bows.** Remove bows (10), (11), and (13) from bow retainer (15).

c. Installation of Bows. The front bow assembly (10) and rear bow assembly (13) contain turnbuttons (8); the front bow assembly (10) has shorter legs. Install bows (10), (11), and (13) into bow retainers (15).

d. Installation of Troop Area Enclosure.

NOTE

- Troop seats and two-man crew area soft top must be installed prior to installation of bows and troop area enclosure.

- To prevent canvas noise and damage, tiedowns can be added to the canvas at the commander's discretion; refer to unit maintenance.

(1) Install troop area soft top (12) over B-pillar (16), front bow (10), intermediate bows (11), and rear bow (13).

(2) Fasten eyelets (9) to turnbuttons (8) on B-pillar (16), front bow (10), and rear bow (13).

(3) Fit troop area soft top (12) evenly to ensure tight fit and secure by installing grommets (18) over footman loops (19). Attach straps (14) to footman loops (19) and pull straps (14) tight.

(4) Secure front flap (7) of troop area soft top (12) to two-man crew area soft top (17).

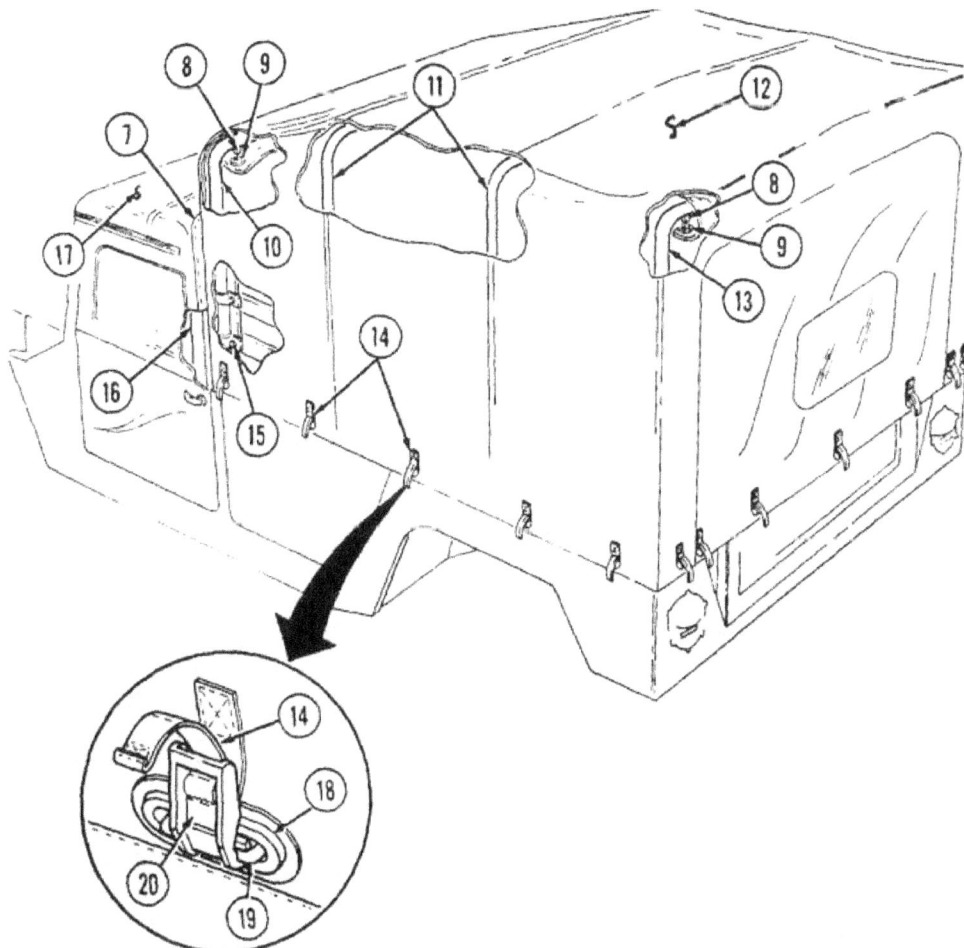

2-54.2. REMOVAL AND INSTALLATION OF FOUR-MAN CREW AREA SOFT TOP AND ARCTIC SOFT TOP ENCLOSURES

NOTE

• For ease of installation, soft top components should be installed when temperatures are above 72°F (22°C).

• To prevent seams from leaking, coat with adhesive as needed. (Refer to appendix D, item 1.)

a. General. The four-man soft top and arctic soft top enclosures consist of four soft doors, rear curtain, cab roof cover, and bow assemblies. For removal and installation of doors, refer to para. 2-46. Para. 2-48 covers removal and installation of four-man cab roof cover, rails, bow assemblies, and rear curtain.

b. Removal of Cab Roof Cover, Rails, and Bow Assemblies.

(1) Unfasten eyelets (2) from turnbuttons (3) on horizontal rails (12) and A-pillar (13). Unfasten hook and loop attachments (5) securing cab roof cover (6) to bow assemblies (4) and B-pillar (11).

(2) Remove cab roof cover (6) from horizontal rails (12).

(3) Unfasten eyelets (8) from turnbuttons (7) on C-pillar (9).

(4) Roll cab roof cover (6) over windshield and slide cab roof cover (6) from channel (1).

(5) Remove eight screws (14) and two bow assemblies (4) from horizontal rails (12).

(6) Remove six screws (15) and two horizontal rails (12) from A-pillar (13), B-pillar (11), and C-pillar (9).

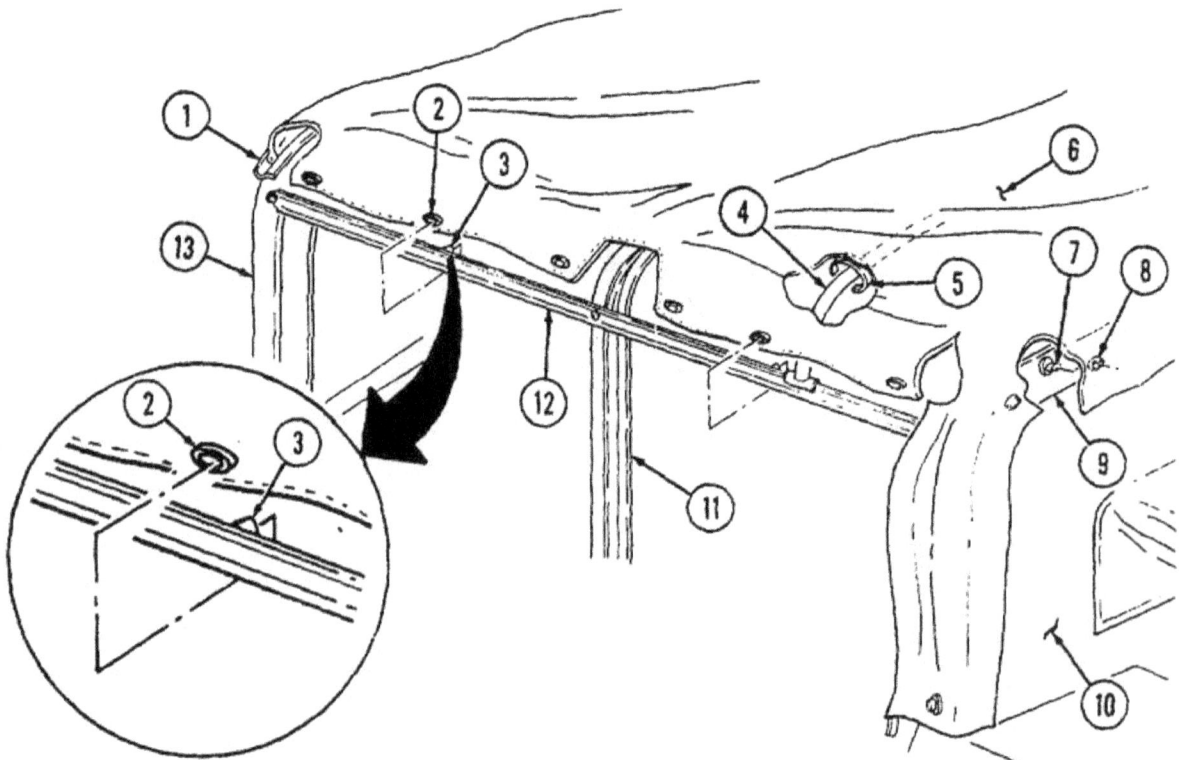

c. Installation of Rails, Bow Assemblies, and Cab Roof Cover.

NOTE

Align horizontal rail holes with holes in A, B, and C-pillars.

(1) Install two horizontal rails (12) on A-pillar (13), B-pillar (11), and C-pillar (9) and secure with six screws (15).

(2) Loosen end bracket screws (16), and install two bow assemblies (4) on horizontal rails (12) with eight screws (14). Tighten bracket screws (16).

(3) Slide cab roof cover (6) into channel (1) on A-pillar (13), roll cab roof cover (6) over cab, and fasten eyelets (2) to turnbuttons (3) on A-pillar (13).

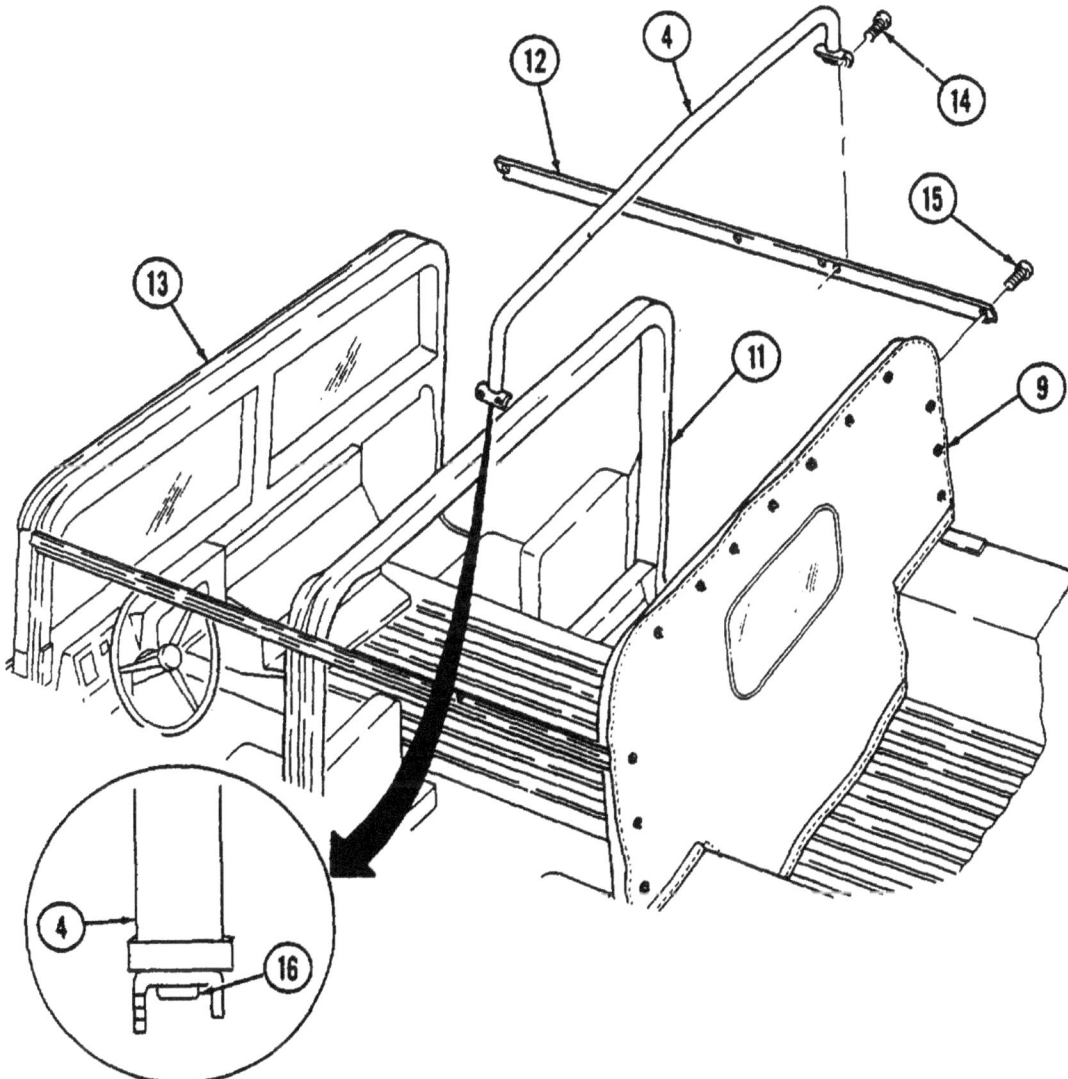

(4) Install cab roof cover (6) over rear curtain (10) at C-pillar (9) and fasten eyelets (8) to turnbuttons (7) on C-pillar (9).

(5) Position cab roof cover (6) around horizontal rails (12) and fasten eyelets (2) to turnbuttons (3). Attach cab roof cover (6) to bow assemblies (4) with hook and loop attachments (5).

d. Removal of Rear Curtain.

(1) Detach four curtain straps (11) from footman loops (13) on wheelhousings (15) and cargo floor (14) by depressing locking tabs (12).

NOTE

Perform step 2 if vehicle is equipped with four-man arctic cab.

(2) Peel curtain (2) back from fastener tape (3).

(3) Unfasten eyelets (10) from turnbuttons (16) on C-pillar (1).

NOTE

Perform step 4 for vehicles equipped with three-point seatbelts.

(4) Unfasten eyelets (7) from turnbuttons (8) on rear seatbelt bracket (9).

(5) Remove curtain (4).

e. Installation of Rear Curtain.

(1) Install curtain (4) on C-pillar (1) by fastening eyelets (10) to turnbuttons (16).

NOTE

• Perform steps 2 and 3 when installing new curtain on vehicles with three-point seatbelts.

• Ensure P/N 12342475 is used for replacement of rear curtain on vehicles equipped with three-point seatbelt.

(2) Cut away inside curtain panel (17) and remove curtain panel (17) from curtain (4).

(3) Cut away outside curtain panel (6) and remove curtain panel (6) from curtain (4).

(4) Extend rear seatbelt bracket boot (5) and install over rear seatbelt bracket (9).

(5) Fasten eyelets (7) to turnbuttons (8) on rear seatbelt bracket (9).

NOTE

Perform step 6 if vehicle is equipped with four-man arctic cab.

(6) Attach curtain (2) to fastener tape (3).

(7) Attach four curtain straps (11) to footman loops (13) on wheelhousings (15) and cargo floor (14). Tighten straps (11) equally, but do not overtighten.

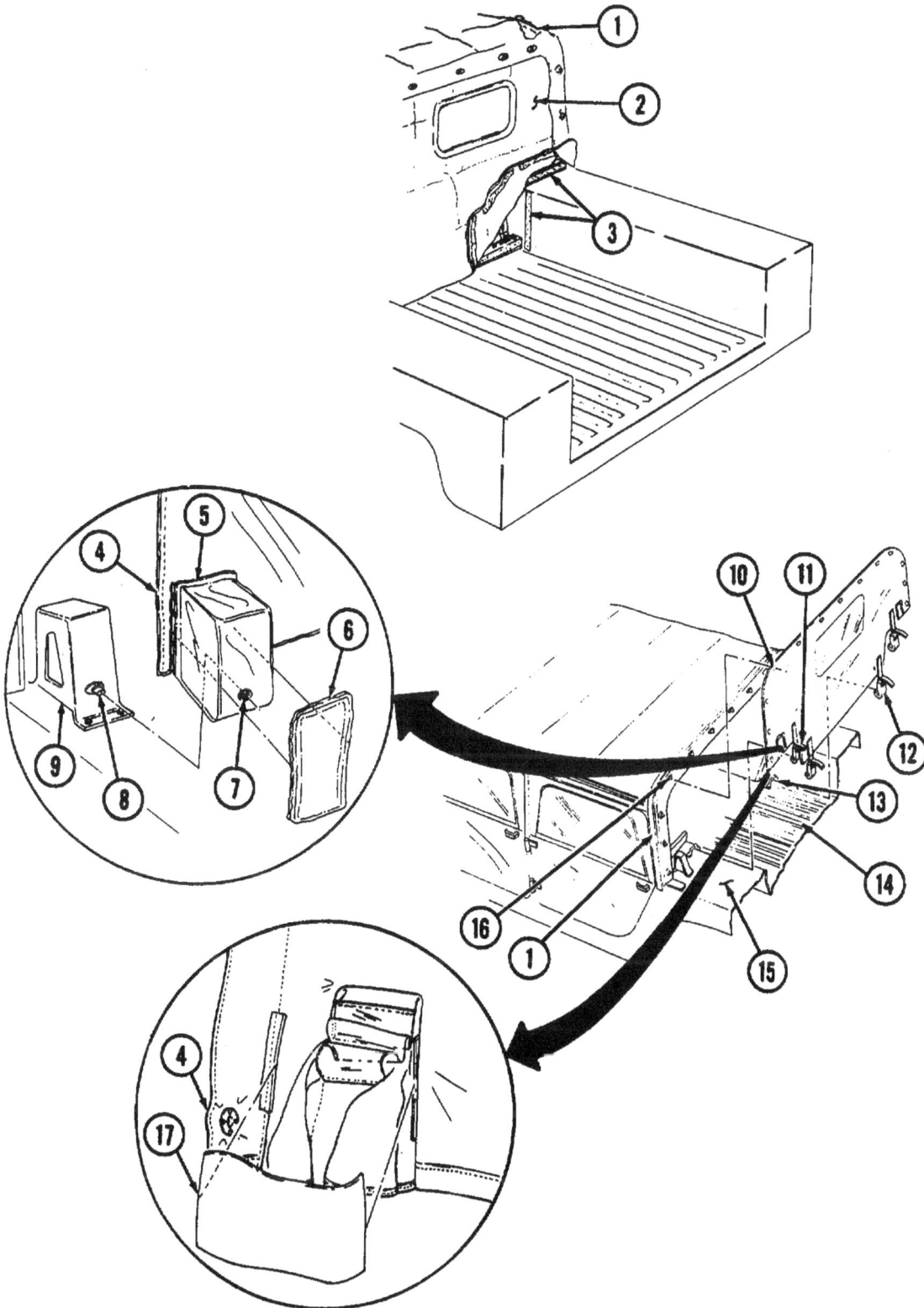

2-54.3. REMOVAL AND INSTALLATION OF FOUR-MAN CARGO AREA SOFT TOP ENCLOSURE

CAUTION

Remove any accumulation of rain, snow, and ice from the cargo cover as soon as possible. Failure to do so could result in damage to the cargo cover assembly. If the vehicle is to be parked for a long period of time, the cargo cover can be removed. The decision whether or not to remove the cover should be based on the length of time the vehicle is to be parked and the expected weather conditions.

NOTE

- For ease of installation, soft top components should be installed when temperatures are above 72°F (22°C).
- To prevent seams from leaking, coat with adhesive as needed. (Refer to appendix D, item 1.)

a. Removal of Cargo Enclosure.

   (1) Unfasten hook and loop attachment (3) securing cargo cover (6) to cab roof (2) at C-pillar (8).

   (2) Unfasten eyelets (5) from turnbuttons (4) on C-pillar (8), front bow (1), and rear bow (7).

   (3) Detach straps (9) from footman loops (11) by depressing locking tabs (12) and loosening straps (9). Remove grommets (10) from footman loops (11).

   (4) Remove cargo cover (6).

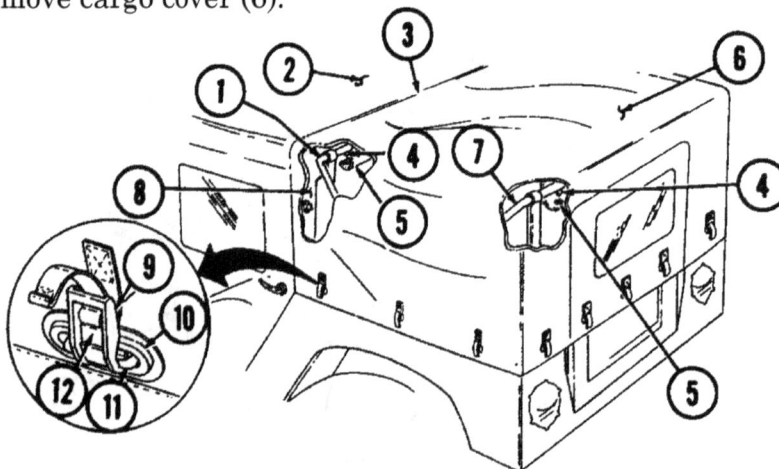

b. Removal of Bows.

NOTE

If bows are broken or bent, reinforce 3/64 in. (1.2 mm) thick bow walls with steel rod NSN 9510-00-596-2063. Reinforce 3/32 in. (2.4 mm) bows with steel rod NSN 9510-00-596-2066.

   (1) Depress locking tabs (14) and unhook straps (13) from footman loops (15).

   (2) Remove capscrews (19) securing longitudinal bow (20) to front bow (1) and rear bow (7). Remove longitudinal bow (20).

   (3) Remove four snaprings (17) and pivot pins (16) securing front bow (1) and rear bow (7) to pivot brackets (18). Remove bows (1) and (7).

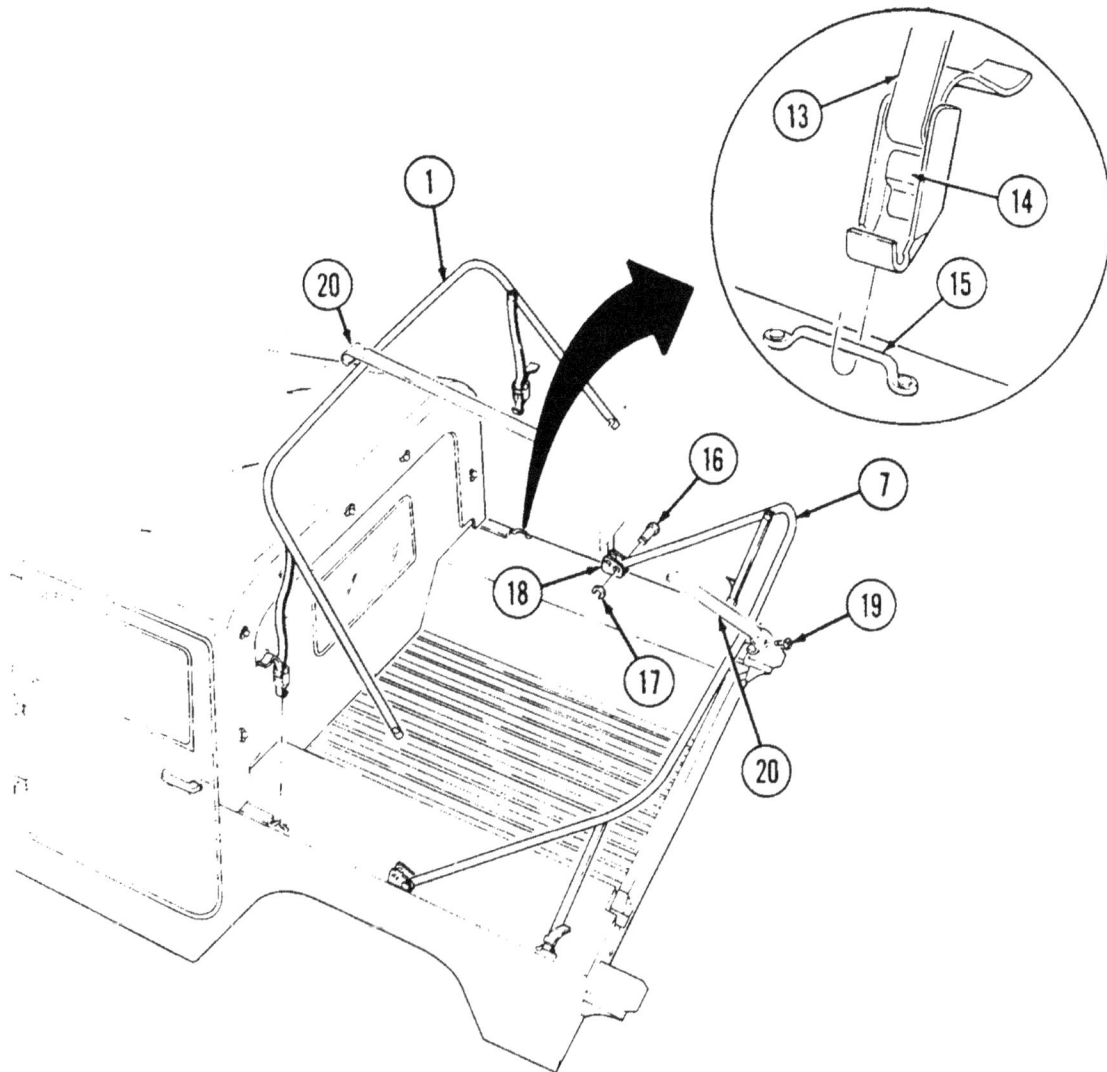

c. Installation of Bows.

NOTE

Four-man crew area soft top must be installed before installation of bows and cargo enclosure.

(1) Install bows (1) and (7) on pivot brackets (18) and secure with four pivot pins (16) and snaprings (17).

(2) Install longitudinal bow (20) on front bow (1) and rear bow (7) and secure with two capscrews (19).

(3) Install straps (13) on footman loops (15) and pull straps (13) tight.

d. Installation of Cargo Enclosure.

(1) Install cargo cover (6) over C-pillar (8), front bow (1), and rear bow (7).

(2) Fasten eyelets (5) to turnbuttons (4) on C-pillar (8), front bow (1), and rear bow (7).

(3) Secure cargo (6) to cab roof (2) at C-pillar (8) with hook and loop attachment (3).

(4) Install grommets (10) on footman loops (11) and attach straps (9) to footman loops (11). Tighten straps (9) equally, but do not overtighten.

2-55. REMOVAL AND INSTALLATION OF S250 SHELTER CARRIER REAR SUSPENSION TIEDOWN KIT

CAUTION

The M1113 shelter carriers are specifically designed to be operated with the S250 shelter installed. The vehicles can be driven safely without the shelter installed, or equivalent payload of 1,500 lb (681 kg), for short distances (e.g., to and from maintenance or from the rail head when being delivered), but this should not be done often or for long distances. Driving for long distances without the shelter installed, or equivalent payload of 1,500 lb (681 kg) evenly distributed in center of cargo area, will cause damage to equipment.

NOTE

Rear suspension tiedown kit is to be used only for shipment and should be removed promptly after shipment.

a. Removal of S250 Rear Suspension Tiedown Kit.

(1) Using wrench provided, turn out turnbuckle assembly (6). Unhook from rear-rear tiedown bracket (1) on frame rail (2) and suspension tiedown bar (5).

(2) Remove suspension tiedown bar (5) from lower control arm (3) and shock mount bracket (4).

b. Installation of S250 Rear Suspension Tiedown Kit.

(1) Install suspension tiedown bar (5) into lower control arm (3) and shock mount bracket (4).

(2) Grease threads on turnbuckle assembly (6). Hook turnbuckle assembly (6) to suspension tiedown bar (5) and to rear-rear tiedown bracket (1) on frame rail (2).

(3) With wrench provided, tighten until turnbuckle assembly (6) is completely turned in.

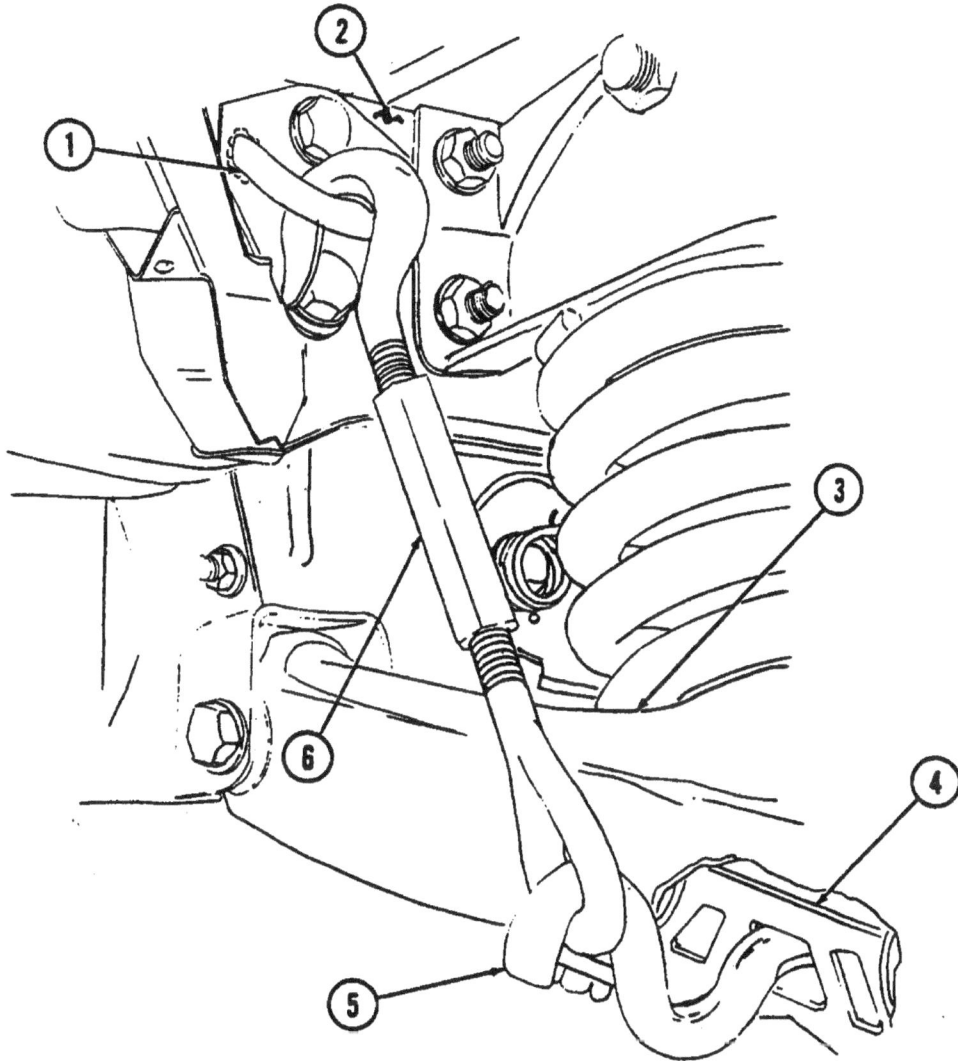

Section VI. M1114 UP-ARMORED CARRIER STOWAGE NET OPERATION

2-56. GENERAL

This section provides instructions for the removal and installation of stowage nets used on the M1114 up-armored carrier.

2-57. M1114 UP-ARMORED CARRIER STOWAGE NET OPERATION REFERENCE INDEX

2-58. REMOVAL AND INSTALLATION OF STOWAGE COMPARTMENT NET

a. Removal.

Remove net (7) by unhooking three J-hooks (4) from D-rings (5) on cargo floor (6) and three J-hooks (2) from D-rings (3) at C-partition (1).

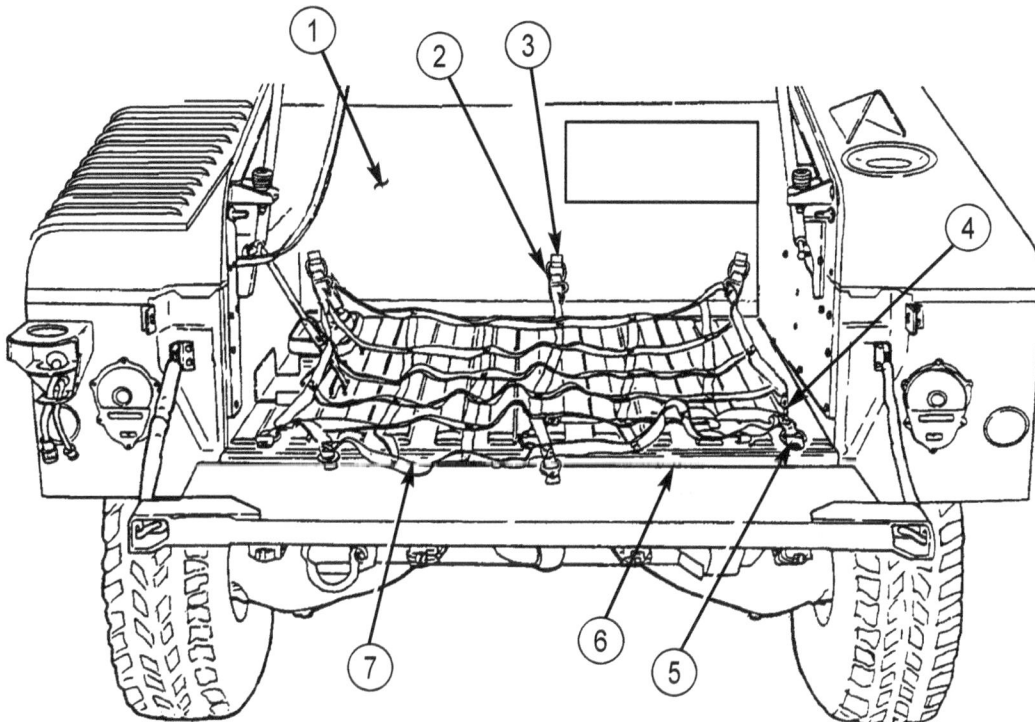

b. Installation.

Position net (7) on cargo floor (6), and install three J-hooks (4) to D-rings (5) on cargo floor (6), and three J-hooks (2) to D-rings (3) on C-partition (1).

2-59. REMOVAL AND INSTALLATION OF HATCH STOWAGE NET

a. **Removal.**

   (1) Detach hatch stowage net (1) from side mounted footman loops (3).

   (2) Detach net (1) from top and bottom mounted footman loops (2). Remove net (1).

b. **Installation.**

   (1) Attach net (1) to top and bottom mounted footman loops (2).

   (2) Attach net (1) to side mounted footman loops (3).

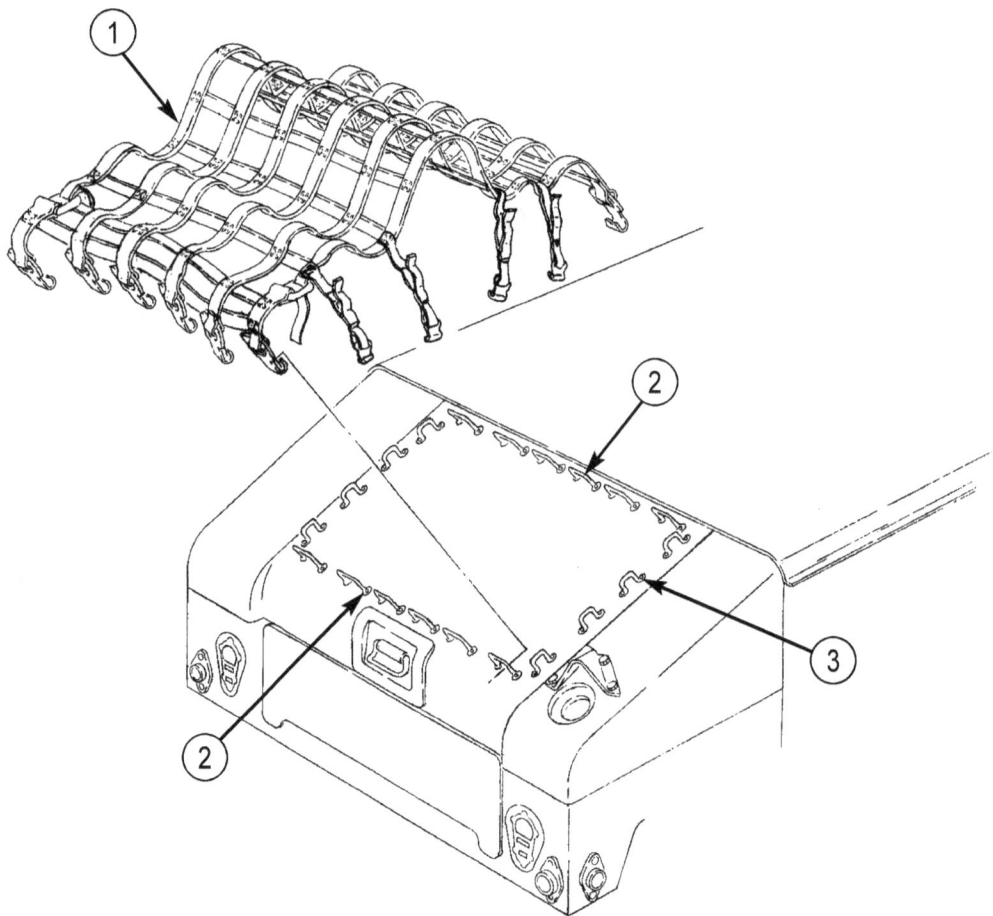

2-60. REMOVAL AND INSTALLATION OF REAR SEAT STOWAGE COMPARTMENT NET

a. Removal.

Detach locking tabs (1) and unhook compartment net (4) from D-rings (5) and mounting brackets (2) and (3). Remove net (4).

b. Installation.

Attach compartment net (4) to D-rings (5) and mounting brackets (2) and (3).

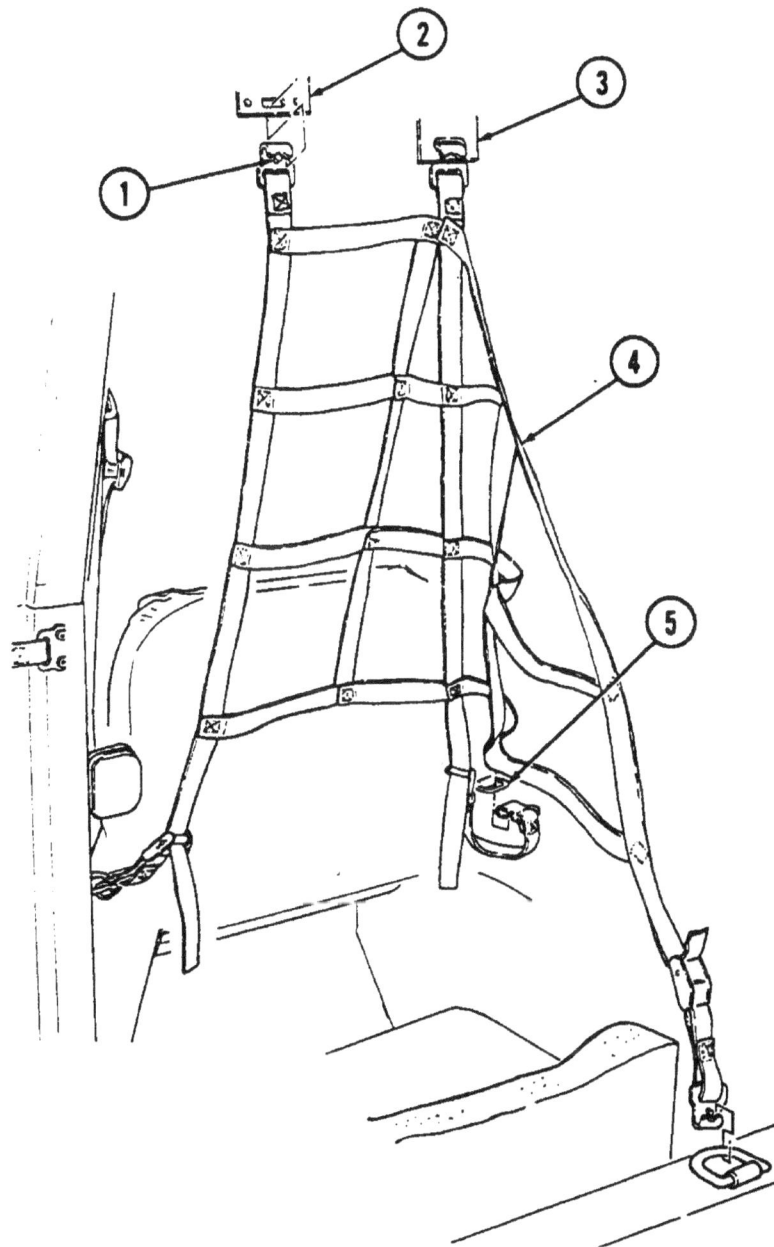

Section VII. REMOVAL AND INSTALLATION OF TOW GUNNER'S PROTECTION KIT (T-GPK) IN PREPARATION FOR AIR DROP.

2-61. GENERAL

Refer to FM 3-22.32 for removal and installation of TOW Gunner's Protection Kit (T-GPK) in preparation for air drop.

CHAPTER 3
MAINTENANCE INSTRUCTIONS

Section I. LUBRICATION

3-1. LUBRICATION INSTRUCTIONS

Lubrication instructions are in appendix G of this manual. All lubrication instructions are mandatory.

3-2. GENERAL LUBRICATION INSTRUCTIONS

a. Service Intervals. Service intervals on the lubrication instructions are for normal operation in moderate temperatures, humidity, and atmospheric conditions.

b. Application Points. Wipe lubricating points and surrounding surfaces before and after applying lubricant.

c. Reports and Records. Report unsatisfactory performance of lubricant or preserving materials on Product Quality Deficiency Report, SF 368, as stated in para. 1-4.

3-3. GENERAL LUBRICATING INSTRUCTIONS UNDER UNUSUAL CONDITIONS

a. Service Intervals. Increase frequency of lubricating service intervals when operating under abnormal conditions such as high or low temperatures, prolonged high-speed driving, or extended cross-country operations. Such operations can destroy lubricant's protective qualities. More frequent lubricating service intervals are necessary to maintain vehicle readiness when operating under abnormal conditions. During inactive periods, with adequate preservation, service intervals can be extended.

b. Changes in Lubricant Grades. Lubricant grades change with weather conditions. Refer to appendix G for lubricant grade changes.

c. Maintaining Lubricant Levels. Lubricant levels must be checked as specified in appendix G. Steps must be taken to replenish and maintain lubricant levels.

3-4. LUBRICATION FOR CONTINUED OPERATION BELOW 0°F (-18°C)

Refer to FM 9-207, Operation and Maintenance of Ordnance Materiel in Cold Weather (0°F to -65°F) (-18°C to -54°C), or appendix G.

Section II. TROUBLESHOOTING

3-5. GENERAL

Troubleshooting, table 3-1, contains instructions that will help the operator identify and correct simple vehicle malfunctions. The table also helps the operator identify major mechanical difficulties that must be referred to unit maintenance. The listing of possible malfunctions come under major vehicle headings. They are:

- Engine
- Heating system
- Transmission
- Transfer case
- Brakes
- Wheels and tires
- Steering
- Winch
- Up-armored carrier

3-6. TROUBLESHOOTING PROCEDURES

   a. Table 3-1 lists the common malfunctions which you may find during the operation or maintenance of the vehicles or their components. You should perform the tests/inspections and corrective actions in the order listed.

   b. This manual cannot list all malfunctions that may occur, nor all tests or inspections and corrective actions. If a malfunction is not listed or is not corrected by listed actions, notify your supervisor.

NOTE

- Hydrostatic lock is caused by the entry of substantial amounts of water into the engine through the air intake system and subsequent contamination of the fuel system. Hydrostatic lock most frequently occurs during or just after fording. Water is forced into the air intake system, is drawn into the engine, and effectively locks up the engine.
- Notify unit maintenance if you suspect hydrostatic lock and they will test the engine.

Table 3-1. Troubleshooting.

```
MALFUNCTION
    TEST OR INSPECTION
        CORRECTIVE ACTION
```

ENGINE

1. ENGINE FAILS TO CRANK

Step 1. Check to see if transmission shift lever is in P (park).

If not, place lever in P (park).

Step 2. Check battery fluid level and check battery cable connections for looseness, damage, or corrosion.

If any of these conditions exist, notify unit maintenance.

Step 3. Attempt to slave-start vehicle (para. 2-23).

Step 4. Other causes.

Notify unit maintenance.

2. ENGINE CRANKS SLOWLY

Step 1. Check battery fluid level and check battery cable connections for looseness, damage, or corrosion.

If any of these conditions exist, notify unit maintenance.

Step 2. Attempt to slave-start vehicle (para. 2-23).

Step 3. Other causes.

Notify unit maintenance.

3. ENGINE CRANKS BUT DOES NOT START

Step 1. Check to see if fuel gauge indicates E (empty).

Fill fuel tank, and start engine.

Step 2. Purge fuel system of air (para. 3-11).

Step 3. Check to see if wait-to-start lamp assembly fails to light or does not go out.

Notify unit maintenance if wait-to-start lamp assembly fails to light or does not go out.

Step 4. Other causes.

Notify unit maintenance.

4. VEHICLE NOT CHARGING ACCORDING TO VOLTMETER

Step 1. Check battery cable connections for looseness, damage, or corrosion.

Notify unit maintenance of any damage to battery cables.

Step 2. Check for broken or missing drivebelt.

Notify unit maintenance if drivebelt is broken or missing.

Step 3. Other causes.

Notify unit maintenance.

Table 3-1. Troubleshooting (Cont'd).

```
MALFUNCTION
    TEST OR INSPECTION
        CORRECTIVE ACTION
```

5. EXCESSIVE EXHAUST SMOKE AFTER ENGINE REACHES NORMAL OPERATING TEMPERATURE 185°-250°F (85°-120°C)

Step 1. Check oil level for overfilling (appendix G).

Notify unit maintenance if fluid level is high.

Step 2. Check for restricted air cleaner.

If emergency situation exists, clean air cleaner element (para. 3-15).

If emergency situation does not exist, notify unit maintenance.

Step 3. Other causes.

Notify unit maintenance.

6. ENGINE STARTS BUT MISFIRES, RUNS ROUGH, OR LACKS POWER

Step 1. Check for water in fuel filter by draining.

Drain fuel filter (para. 3-11).

Step 2. Check for restricted air cleaner.

If emergency situation exists, clean air cleaner element (para. 3-15).

If emergency situation does not exist, notify unit maintenance.

Step 3. Other causes.

Notify unit maintenance.

7. ENGINE OVERHEATS ACCORDING TO ENGINE COOLANT TEMPERATURE GAUGE

Step 1. Check to see if fan is running.

If fan is not running, perform emergency fan clutch override procedure (para. 3-23).

Step 2. Allow engine to cool and check for low coolant level.

Add coolant as necessary (para. 3-10).

Step 3. Check for debris blocking radiator fins.

Remove debris (para. 3-26).

Step 4. Check for broken or missing drivebelt.

Notify unit maintenance if drivebelt is broken or missing.

Step 5. Other causes.

Notify unit maintenance.

8. LOW ENGINE OIL PRESSURE ACCORDING TO OIL PRESSURE GAUGE

Step 1. Check for low oil level (para. 3-18).

Add oil (appendix G).

Step 2. Other causes.

Notify unit maintenance.

Table 3-1. Troubleshooting (Cont'd).

MALFUNCTION
TEST OR INSPECTION
CORRECTIVE ACTION

HEATING SYSTEM

9. HOT WATER PERSONNEL HEATER FAILS TO PRODUCE HEAT AFTER ENGINE REACHES OPERATING TEMPERATURE

Step 1. Check operating controls for correct settings.

Step 2. Other causes.

Notify unit maintenance.

TRANSMISSION

10. NO RESPONSE TO SHIFT LEVER MOVEMENT

Step 1. Check to see if transmission lever is in P (park).

Place transmission shift lever in P and select transfer gear range.

Step 2. Other causes.

Notify unit maintenance.

11. ROUGH SHIFTING

All causes.

Notify unit maintenance.

12. FLUID THROWN FROM TRANSMISSION FILL TUBE

Step 1. Check to see if transmission dipstick is loose.

Secure dipstick.

Step 2. Check fluid level for overfilling (para. 3-19).

Notify unit maintenance if fluid level is high.

Step 3. Other causes.

Notify unit maintenance.

13. SLIPPAGE IN ALL RANGES

Step 1. Check for low fluid level (para. 3-19).

Add fluid (appendix G).

Step 2. Other causes.

Notify unit maintenance.

13.1. TRANSMISSION LIGHT ON

Check to see if transmission light stays on. (Refer to para. 2-2).

If light stays on, notify unit maintenance.

Table 3-1. Troubleshooting (Cont'd).

MALFUNCTION
TEST OR INSPECTION
CORRECTIVE ACTION

TRANSFER CASE

14. TRANSFER CASE SHIFT LEVER WILL NOT SHIFT

Step 1. Check for proper shifting sequence.

Ensure proper shifting sequence is used (para. 2-13).

Step 2. Other causes.

Notify unit maintenance.

BRAKES

15. POOR SERVICE BRAKING ACTION

All causes.

Notify unit maintenance.

16. SERVICE BRAKES DRAGGING

All causes.

Notify unit maintenance.

17. BRAKE WARNING LAMP ASSEMBLY ON

Step 1. Check to see if parking brake is partially applied.

Disengage parking brake.

Step 2. Other causes.

Notify unit maintenance.

18. PARKING BRAKE FAILS TO HOLD VEHICLE

Clean and adjust parking brake (para. 3-13).

If parking brake still fails to hold vehicle, notify unit maintenance.

WHEELS AND TIRES

19. WHEELS WOBBLE OR SHIMMY

Step 1. Check to see if wheel lug nuts are loose.

Tighten loose lug nuts (para. 3-25) and notify unit maintenance to properly torque lug nuts.

Step 2. Inspect for mud or dirt buildup inside the rim.

Remove any mud or dirt buildup.

Step 3. Check tires for proper inflation pressure.

Step 4. Other causes.

Notify unit maintenance.

Table 3-1. Troubleshooting (Cont'd).

MALFUNCTION
TEST OR INSPECTION
CORRECTIVE ACTION

20. EXCESSIVE OR UNEVEN TIRE WEAR

Step 1. Check tires for proper inflation pressure.
Step 2. Other causes.
   Notify unit maintenance.

21. VEHICLE WANDERS TO ONE SIDE ON LEVEL PAVEMENT

Step 1. Check tires for proper inflation pressure.
Step 2. Other causes.
   Notify unit maintenance.

STEERING

22. HARD STEERING

Step 1. Check tires for proper inflation pressure.
Step 2. Check power steering reservoir for low fluid level (para. 3-20).
   Add steering fluid (appendix G).
Step 3. Check for broken or missing drivebelt.
   Notify unit maintenance if drivebelt is broken or missing.
Step 4. Check power steering oil cooler for bent fins or any other air flow obstructions.
   Remove obstructions if fins are not damaged.
   Notify unit maintenance if power steering oil cooler is damaged.
Step 5. Other causes.
   Notify unit maintenance.

WINCH

23. ELECTRIC WINCH STOPS DURING NORMAL OPERATION

Step 1. Wait 2 minutes and attempt winch operation again.
   Refer to para. 2-28 for winch operation.
Step 2. Check to see if clutch lever is engaged (para. 2-28).
   If not, engage clutch lever.
Step 3. Check to see if voltmeter is in red or yellow (engine not running).
   Start engine and charge batteries.
Step 4. Other causes.
   Notify unit maintenance.

Table 3-1. Troubleshooting (Cont'd).

MALFUNCTION
TEST OR INSPECTION
CORRECTIVE ACTION

23.1. HYDRAULIC WINCH STOPS DURING NORMAL OPERATION

Step 1. Check power steering fluid level (para. 3-20).

If fluid level is low add fluid.

Step 2. Ensure winch selector levers are in LOCK LOW GEAR position (para. 2-28.1).

If not, move winch selector levers to LOCK LOW GEAR position.

WARNING

• Wear leather gloves when handling winch cable. Do not handle cable with bare hands.

• When fully extending winch cable, ensure that four wraps of winch cable remain on drum at all times. Failure to do so may cause damage to equipment or injury or death to personnel.

• Direct all personnel to stand clear of winch cable during winch operation. Failure to do so may cause damage to equipment and injury or death to personnel.

• Do not even slide gloved hands across winch cable. Injury can result.

Step 3. Ensure winch cable is not stacked or binding against tiebars or winch housing.

If winch cable is stacked or binding, remove load from cable, move winch control levers to FREESPOOL and pay out winch cable by hand. If unable to pay out winch cable by hand, notify unit maintenance.

Step 4. Other causes.

Notify unit maintenance.

■ UP-ARMORED AND TOW ITAS/ARMAMENT CARRIER

24. WEAPON STATION WILL NOT ROTATE OR LOCK

Step 1. Check for obstructions that may be restricting weapon station movement.

Remove obstructions.

Step 2. Other causes.

Notify unit maintenance.

25. CARGO SHELL DOOR WILL NOT SEAL PROPERLY

All causes.

Notify unit maintenance.

Section III. MAINTENANCE PROCEDURES

3-7. GENERAL

The operator/crew is responsible for daily, weekly, and monthly preventive maintenance checks and services listed in table 2-2. Other maintenance services, also the responsibility of the operator/crew, are explained in this section.

3-8. MAINTENANCE PROCEDURES REFERENCE INDEX

3-9. RAISING AND SECURING HOOD

WARNING

To ensure opening of the hood assembly is accomplished safely and effectively, always maintain the proper lifting posture, legs bent with back straight. Failure to do so may cause damage to equipment or injury to personnel.

NOTE

Due to the inherent weight of the hood assembly, hood may flex when opening, possibly causing interference between the right side of hood assembly and body. This interference can be eliminated by pushing hood assembly laterally away from individual prior to lifting.

a. Raising Hood.

(1) Apply parking brake.

(2) Release left and right hood latches (1).

(3) Facing driver's side of hood, position one hand at the rear area of hood and other at rear area of wheel well.

WARNING

When raising and securing hood, ensure hood prop rod is secured to hood support bracket. Damage to equipment or injury to personnel may occur if hood is not properly secured in raised position.

(4) Push the hood toward the passenger side and lift at the same time, moving your hands toward the front of the hood as it opens. The prop rod (2) should automatically engage the support bracket (3) when hood is raised.

b. Lowering Hood.

WARNING

When releasing hood prop rod, do not pull rod at hook end. Injury to fingers may occur.

CAUTION

Lower hood slowly. Damage to hood and/or headlights can occur if hood is dropped.

(1) While supporting and slightly raising hood, grasp prop rod (2) above retaining ring (5), pull out, and release hood prop rod (2).

(2) Once prop rod hook (4) is clear of support bracket (3), slowly lower hood and secure left and right hood latches (1).

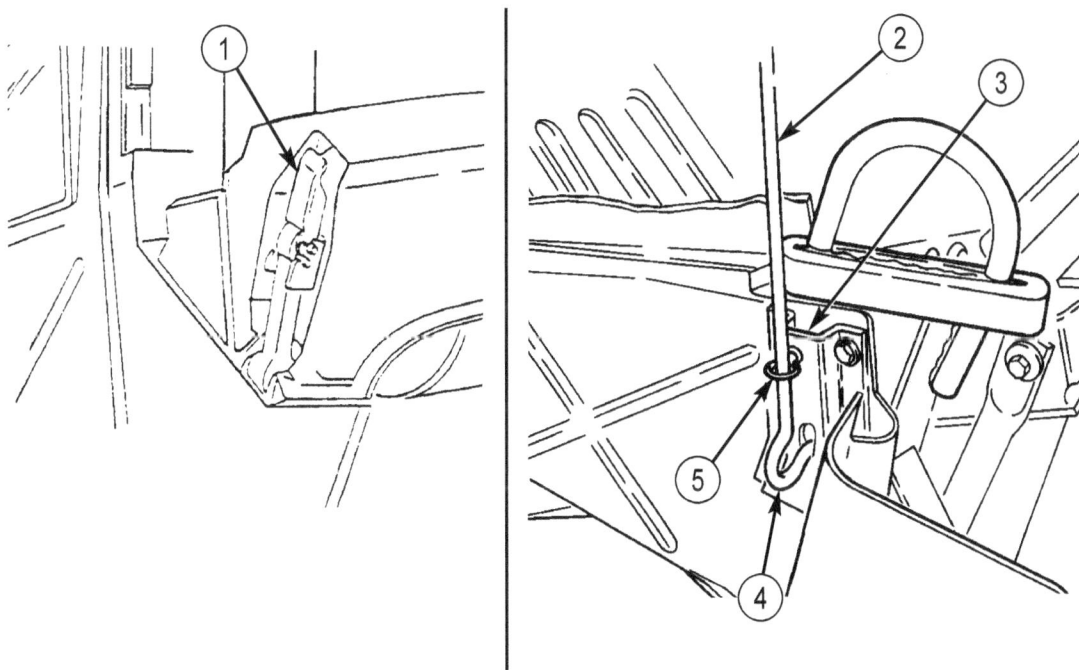

3-10. SERVICING COOLANT SURGE TANK

WARNING

Extreme care should be taken when removing surge tank filler cap if temperature gauge reads above 165°F (74°C). Do not add coolant to cooling system when engine is hot unless engine is running. Add coolant slowly. Steam or hot coolant under pressure may cause injury.

CAUTION

- Type 1, ethylene glycol (green), and Type 2, propylene glycol (purple), should never be mixed due to their difference in toxic properties. Failure to comply may result in damage to equipment.
- Using antifreeze without mixing it with water can cause high operating temperatures, blockage of cooling system passages, and damage to water pump seals.

NOTE

Type 1 antifreeze is an ethylene glycol based coolant, green in color. Type 1 can be added to factory-filled pink coolant. When it becomes necessary to flush factory coolant, Type 1, ethylene glycol, will be used. When mixing Type 1 antifreeze with water, distilled water is recommended. Tap water should only be used in emergency situations.

   a. Raise and secure hood (para. 3-9).

   b. Visually check coolant level. Surge tank level should be at COLD FILL LINE before operation and slightly above COLD FILL LINE after operation. If coolant is low, perform steps c through h.

   c. Place a thick cloth over surge tank filler cap (1). Carefully turn cap (1) counterclockwise to its first stop to allow cooling system pressure to escape.

   d. After cooling system pressure is vented, push down and turn cap (1) counter-clockwise to remove. Add coolant until surge tank level is at COLD FILL LINE.

   e. Start engine (para. 2-12) and run for one minute.

   f. Stop engine (para. 2-14) and recheck coolant level. If coolant level is low, add coolant until surge tank level is at COLD FILL LINE.

   g. Repeat steps e and f until surge tank level remains at COLD FILL LINE.

   h. Install cap (1). Lower and secure hood (para. 3-9).

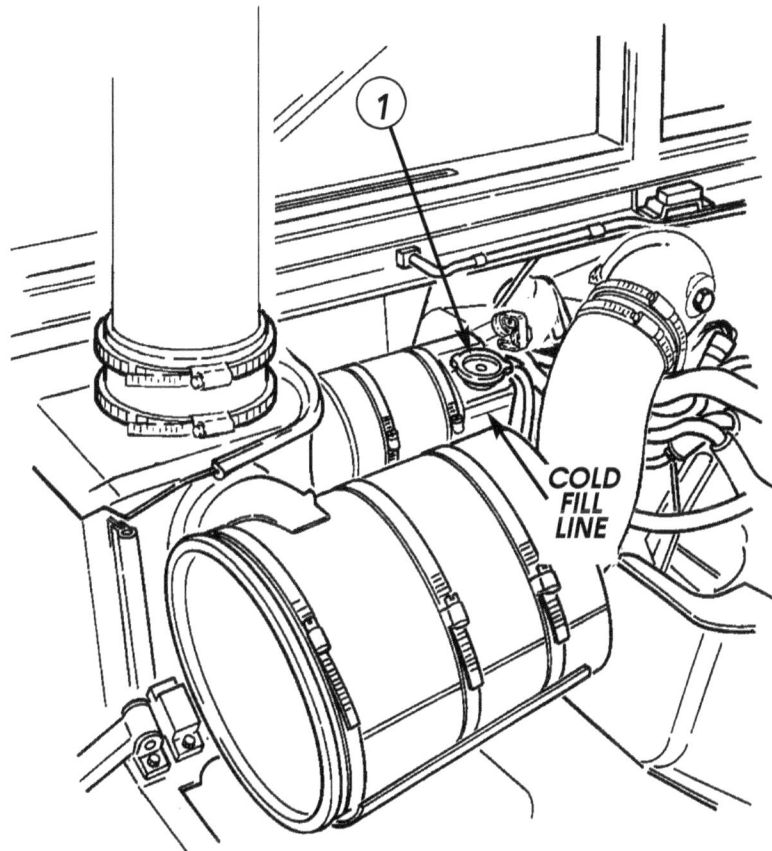

COLD
FILL
LINE

3-11. FUEL FILTER MAINTENANCE

WARNING

Do not perform fuel system checks, inspection, or maintenance while smoking or near fire, flames, or sparks. Fuel may ignite, causing damage to vehicle and injury or death to personnel.

a. **Draining Fuel Filter.**

(1) Raise and secure hood (para. 3-9).

(2) Start engine (para. 2-12).

(3) Place toggle fuel switch (3) to the UP position and allow approximately 1 pt (0.47 L) of fuel to drain into a suitable container.

(4) Place toggle fuel switch (3) to the DOWN position when draining is complete.

(5) Stop engine (para. 2-14).

(6) Lower and secure hood (para. 3-9).

b. **Purging Fuel System of Air.**

NOTE

This procedure is used to purge fuel system of air if vehicle has run out of fuel.

(1) Raise and secure hood (para. 3-9).

(2) Disconnect fuel line (1) from fuel filter outlet (2).

(3) Place a rag over fuel filter outlet (2).

(4) Crank engine until rag is wet with fuel.

(5) Connect fuel line (1) to fuel filter outlet (2).

(6) Start engine (para. 2-12) and ensure fuel system has been purged of air.

(7) Lower and secure hood (para. 3-9).

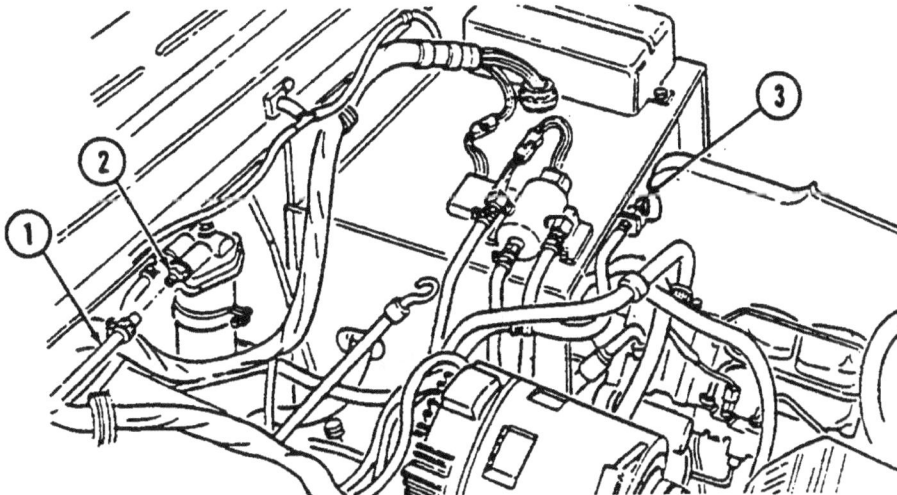

3-12. SERVICING BATTERIES

a. Unhook latches (6) and remove companion seat (1) from battery box (9).

WARNING

Do not perform battery system checks or inspections while smoking or near fire, flames, or sparks, especially if the caps are off. Batteries may explode, causing damage to vehicle and injury or death to personnel.

b. Check electrolyte level.

(1) Remove all battery filler caps (2) and check electrolyte level. If electrolyte level is below ledge in battery filler opening, add distilled water (appendix D, item 7).

(2) A battery that is continually in need of electrolyte may indicate an improperly adjusted charging system. Notify unit maintenance if problem continues.

(3) Inspect vented battery filler caps (2) to ensure that vents are clear, unobstructed, and permit escape of battery gases. Clean vents if obstructed; replace caps (2) if damaged.

(4) Install filler caps (2).

c. Inspect all battery compartment components, including terminal clamps (5), battery cables (3), battery holddowns (8), and shunt (4) for corrosion, damage, or looseness. Inspect terminal boots (7), if installed. Notify unit maintenance if any of these problems exist.

d. Ensure that battery terminal clamps (5) have a light coat of lubricating oil for corrosion protection (appendix D, item 23).

e. Install companion seat (1) and secure to battery box (9) with latches (6).

f. Refer to TM 9-6140-200-14 for additional information.

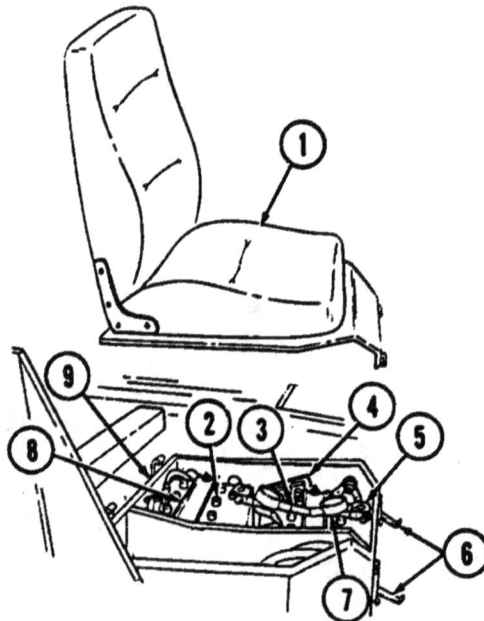

3-13. PARKING BRAKE ADJUSTMENT AND CLEANING

a. Chock wheels and release parking brake handle (3) by depressing safety release button (2).

b. Turn adjusting knob (1) clockwise as tightly as possible by hand.

c. Apply parking brake handle (3).

d. If parking brake cannot be applied, turn adjusting knob (1) counterclockwise until parking brake can be applied.

e. Test parking brake.

    (1) Remove chocks.

    (2) Depress service brake pedal and start engine (para. 2-12).

    (3) Place transfer case shift lever (4) in H (high) and transmission shift lever (5) in ⒟ (overdrive).

    (4) Slowly let up on service brake pedal. Parking brake should hold vehicle stationary.

NOTE

Perform step f. for vehicles with serial numbers 299999 and below only.

f. After operating in mud or sand, use a low-pressure water source to ensure actuating lever (7) and spring (6) are thoroughly cleaned of mud, sand, or other debris. Lubricate lever (7) in accordance with appendix G as soon as possible.

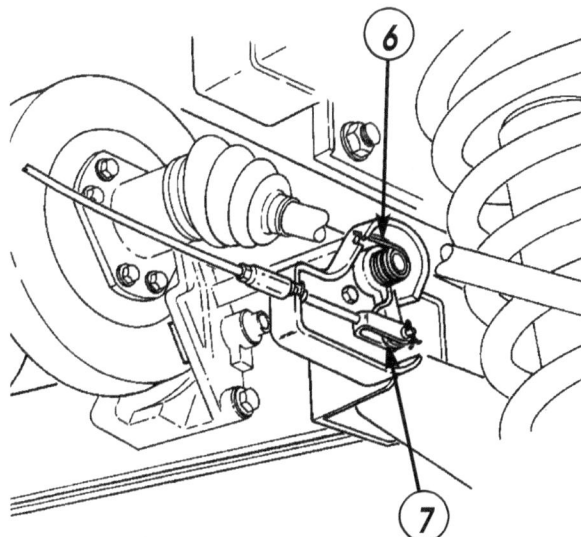

S/N 299999 AND BELOW ONLY

3-14. SERVICING WINDSHIELD WASHER RESERVOIR

a. Raise and secure hood (para. 3-9).

b. Check fluid level in washer reservoir (2). Check frequently under adverse weather conditions.

c. If fluid is required, fill washer reservoir with water and cleaning compound (appendix D, item 4), refer to table 3-2.

Table 3-2. Cleaning Compound-to-Water Ratio.

Temperature Range	Cleaning Compound	to	Water
Above +15°F (-9°C)	1	to	2
+40° to -15°F (+4° to -26°C)	1	to	1
+40° to -65°F (+4° to -54°C)	2	to	1

d. If washer system does not work, check washer nozzles (1) to see if they are blocked with dirt and/or debris. Remove dirt and/or debris with fine wire.

e. Check hoses (3) for leaks or poor condition.

f. Lower and secure hood (para. 3-9).

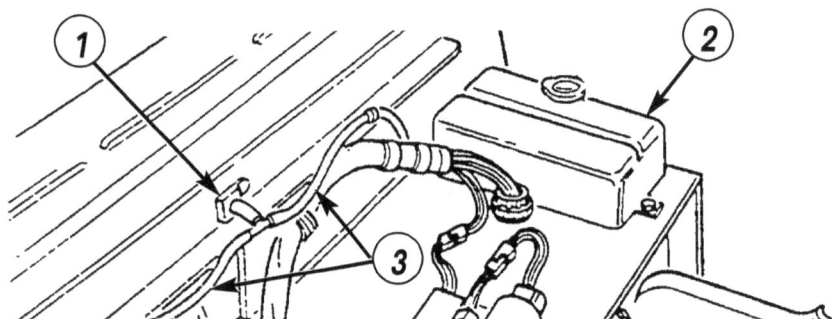

3-15. AIR CLEANER SERVICING (EMERGENCY PROCEDURE)

a. General. Air cleaner service is required when yellow air restriction indicator (3) reaches red zone (2) of gauge (1).

CAUTION

Do not operate engine without air cleaner element. Damage to engine may result.

b. Filter Element Removal.

(1) Raise and secure hood (para. 3-9).

NOTE

Perform step 2.1 for vehicles with over-center clamp configuration only.

(2) Loosen clamp screw (10) and remove clamp (4) and cover (5) from air cleaner assembly (8).

(2.1) Release and remove over-center clamp (4) and cover (5) from air cleaner assembly (8).

WARNING

- NBC contaminated filters must be handled using adequate precautions (FM 3-5) and must be disposed of by trained personnel.
- After Nuclear, Biological, or Chemical (NBC) exposure of this vehicle, all air filters shall be handled with extreme caution. Unprotected personnel may experience injury or death if residual toxic agents or radioactive material are present. Servicing personnel will wear protective overgarments, mask, hood, and chemical-protective gloves and boots. All contaminated air filters will be placed into double-lined plastic bags and moved immediately to a temporary segregation area away from the work site. If contaminated by radioactive dust, the Company NBC team will measure the radiation before removal. The NBC team will determine the extent of safety procedures required. The temporary segregation area will be marked with the appropriate NBC signs. Final disposal of contaminated air filters will be in accordance with local Standard Operating Procedures (SOP).
- Failure to observe above warnings may result in injury or death.

(3) Remove nut and washer assembly (6) securing filter element (7) to stud (9) and pull filter element (7) from air cleaner assembly (8).

(4) Place cover (5) and clamp (4) back on air cleaner assembly (8) to prevent dirt and dust from entering air induction system while cleaning filter element (7).

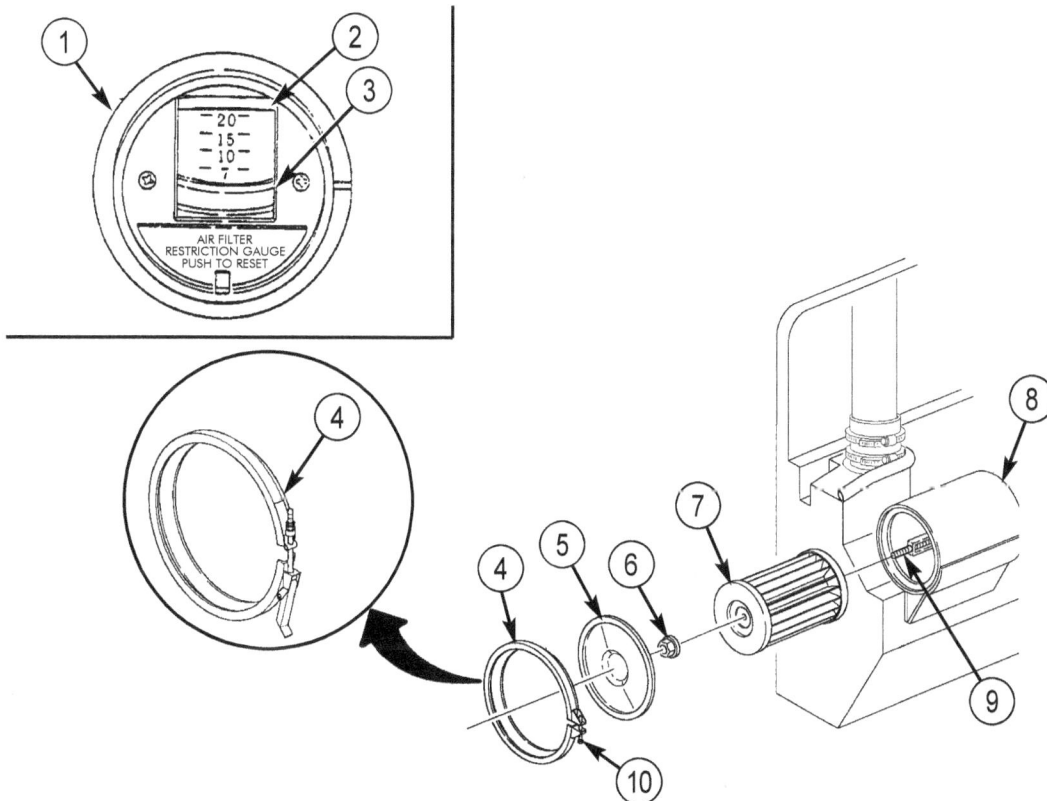

3-15. AIR CLEANER SERVICING (EMERGENCY PROCEDURE) (Cont'd)

c. Air Filter Element Cleaning.

CAUTION

Do not strike ends of filter element on hard surface, or damage to filter element may result.

(1) Hold filter element (1) with open end facing ground.

(2) Gently tap completely around filter element (1) with hand to free trapped dirt.

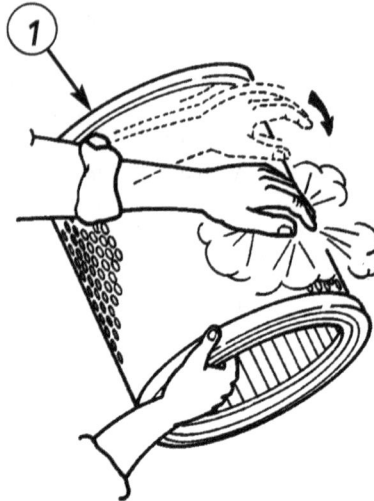

d. Filter Element Installation.

CAUTION

The turbocharged engine requires a 420CFM filter element. Use of any other filter element will damage equipment.

(1) Remove clamp (2) and cover (3) from air cleaner assembly (5).

(2) Position filter element (1) into air cleaner assembly (5) and secure element (1) to stud (6) with nut and washer assembly (4).

CAUTION

When clamp is secured to end of air cleaner, ensure clamp screw is between three and six o'clock positions to prevent damaging hood when hood is closed.

NOTE

Perform step 3.1 for vehicles with over-center clamp configuration only.

(3) Install cover (3) and clamp (2) on end of air cleaner assembly (5) and tighten clamp screw (7).

(3.1) Install cover (3) on air cleaner assembly (5) and secure with over-center clamp (2).

(4) Lower and secure hood (para. 3-9).

(5) Notify unit maintenance to properly torque or set gap of clamp.

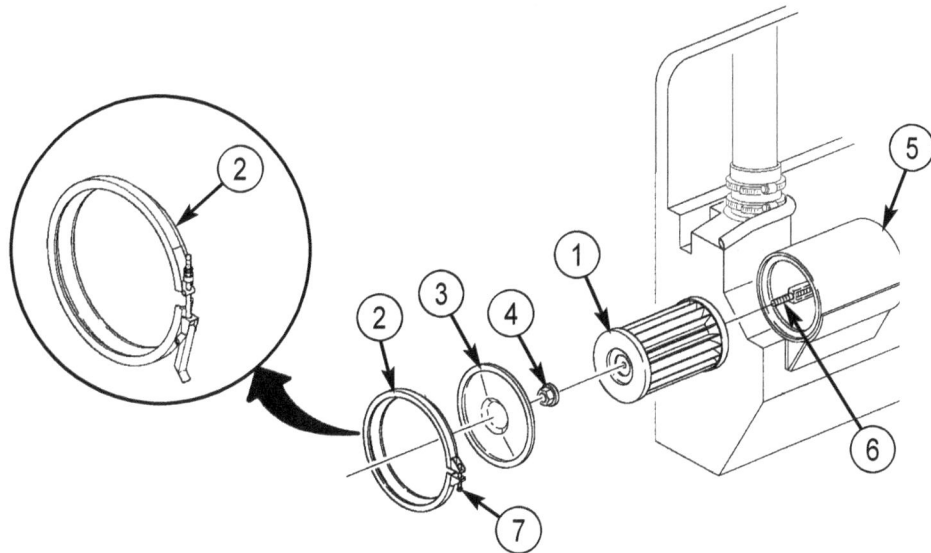

3-16. AIR CLEANER DUMP VALVE SERVICING

NOTE

Air cleaner dump valve should be serviced after any operation through sand, mud, or water.

a. Raise and secure hood (para. 3-9).

b. Squeeze dump valve (2) to clear any sand, mud, or water from air cleaner assembly (1).

c. For vehicles equipped with a deep water fording kit, loosen clamp (3) and remove cap (4). Clean and install cap (4) and tighten clamp (3).

d. Lower and secure hood (para. 3-9).

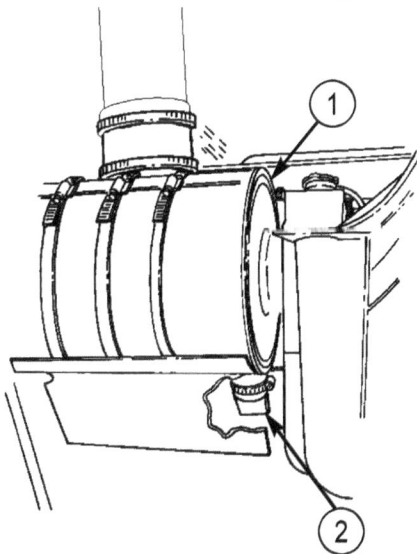
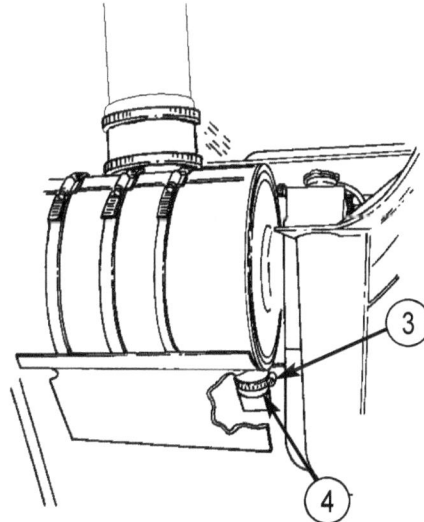

3-17. PLASTIC WINDOW CLEANING

CAUTION

Never clean plastic windows with abrasives. Repeated use of abrasives or failure to follow instructions below will eventually cause damage to windows.

NOTE

This paragraph provides instructions for cleaning soft top plastic windows and ballistic windshield and windows. This procedure is to be used on the inner plastic laminate surfaces of ballistic glass. Clean outer surface of ballistic glass as you would plain glass.

a. Plastic Window Cleaning.

(1) Wash windows using soap, water, and a soft, clean cloth.

(2) Rinse with clean water.

(3) Apply cream cleaner (appendix D, item 5) to plastic windows.

(4) Wipe cream cleaner off with dry cloth. Cream cleaner improves visual clarity after cleaning with soap and water.

a.1. Plastic Window Cleaning - Fungus and Mold.

(1) Wash windows using soap and water and a soft, clean cloth.

(2) Rinse with clean water.

(3) Using a soft, clean cloth and isopropyl alcohol (Appendix D, Item 1.01), clean off fungus and mold.

(4) Wipe area with soft, clean cloth.

b. Ballistic Glass Cleaning.

CAUTION

- Do not clean interior surfaces of ballistic glass by any other means than specified below.
- Do not use a scraper or other objects with sharp edges that may scratch the inside surfaces of ballistic glass.
- Do not apply stickers, labels, solvents, abrasive materials, or cleaners.

(1) Remove dust and loose abrasive particles using clean, filtered air at 20 psi (138 kPa) maximum.

(2) Wash with mild detergent (appendix D, item 6) and warm water. Dry using a soft, clean cloth.

(3) Remove stubborn marks and stains using a soft, clean cloth and equal parts by volume of isopropyl alcohol (appendix D, item 18) in water.

(4) Repeat step (2).

3-18. ENGINE OIL SERVICING

a. Raise and secure hood (para. 3-9).

CAUTION

Do not permit dirt, dust, or grit to enter engine oil dipstick tube. Internal engine damage may result if engine oil becomes contaminated.

NOTE

• Engine oil level is checked with engine off.

• If oil level checks above FULL, it may be due to oil cooler drain-back. Operate engine for one minute, shut down, wait one minute, and recheck oil level. If it still checks above full, drain oil to obtain correct level.

• Vehicles equipped with deep water fording kit will have a sealed dipstick.

b. Pull out dipstick (1) and check for proper oil level. Level should be at crosshatch marks (2) between FULL and ADD 1 QT.

CAUTION

Do not overfill engine crankcase. Damage to engine may result.

c. If engine oil is low, remove oil filler cap (3) and add engine oil (appendix G).

CAUTION

Use a non-vented filler cap only. An incorrect filler cap will not seal properly, causing water to enter and damage engine.

d. Replace oil filler cap (3), tighten securely, and wipe away any spilled oil.

e. Lower and secure hood (para. 3-9).

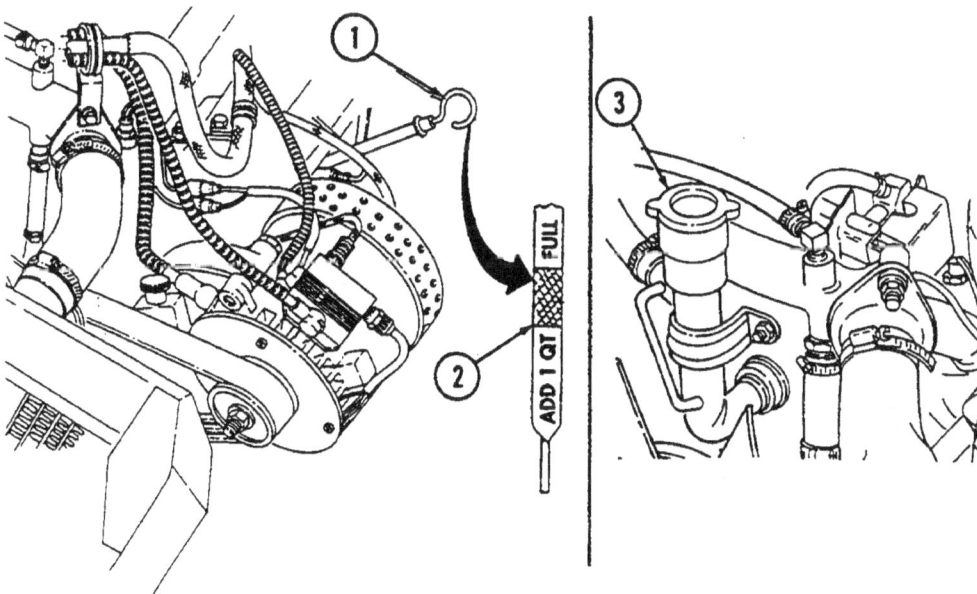

3-19. TRANSMISSION FLUID SERVICING

  a. Raise and secure hood (para. 3-9).

  b. Start engine (para. 2-12).

  c. While depressing service brake pedal, move transmission shift lever through all operating ranges and R (reverse) before checking fluid level in N (neutral) with parking brake applied.

CAUTION

Do not permit dirt, dust, or grit to enter transmission oil dipstick tube. Internal transmission damage may result if transmission fluid becomes contaminated.

  d. Pull out dipstick (2) and check for proper transmission fluid level. At normal operating temperature, fluid level should be at crosshatch marks (3).

CAUTION

Do not overfill transmission. Damage to transmission may result.

  e. If transmission fluid level is low, add fluid (appendix G) through fill pipe (1), insert dipstick (2), and wipe away any spilled fluid.

  f. Shut off engine (para. 2-14).

  g. Lower and secure hood (para. 3-9).

3-20. POWER STEERING FLUID SERVICING (P/N RCSK 18330)

a. Raise and secure hood (para. 3-9).

CAUTION

Do not permit dirt, dust, or grit to enter power steering reservoir. Damage to power steering system may result if power steering fluid becomes contaminated.

b. Remove cap/dipstick (2) and check steering fluid level. If engine is warm, level should be between HOT and COLD marks. If engine is cool, level should be between ADD and COLD marks. In either condition, fluid must be added if level reads below ADD mark.

CAUTION

Do not overfill power steering reservoir. Damage to power steering system will result.

c. If fluid level is low, add fluid (appendix G) to power steering reservoir (1) and wipe away any fluid spilled.

d. Install cap/dipstick (2).

e. Lower and secure hood (para. 3-9).

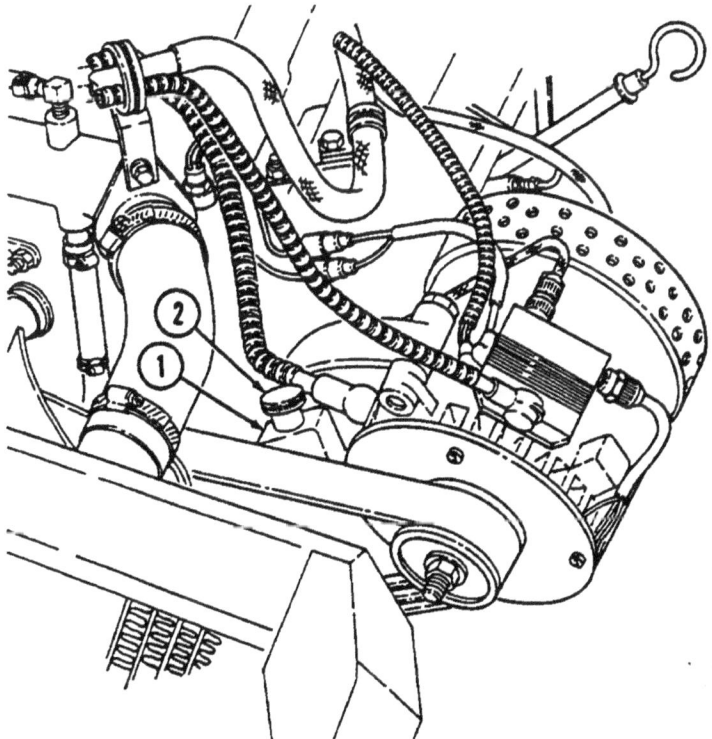 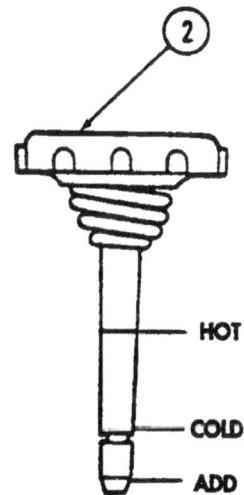

3-20.1. POWER STEERING FLUID SERVICING (P/N 94252A)

a. Raise and secure hood (para. 3-9).

b. Check sightglass (3) for proper fluid level. If fluid is HOT (at engine operating temperature), level should be at top of sightglass (3). If fluid is COLD, fluid should be in center of sightglass (3).

c. If fluid level is low, remove fill cap (1) on power steering reservoir (2) and fill with Dexron VI hydraulic fluid (appendix G) until fluid is in center of sightglass (3) and replace fill cap (1) on power steering reservoir (2).

CAUTION

Do not permit dirt, dust, or grit to enter power steering reservoir. Damage to power steering system may result if power steering fluid becomes contaminated.

NOTE

Inspect power steering fluid in sightglass for bubbles or foam. If fluid or bubbles are visible, return vehicle to unit maintenance.

d. Run engine until fluid level is below sightglass (3), shut off engine.

e. Add Dexron VI hydraulic fluid (appendix G) until fluid level is at center of sightglass (3) when cold and top of sightglass (3) when hot. Do not overfill.

f. Start engine and check power steering system for leaks.

g. Inspect fluid in sightglass to ensure fluid is free of air. If fluid does not stabilize at proper fluid level, return vehicle to unit maintenance.

h. Lower and secure hood (para. 3-9).

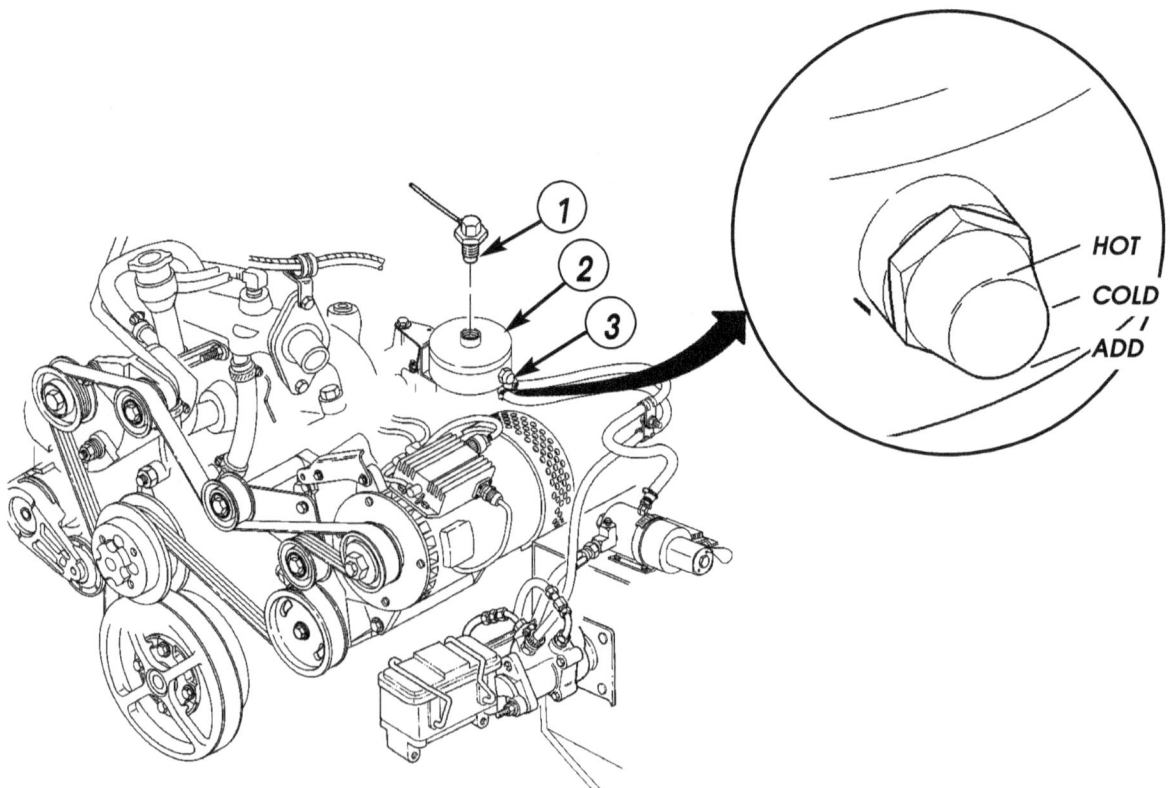

3-21. FUEL TANK SERVICING

a. Shut off engine (para. 2-14).

WARNING

Do not perform fuel system checks, inspection, or maintenance while smoking or near fire, flames, or sparks. Fuel may ignite, causing damage to vehicle and injury or death to personnel.

CAUTION

Do not turn fuel cap handle more than necessary to remove or seal fuel cap. Fuel cap chain links may separate or become damaged.

NOTE

For M1114 vehicles, the fuel door must be open before removing fuel cap (para. 2-34).

b. Turn fuel cap handle (1) on fuel cap (2) counterclockwise and remove fuel cap (2).

CAUTION

When refueling vehicle, be careful that nozzle does not slam into filler pipe. Damage to strainer may occur, allowing dirty fuel to clog up filters.

NOTE

To help prevent spills and overflows, pay close attention to the dispensing nozzle while refueling. Do not exceed a safe refueling rate. If fuel backup and spillage is being experienced, reduce flow rate. Also, when topping off the tank, a reduced flow rate should be used.

c. Insert fuel filler nozzle into filler neck rubber cone, using caution not to puncture fuel filler neck strainer. Dispense fuel into fuel tank.

d. If fuel cap handle (1) is turned more than necessary to remove fuel cap (2), fuel cap backing plate (3) may become jammed on staked threads (4). Correct the problem by holding backing plate (3) and turning the fuel cap handle (1) clockwise until backing plate (3) turns freely.

e. When fueling operation is complete, install fuel cap (2) and turn fuel cap handle (1) clockwise to seal fuel cap (2).

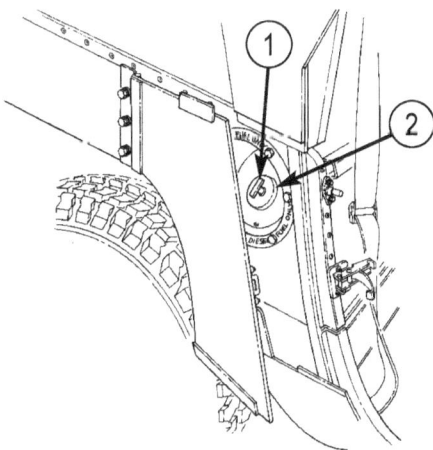

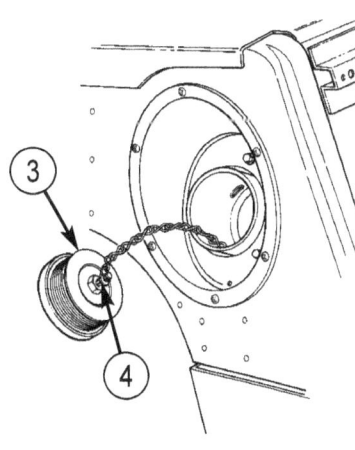

3-22. TIRE CHAIN INSTALLATION AND REMOVAL

CAUTION

Tire chains are only used when extra traction is required and must be used as an axle set. Any other combination may cause damage to the drivetrain.

a. Tire Chain Installation.

(1) Spread out tire chain assembly (1) and line up with tire.

(2) Cautiously move or drive vehicle over tire chain assembly (1) until wheel is positioned at either end of chain assembly (1), allowing tire chain assembly (1) to be draped up and over tire.

(3) Maneuver tire chain assembly (1) until cross-link sections are evenly spaced around tire. Secure one side of tire chain assembly (1) to tire by hooking inside fastener (2) to chain assembly (1). Tighten chain assembly (1) as much as possible.

(4) Repeat steps 1 through 3 until all tire chain assemblies have been properly installed.

(5) Hook end fastener (3) to chain assembly (1) and secure with locking retainer (4) to tighten chain assembly (1). Ensure as many chain links as possible lay between the sidewall head lugs (5) on both sides of tires.

(6) Move vehicle forward a few feet and retighten chain assembly (1) to remove any slack from where tire was resting on chain assembly (1). Secure loose chain linkage to chain assembly (1) with wire or other field expedient method.

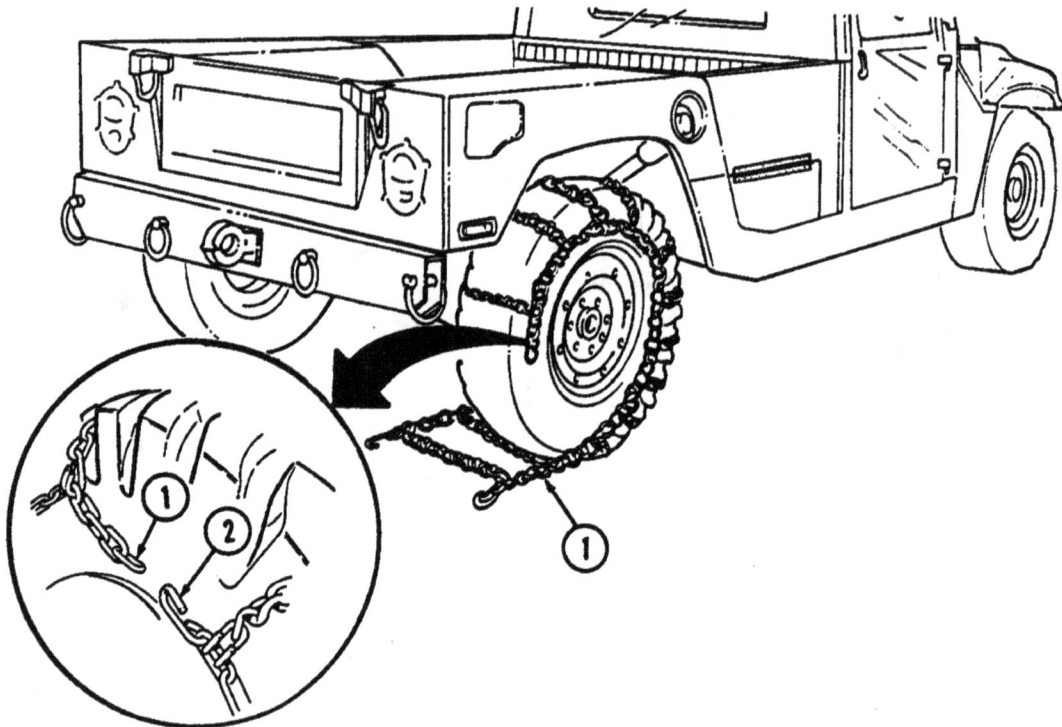

(7) After vehicle is driven one or two miles, stop and retighten tire chains. Ensure as many chain links as possible lie between sidewall head lugs (5) on both sides of tires.

(8) After final tightening, secure loose chain linkage to chain assembly (1) with wire or other field expedient method.

(9) Occasionally check tire chains (1) during operations to ensure tire chains (1) have not slipped.

b. **Tire Chain Removal.**

CAUTION

Remove tire chains from tires as soon as possible after leaving area requiring their use. Prolonged use of tire chains may damage drivetrain.

(1) Detach locking retainer (4) from end fastener (3) and unhook end fastener (3) from chain assembly (1).

(2) Unhook inside fastener (2) from chain assembly (1) and remove chain assembly (1) from tire.

(3) Drive vehicle off chain assembly (1).

(4) Repeat steps 1 through 3 until all tire chain assemblies (1) have been removed from tires.

(5) Stow tire chain assemblies (1).

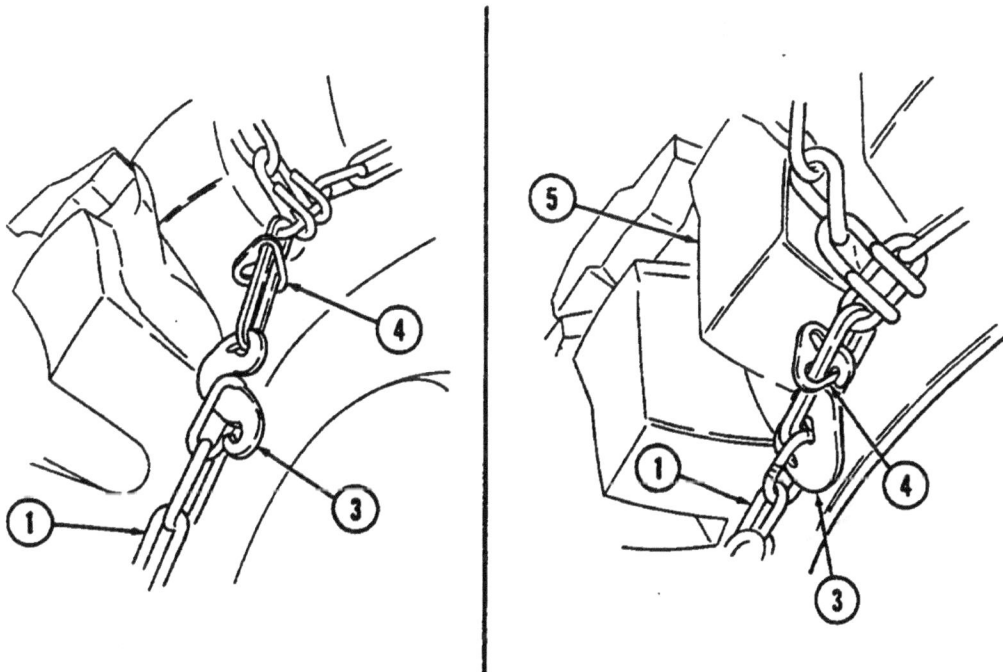

3-23. EMERGENCY FAN CLUTCH OVERRIDE

a. The radiator fan normally activates when engine temperature exceeds 220°F (104°C) and deactivates when engine temperature drops to 190°F (88°C). If overheating occurs in an emergency situation, this procedure will ensure continuous fan operation.

b. Raise and secure hood (para. 3-9).

c. Stop engine (para. 2-14) prior to disconnecting time delay module.

d. Disconnect time delay module connector (1) from control valve connector (2).

e. Start engine (para. 2-12).

f. Check fan for continuous operation. If fan is not operating continuously, stop engine and notify unit maintenance.

g. Lower and secure hood (para. 3-9).

h. Allow engine to cool at idle until engine temperature lowers to normal operating temperature of 185°-250°F (85°-120°C).

i. Proceed to unit maintenance with vehicle. Ensure unit maintenance is notified of emergency service performed on vehicle.

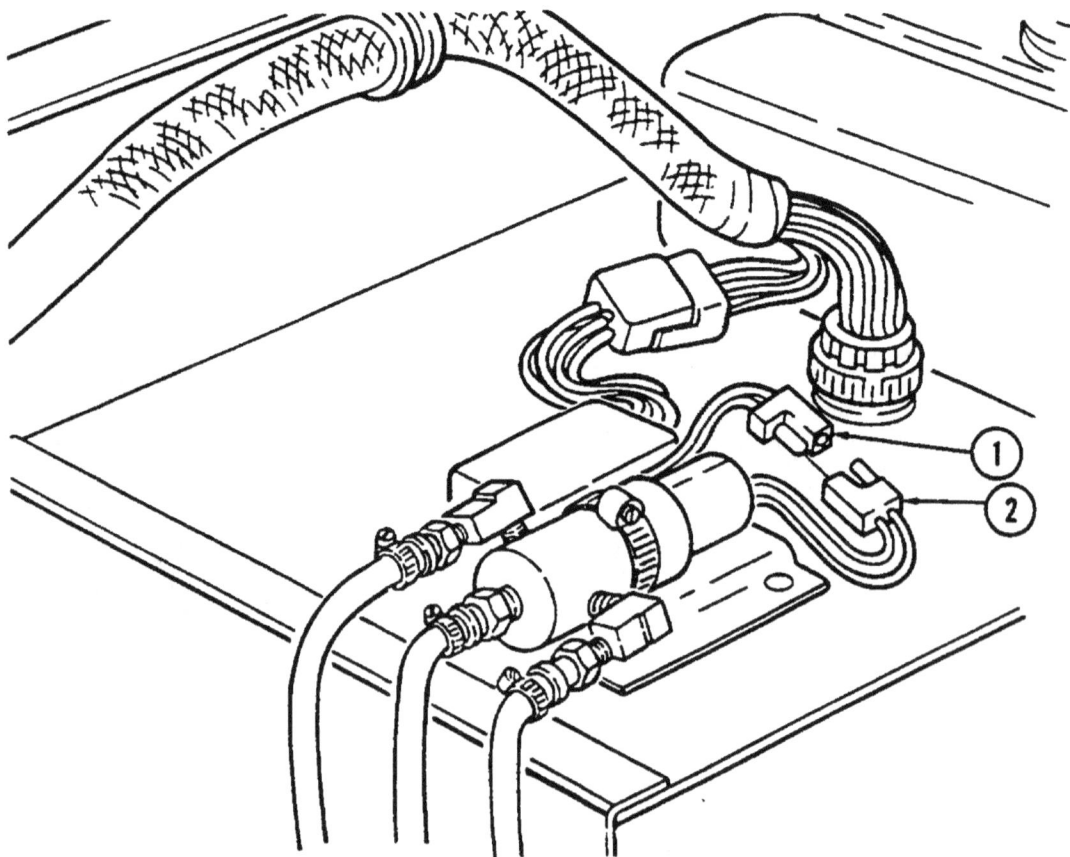

3-24. REARVIEW MIRROR ADJUSTMENT

CAUTION
Visibility from the right mirror may be impaired.

NOTE
Prior to mirror adjustment, move mirror arm assembly until assembly locks in place.

a. Left Mirror (Driver's Side). Loosen top nut (5) and position mirror arm (7) forward of windshield frame (4) so that center line of mirror arm (7) is approximately 9-1/2 in. (24.1 cm) (M1113, M1151, and M1152), 10-1/2 in. (26.7 cm) (M1114) from vehicle (1). Tighten top nut (5). Adjust mirror head (6) for maximum visibility.

b. Right Mirror (Passenger's Side). Loosen top nut (5) and position mirror arm (3) forward of windshield frame (4) so that center line of mirror arm (3) is approximately 7-3/4 in. (19.7 cm) (M1113, M1151, and M1152), 7-1/2 in. (19.1 cm) (M1114) from vehicle (1). Tighten top nut (5). Adjust mirror head (2) for maximum visibility.

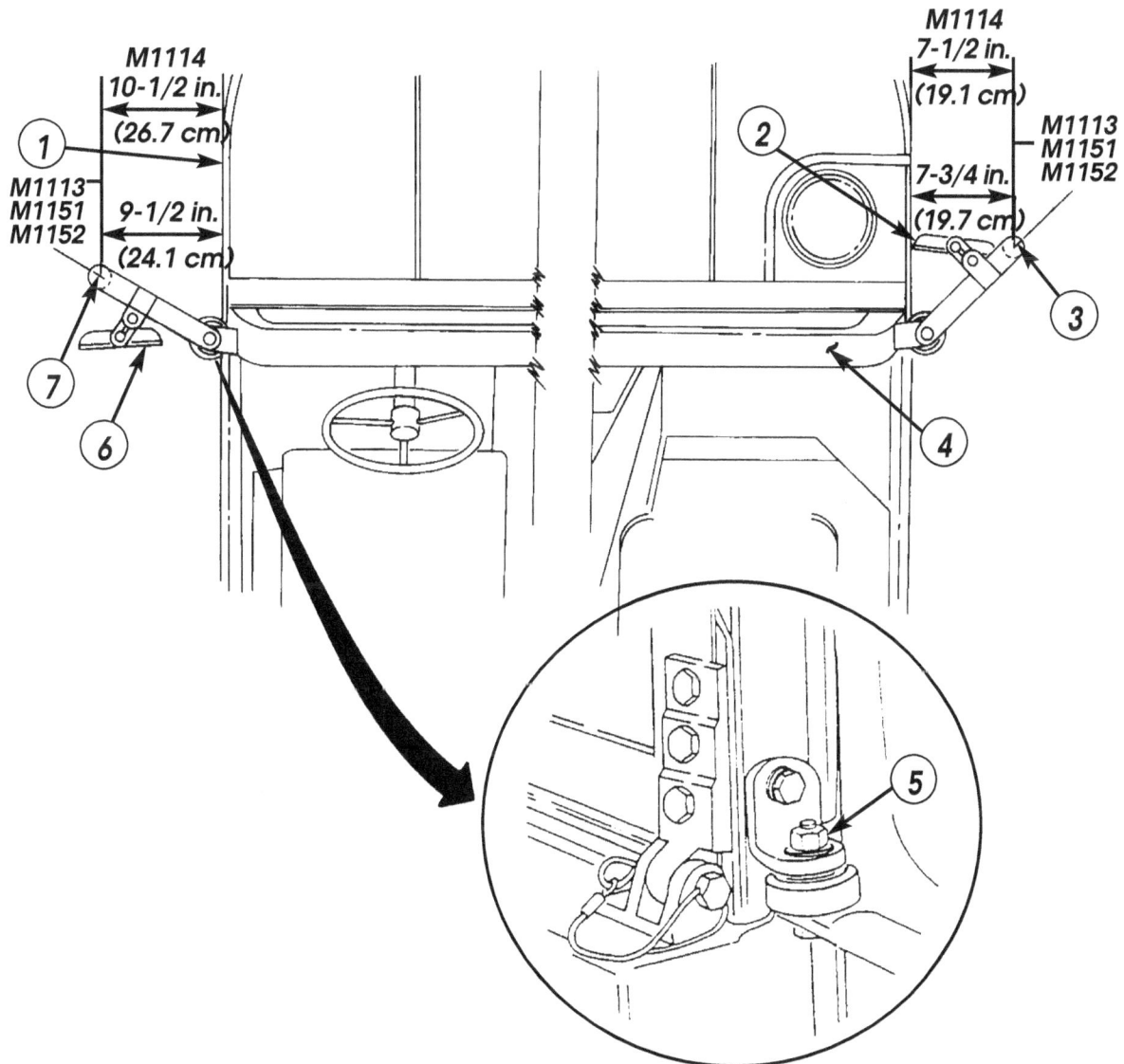

3-25. WHEEL ASSEMBLY REPLACEMENT

a. Wheel Assembly Removal.

WARNING

- Always apply parking brake and block opposite wheel before removing wheel assembly. Avoid removing wheel assembly when vehicle is on sloping terrain. Injury to personnel or damage to equipment may result.
- Remove only the inner group of nuts when removing a wheel from the vehicle. Removing the outer nuts which hold the rim together while the wheel assembly is inflated could result in serious injury or death.
- Ensure scissors jack is positioned directly under lower control arm next to wheel being replaced. Do not place at any other location such as frame rails. Injury to personnel or damage to equipment may result.

NOTE

- Radial tires are non-directional. They provide equal traction and performance when installed in either direction. For further information, refer to TM 9-2610-200-14.
- Perform step 1 for old style control arms only.

(1) Place jack (3) in position (2) on lower control arm (1) next to wheel being replaced. Center jack (3) squarely under point of contact.

NOTE

Perform step 2 for new style control arms only.

(2) Place jack (3) in position (5) on lower control arm (4) next to wheel being replaced. Center jack (3) under point of contact.

(3) Loosen eight lug nuts (7), but do not remove.

(4) Raise vehicle high enough to allow removal of wheel assembly (6).

(5) Remove eight lug nuts (7) and wheel assembly (6) from geared hub (8).

b. Wheel Assembly Installation.

(1) Install wheel assembly (6) on geared hub (8) and secure with eight lug nuts (7). Tighten lug nuts (7) to full engagement of wheel assembly (6) to geared hub (8).

(2) Slowly lower vehicle and remove jack (3).

(3) Tighten eight lug nuts (7) in sequence indicated.

(4) Notify unit maintenance to tighten lug nuts (7) to proper torque.

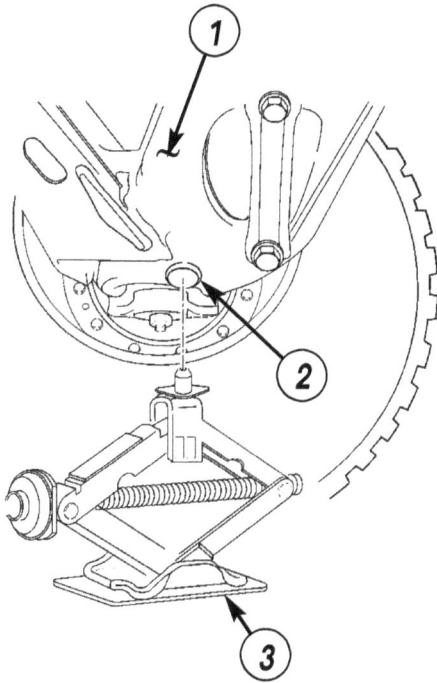

OLD STYLE

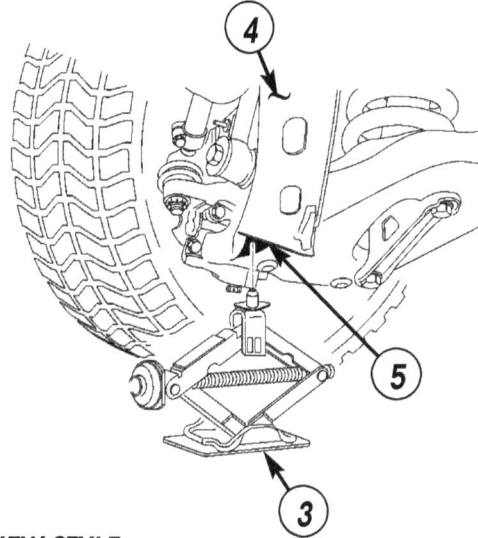

NEW STYLE

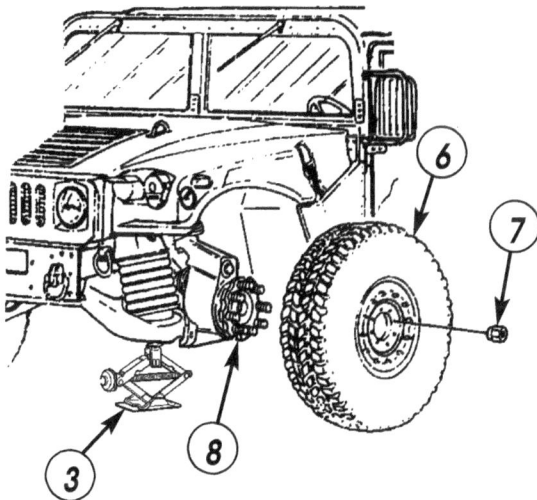

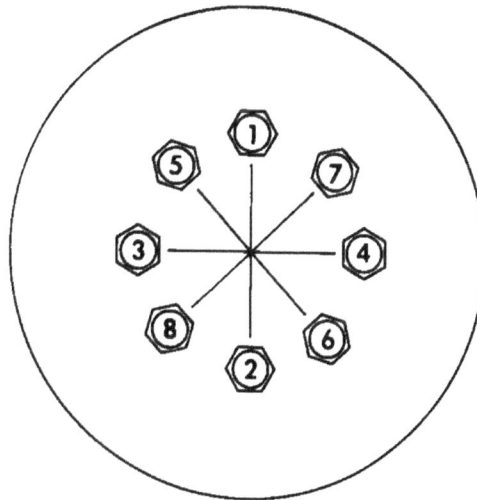

TIGHTENING SEQUENCE

APPENDIX A
REFERENCES

A-1. SCOPE

This appendix lists all forms, field manuals, and technical manuals for use with these vehicles.

A-2. DEPARTMENT OF THE ARMY PAMPHLETS

Consolidated Index of Army Publications and Blank Forms . . . . . . D A P a m 25-30

The Army Maintenance Management System (TAMMS)
User's Manual . . . . D A P a m 750-8

A-3. FORMS

Recommended Changes to Publications and Blank Forms . . . . D A Form 2028

Recommended Changes to Equipment Technical Publications . . DA Form 2028-2

Hand Receipt/Annex Number . . . . D A Form 2062

Exchange Tag . . . . D A Form 2402

Equipment Inspection and Maintenance Worksheet . . . . D A Form 2404

Maintenance Request . . . . D A Form 2407

Preventive Maintenance Schedule and Record . . . . D D Form 314

Product Quality Deficiency Report . . . . S F 3 6 8

Equipment Inspection and Maintenance Worksheet
(Automated) . . . . D A Form 5988-E

Maintenance Request (Automated) . . . . D A Form 5990-E

Preventive Maintenance Schedule and Record (Automated) . . . . D A Form 5986-E

A-4. FIELD MANUALS

NBC Decontamination . . . . F M 3 - 5

Operation and Maintenance of Ordnance Materiel in
Cold Weather (0° to -65°F) . . . . F M9-207

Vehicle Recovery Operations . . . . F M 9 - 4 3 - 2

First Aid for Soldiers . . . . F M21-11

Manual for the Wheeled Vehicle Driver . . . . F M21-305

Browning Machine Gun, Caliber .50 HB, M2 . . . . F M23-65

Basic Cold Weather Manual . . . . F M31-70

Northern Operations . . . . F M31-71

Army Motor Transport Units and Operations . . . . F M55-30

Desert Operations . . . . F M90-3

Jungle Operations . . . . F M90-5

Mountain Operations . . . . F M90-6

Improved Target Acquisition System, M41 . . . . F M3-22.32

A-5. TECHNICAL MANUALS

Machine Gun, Caliber .50 Browning, M2 . . . . T M9-1005-213-10

Operator's, Unit, Direct Support, and General Support
  Maintenance Manual for Care, Maintenance,
  Repair, and Inspection of Pneumatic Tires
  and Inner Tubes. . . . . T M9-2610-200-14

Operator's, Unit, Intermediate Direct Support, and
  Intermediate General Support Maintenance Manual for
  Lead-Acid Storage Batteries. . . . . T M9-6140-200-14

Operating Instructions for Computer Truck Wheel Balancer . . TM 9-4910-785-10

A-6. TECHNICAL BULLETINS

Security of Tactical-Wheeled Vehicles . . . . T B9-2300-422-20

Equipment Improvement Report and Maintenance Digest . . . . . . T B43-0001-62

Use of Antifreeze Solutions, Antifreeze Extender, and
  Cleaning Compounds Test Kit in Engine Cooling Systems . . . . T B750-651

A-7. MISCELLANEOUS PUBLICATIONS

Hand Receipt . . . . T M9-2320-387-10-HR

A-8. ARMY REGULATIONS

The Army Integrated Publishing and Printing Program . . . . A R25-30

Dictionary of United States Army Terms . . . . A R310-25

Army Materiel Policy and Retail Maintenance Operation . . . . A R750-1

Army Acquisition Policy. . . . . A R - 7 0 - 1

Prevention of Motor Vehicle Accidents . . . . A R385-55

Expendable/Durable Items (Except: Medical, Class V,
  Repair Parts and Heraldic Items) . . . . C T A50-970

APPENDIX B
COMPONENTS OF END ITEM (COEI)
AND BASIC ISSUE ITEMS (BII) LISTS

Section I. INTRODUCTION

B-1. SCOPE

This appendix lists components of the end item and basic issue items for the EVC
vehicles to help you inventory items required for safe and efficient operation of the
equipment.

B-2. GENERAL

The Components of End Item (COEI) and Basic Issue Items (BII) lists are divided
into the following sections:

a. Section II. Components of End Item (COEI). This listing is for
informational purposes only and is not authority to requisition replacements.These
items are part of the end item, but are removed and separately packaged for
transportation or shipment.As part of the end item, these items must be with the
end item whenever it is issued or transferred between property accounts.
Illustrations are furnished to assist in identifying the items.

b. Section III. Basic Issue Items (BII). These essential items are required to
place EVC vehicles in operation, to operate them, and to perform emergency repairs.
Although shipped separately packaged, BII must be with the vehicle during
operation and whenever it is transferred between property accounts.This list is
your authority to request/requisition replacement BII, based on TOE/MTOE
authorization of the end item.The illustrations will assist you with hard-to-identify
items.

B-3. EXPLANATION OF COLUMNS

The following provides an explanation of columns found in the tabular listings:

a. Column (1) - Illustration Number (Illus Number). This column indicates
the number of the illustration in which the item is shown.

b. Column (2) - National Stock Number. Indicates the National Stock
Number assigned to the item and will be used for requisitioning purposes.

c. Column (3) - Description. Indicates the Federal item name and, if required,
a minimum description to identify the item.The last line for each item indicates the
Commercial and Government Entity Code (CAGEC) in parentheses followed by the
part number.

d. Column (4) - Unit of Issue (U/I). Indicates how the item is issued for the
National Stock Number shown in column two.This unit is expressed by a two-
character alphabetical abbreviation; i.e., (ea, pr).

e. Column (5) - Quantity Required (Qty rqr). Indicates the quantity of the
item authorized to be used with/on the vehicle.

Section II. COMPONENTS OF END ITEM

These items are installed in the vehicle at the time of manufacture or rebuild.

(1) Illus Number	(2) National Stock Number	(3) Description CAGEC and Part Number	Usable On Code	(4) U/I	(5) Qty rqr
1	3120-01-188-5082	BEARING, SLEEVE: Pintle Adapter (located in the armament mounting assembly) (19207) 12340310	XBB	EA	1
		consisting of:			
	5305-01-204-4190	a. Setscrew 3/8 - 24 (7X677) 9428747		EA	4
	N/A	b. Pintle Adapter (19207) 12340310-1		EA	1
2	2510-01-498-4996	GUNNERSHIELD ASSEMBLY: (19207) 57K4470	XBB	EA	1
		consisting of:			
		a. Screw (19207) 12484861-054		EA	4
	5310-01-442-1109	b. Washer (24617) 9421888		EA	8
	2540-01-500-3446	c. Shield, gunner (19207) 12484846		EA	1
	1005-01-500-3744	d. Bracket, gun shield support (19207) 12484852		EA	1
	5310-01-417-7273	e. Lockwasher (19207) 12484862-008		EA	8
		f. Nut (19207) 12484864-005		EA	8

Section II. COMPONENTS OF END ITEM (Cont'd)

(1) Illus Number	(2) National Stock Number	(3) Description CAGEC and Part Number		(4) U/I	(5) Qty rqr
2 (Cont'd)	1010-01-447-2983	g. Handle assembly, locking (19200) 12012065		EA	1
	5305-00-514-0237	h. Screw (19205) 5140237		EA	1
	5325-01-447-2281	i. Sleeve, split (19200) 12012071		EA	1
	5315-01-513-0027	j. Pawl, clamping (19200) 12012067		EA	1
	2540-01-500-0309	k. Adapter, pintle (19207) 12484847		EA	1
		l. Screw (19207) 12484861-055		EA	4

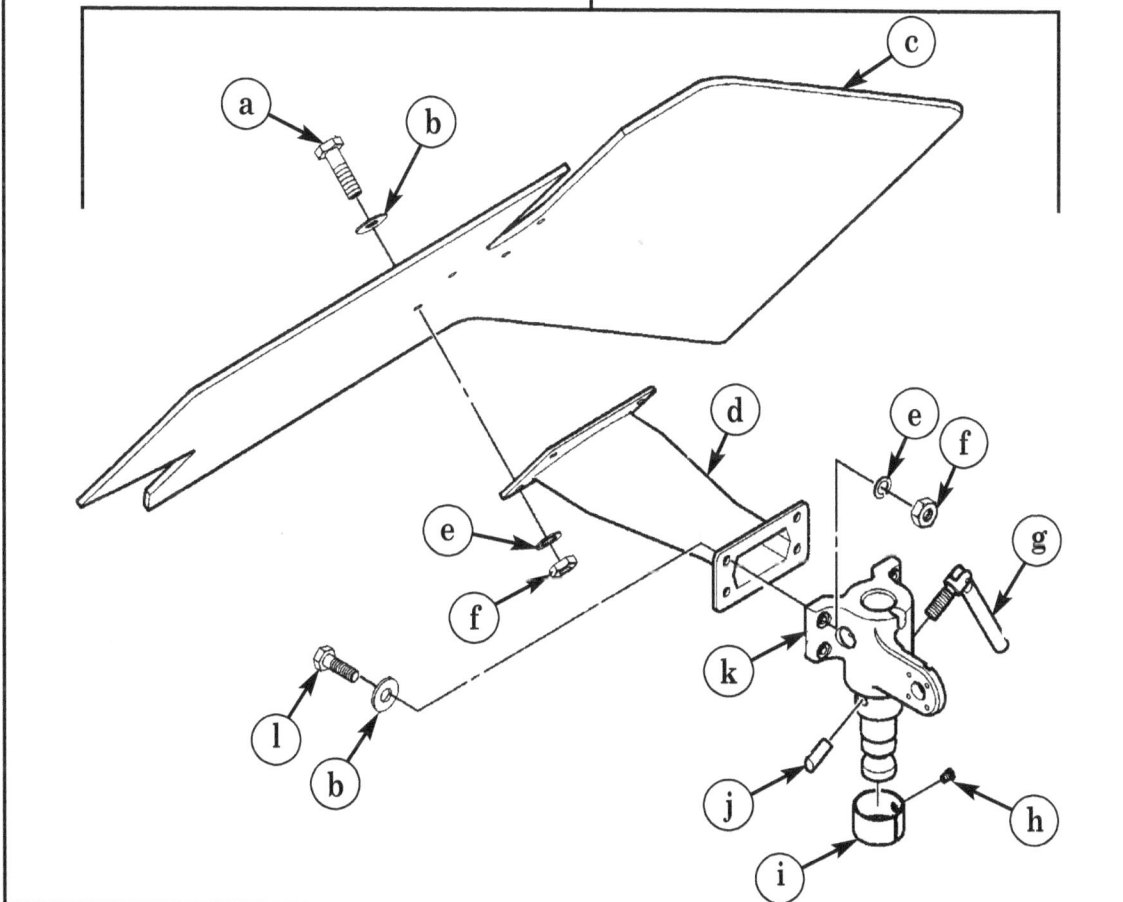

Section III. BASIC ISSUE ITEMS (Cont'd)

(1) Illus Number	(2) National Stock Number	(3) Description CAGEC and Part Number	(4) U/I	(5) Qty rqr
1		TM 9-2320-387-10 [in cotton duck pamphlet bag]	EA	1
		TM 9-2320-387-10-HR [in cotton duck pamphlet bag]	EA	1
2	5140-01-429-6945	BAG: jack and tools stowage, cotton duck, 13 in. x 23 in. x 9-1/2 in. (folded) [in footwell area M1113; shell M1114; behind driver's seat M1151, M1152] (19207) 12447043	EA	1
3	2540-00-670-2459	BAG ASSEMBLY, PAMPHLET: cotton duck, 3 in. x 9-1/4 in. x 11-1/4 in. [behind driver's seat area] (19207) 11676920	EA	1
4	5140-00-473-6256	BAG, TOOL: cotton duck, 10 in. x 20 in., w/flap [behind driver's seat area] (19207) 11655979	EA	1

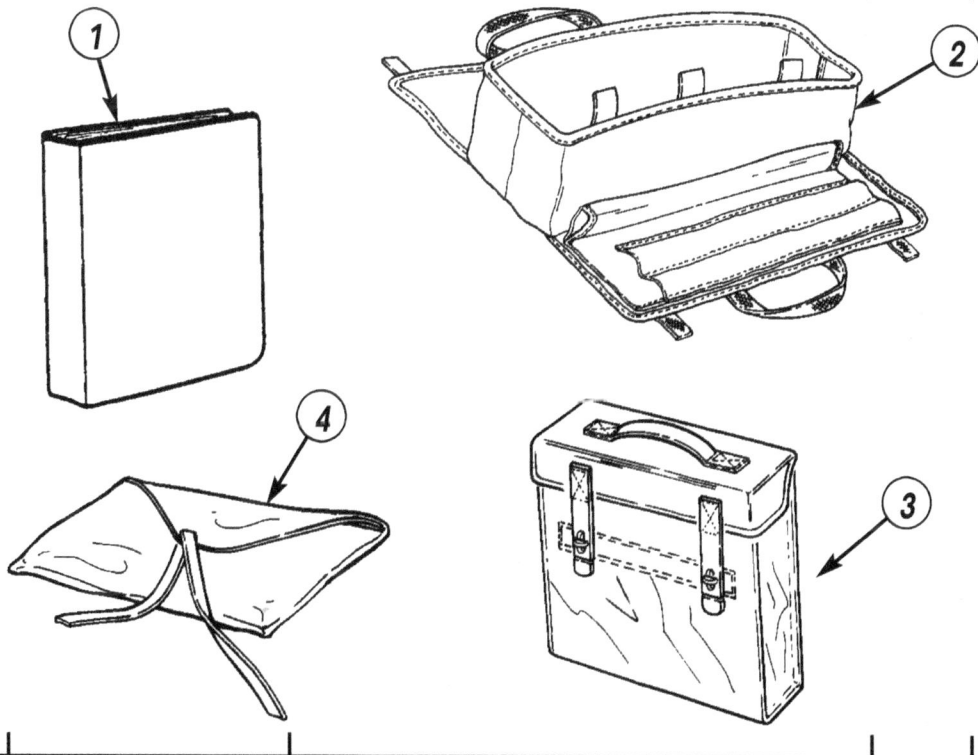

Section III. BASIC ISSUE ITEMS (Cont'd)

(1) Illus Number	(2) National Stock Number	(3) Description CAGEC and Part Number	(4) U/I	(5) Qty rqr
5	5120-01-429-6065	EXTENSION, JACK HANDLE: 15-1/2 in. long [in jack and tools bag] (19207) 12447040	EA	1
6	6545-00-922-1200	FIRST AID KIT, GENERAL PURPOSE: 3 in. x 5-13/16 in. x 8-7/32 in. [under driver's seat area M1113, M1151, and M1152; under passenger's seat area M1114] (19207) 11677011	EA	1
7	4210-01-481-3875	FIRE EXTINGUISHER: hand, type 1, class 2, size 5 [driver's seat area] (54905) 14091	EA	1
8	7510-01-065-0166	FOLDER, EQUIPMENT RECORD: 2-1/2 in. x 8 in. x 10 in. [in pamphlet bag] (72094) 43986-1	EA	1

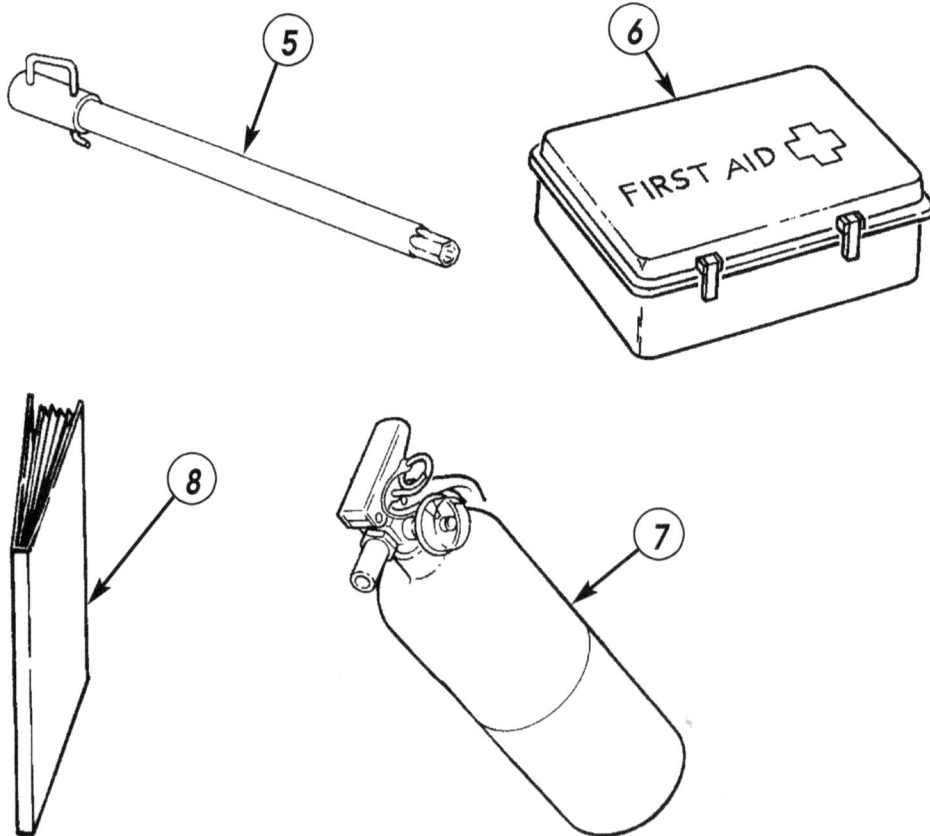

Section III. BASIC ISSUE ITEMS (Cont'd)

(1) Illus Number	(2) National Stock Number	(3) Description CAGEC and Part Number	(4) U/I	(5) Qty rqr
9	5120-01-429-8137	HANDLE, MECHANICAL JACK: 18 in. long [in jack and tools bag] (19207) 12447041	EA	1
10	5120-01-430-3123	JACK, SCISSORS, HAND: mechanical, 3.5-ton cap., 6.33 in. closed to 18.70 in. (max) open [in jack and tools bag] (19207) 12447042	EA	1

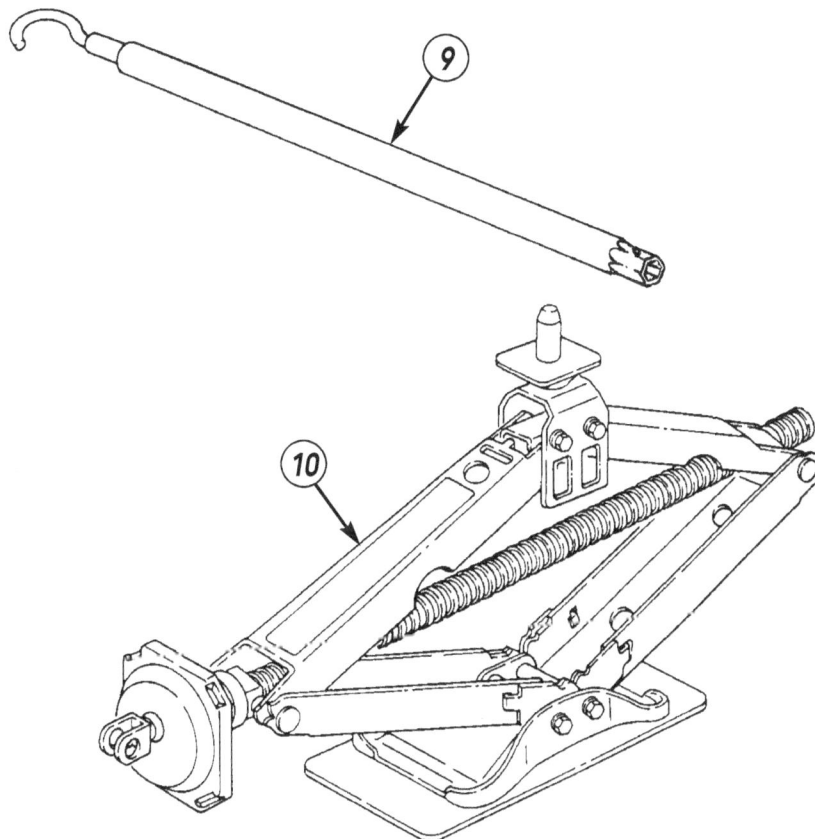

Section III. BASIC ISSUE ITEMS (Cont'd)

(1) Illus Number	(2) National Stock Number	(3) Description CAGEC and Part Number	(4) U/I	(5) Qty rqr
11	5120-01-416-8568	MAX TOOL KIT, COMBINATION TOOL, HAND: [in footwell area on M1113; rear tailgate area on M1114 and M1151; on tunnel on M1152] (19207) 57K3528	EA	1
	5140-01-416-8569	a. Carrying case (0T9K4) 595-030		
	5110-01-416-7827	b. Ax (0T9K4) 595-010		
	5110-01-416-7830	c. Ax sheath (0T9K4) 595-020		
	5120-01-416-8570	d. Shovel attachment (0T9K4) 595-040		
	5120-01-416-8571	e. Mattock attachment (0T9K4) 595-050		
	5120-01-416-8573	f. Pick attachment (0T9K4) 595-060		
	5120-01-416-8572	g. Broad pick attachment (0T9K4) 595-070		
	5120-01-416-8577	h. Rake hoe attachment (0T9K4) (595-080)		
	5120-01-416-8574	i. Rake hoe fastener (0T9K4) 595-090		
	5120-01-416-8575	j. Safety locking pin (0T9K4) 595-999		

Section III. BASIC ISSUE ITEMS (Cont'd)

(1) Illus Number	(2) National Stock Number	(3) Description CAGEC and Part Number	(4) U/I	(5) Qty rqr

Section III.   BASIC ISSUE ITEMS (Cont'd)

(1) Illus Number	(2) National Stock Number	(3) Description CAGEC and Part Number	(4) U/I	(5) Qty rqr
12	5120-00-223-7397	PLIERS, SLIP JOINT: combination, slip joint, straight nose w/cutter, 8 in. long, phosphate finish [in tool bag] (19207) 11655775-3	EA	1
13	5120-00-234-8913	SCREWDRIVER, CROSS TIP: Phillips type, plastic handle, point no. 2, 7-1/2 in. long [in tool bag] (19207) 11655777-12	EA	1
14	5120-00-227-7356	SCREWDRIVER, FLAT TIP: flared sides, plastic handle, round blade, 3/16 in. wide tip, 7-3/4 in. long [in tool bag] (55719) SSDE-66	EA	1

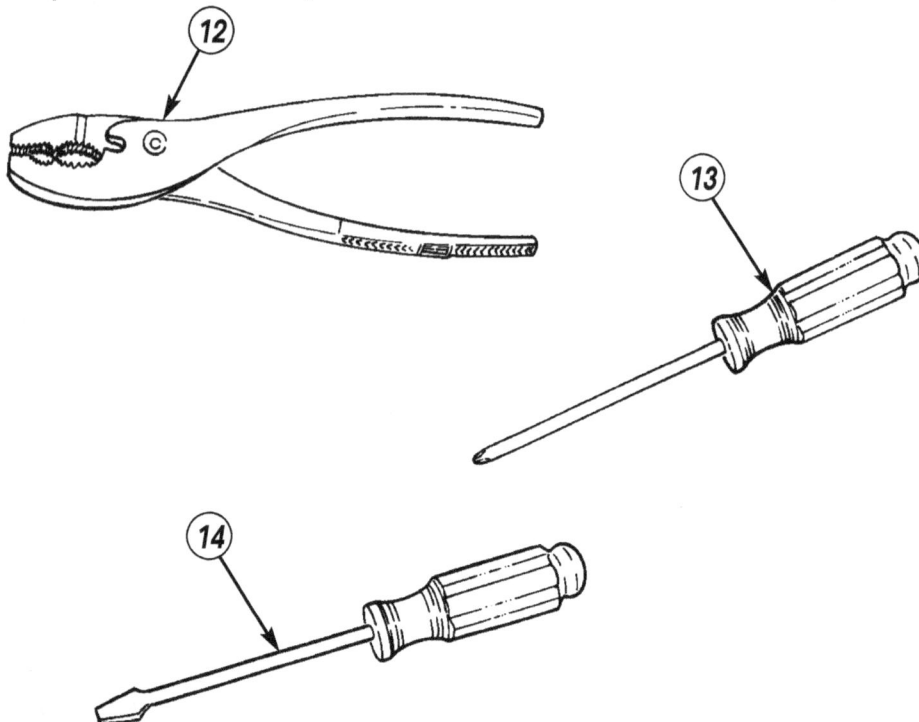

Section III. BASIC ISSUE ITEMS (Cont'd)

(1) Illus Number	(2) National Stock Number	(3) Description CAGEC and Part Number	(4) U/I	(5) Qty rqr
15	4710-01-475-3753	TUBE ASSEMBLY: U-end, allows operation of vehicle with hydraulic winch or controller valve assembly removed. [M1114 - stowage box under commander's seat; M1113, M1151, and M1152 - also, if 10,500 lb. hydraulic winch kit has been installed]. (OGBZ7) 983-90-50201	EA	1
16	5120-00-240-5328	WRENCH, ADJUSTABLE: open end, 8 in. long [in tool bag] (96906) MS15461-3	EA	1
17	5120-01-429-6964	WRENCH, RATCHET: 5/8 in., 3/4 in. hex socket, 1-1/8 in. [in jack and tools bag] (19207) 12447039	EA	1
18	5120-01-156-7296	WRENCH, WHEEL LUG: 1-15/32 - 1-19/64 in. socket dia., 1-3/16 in. socket depth, 17-1/2 in. long [in jack and tools bag] (11862) 14009303	EA	1

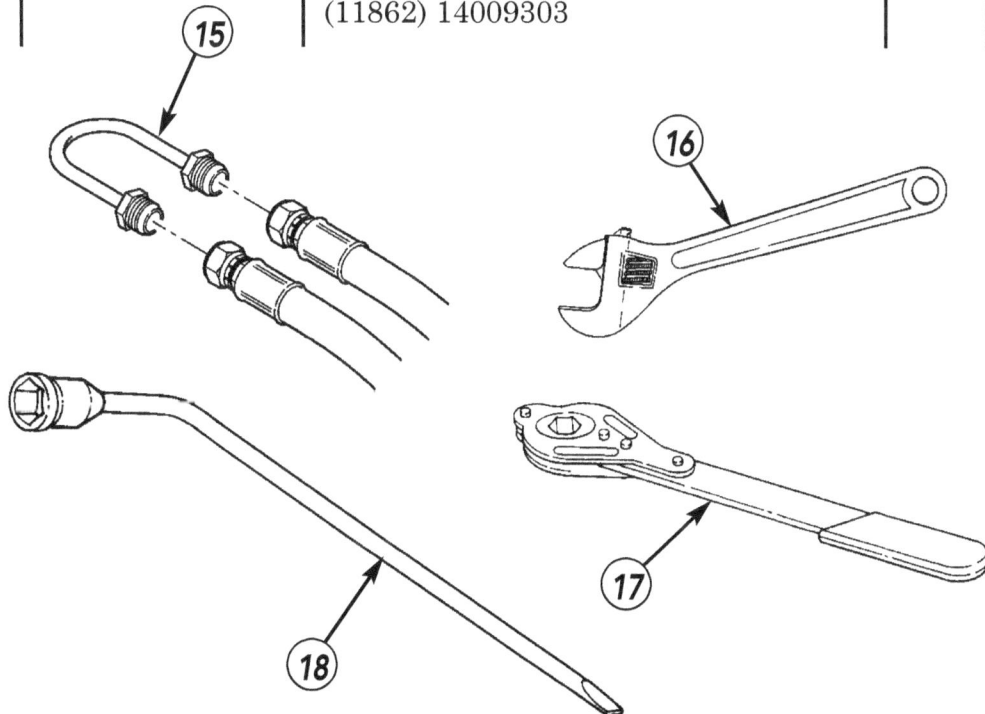

APPENDIX C
ADDITIONAL AUTHORIZATION LIST (AAL)

Section I. INTRODUCTION

C-1. SCOPE

This appendix lists additional items authorized for support of ECV vehicles.

C-2. GENERAL

This list identifies items that do not have to accompany the vehicle and do not have to be turned in with it.These items are all authorized to you by CTA, MTOE,TDA, or JTA.

C-3. EXPLANATION OF LISTING

National Stock Numbers, descriptions, and quantities are provided to help you identify and request the additional items you require to support the vehicles.The items are listed in alphabetical sequence by item name. If item required differs for different models of this equipment, the model is shown under the "Usable On Code" heading in the description column.These codes are identified as:

Code	Used On
(Blank)	All
XAA	M1113
XBB	M1114
TTT	M1151
TTA	M1151A1
UUU	M1152
UUA	M1152A1
UU1	M1165
UU2	M1165A1
TT1	M1167

Section II. ADDITIONAL AUTHORIZATION LIST

(1) NATIONAL STOCK NUMBER	(2) DESCRIPTION		(3)	(4)
	CAGEC AND PART NUMBER	USABLE ON CODE	U/I	QTY AUTH
6665-01-438-6963	ALARM: chemical agent, M8A1 (81361) EA-PRF-2058		EA	1
2540-00-670-2459	BAG: pamphlet, cotton duck, 3 in. x 9-1/4 in. x 11-1/4 in. (19207) 11676920		EA	1
2510-01-276-9249	BAR: suspension tiedown [in jack and tools bag in footwell area] (34623) 5598373	XAA, UUU, UUA	EA	2
2540-00-678-3469	BLOCK: chock (96906) MS52127-3		EA	2
3940-00-151-6769	BLOCK: tackle (19207) 11676932		EA	1
2510-01-050-9770	BRACKET: support, decontaminating apparatus (19207) 11644841		EA	1
2540-01-276-9250	BRACKET: suspension tiedown [in tool bag behind driver's door] (34623) 5598408	XAA, UUU UUA	EA	2
2590-00-273-6331	BRACKET: water can (96906) MS53052-1	TTT,TTA, UUU UUA, UU1, UU2 TT1	EA	2
2590-01-222-7946	CABLE, NATO SLAVE: intervehicular power cable, 12 ft long (19207) 11682379-4		AY	1
7240-00-089-3827	CAN: water, military, plastic, 5 gal. (81349) MIL-C-43613		EA	1
3940-01-509-9096	CARGO NET: (098P0) B9154-4854-2R-1855	XBB	EA	1
2540-01-214-1264	CHAIN: tire, 9/32 side x 1/4 cross link (34623) 5569255		SET	2
4010-00-443-4845	CHAIN: tow, 9/32 side x 12 ft long (19207) 10944642-2		EA	1

Section II. ADDITIONAL AUTHORIZATION LIST (Cont'd)

(1) NATIONAL STOCK NUMBER	(2) DESCRIPTION		(3)	(4)
	CAGEC AND PART NUMBER	USABLE ON CODE	U/I	QTY AUTH
4230-01-133-4124	DECONTAMINATING APPARATUS: M-13 (81361) E5-51-527		EA	1
4230-00-720-1618	DECONTAMINATION APPARATUS: portable, DS-2, 1-1/2 UT,ABC-M11 w/bracket (81361) D5-51-269	TTT,TTA UUU, UUA UU1, UU2 TT1	EA	1
6230-00-163-1856	FLASHLIGHT: 2 x 3-1/2, GFST x 8 in. (82214) 1259		EA	1
4910-00-204-3170	GAUGE: tire pressure (19207) 7081758		EA	1
5120-01-355-1901	HANDLE, SOCKET WRENCH: 1/2 in. square drive reversible, 9 in. long (05047) B107.10M		EA	1
5315-00-732-1019	KEY, MACHINE: geared hub oil drainplug, straight bar key, 1/2 in. square, 2-1/2 in. long (96906) MS20066-543		EA	1
5340-00-158-3805	PADLOCK SET: (96906) MS35647-10		EA	1
5820-00-223-7473	RADIO SET: 9 x 14 x 16 in. (80058) AN/GRC-160		EA	1
5120-00-222-8852	SCREWDRIVER: flat tip, flared sides, plastic handle, round blade, 1/4 in. wide tip (80063) SCC539502-2		EA	1
5120-01-430-3096	SOCKET, SOCKET WRENCH: 1/2 in. square drive, 9/16 in. wrenching size, standard length, 1/2 point, steel (2K880) 40218		EA	1
7240-00-177-6154	SPOUT: can, gas, flexible with filter screen, 16 in. long (19207) 11677020		EA	1
2510-01-197-8572	TRAY: water can (19207) 12340155	XBB	EA	1

Section II. ADDITIONAL AUTHORIZATION LIST (Cont'd)

(1) NATIONAL STOCK NUMBER	(2) DESCRIPTION		(3) U/I	(4) QTY AUTH
	CAGEC AND PART NUMBER	USABLE ON CODE		
2610-01-333-7632	TIRE: radial, 37 x 12.50R16.5LT (34623) 5935336		EA	1
9905-00-148-9546	TRIANGLES: folding, reflective (81348) RR-W-1817		SET	1
1005-00-322-9716	TRIPOD MOUNT: w/cover, M3, mach. gun (19204) 8403398	TTT,TTA, UU1, UU2 TT1	EA	1
5340-01-277-2460	TURNBUCKLE ASSY: suspension tiedown (34623) 5598406	XAA UUU UUA	EA	2
N/A	WHEEL AND RUNFLAT, RADIAL: (19207) 12460176		EA	1
5120-01-279-4788	WRENCH: open end, 1-1/8 x 18 in. long (34623) 5598407	XAA UUU UUA	EA	1

APPENDIX D
EXPENDABLE/DURABLE SUPPLIES AND MATERIALS LIST

Section I. INTRODUCTION

D-1. SCOPE

This appendix lists expendable/durable supplies and materials you will need
to operate and maintain the ECV vehicles. These items are authorized to you by
CTA 50-970, Expendable Items.

D-2. EXPLANATION OF COLUMNS

a. Column (1) - Item Number. This number is assigned to each entry in the
listing.

b. Column (2) - Level. This column identifies the lowest level of maintenance
that requires the listed item:

C – Operator/Crew

c. Column (3) - National Stock Number. This is the National Stock Number
assigned to the item; use it to request or requisition the item.

d. Column (4) - Description. Indicates the Federal item name and, if required,
a description to identify the item. The last line for each item listing indicates the
Commercial and Government Entity Code (CAGEC) in parentheses followed by the
part number.

e. Column (5) - Unit of Measure (U/M). Indicates the measure used in
performing the actual maintenance function. This measure is expressed by an
alphabetical abbreviation (QT, GAL.). If the unit of measure differs from the unit of
issue, requisition the lowest unit of issue that will satisfy your requirements.

Section II. EXPENDABLE/DURABLE SUPPLIES AND MATERIALS LIST

(1) ITEM NUMBER	(2) LEVEL	(3) NATIONAL STOCK NUMBER	(4) DESCRIPTION	(5) U/M
1	C		ADHESIVE: silicone rubber (81349) MIL-A-46106	
		8040-01-010-8758	1 Kit	KT
2	C		ANTIFREEZE: arctic-type (58536) A-A-52624	
		6850-01-464-9096	55 Gallon Drum	GAL.
3	C		ANTIFREEZE: ethylene glycol, inhibited, heavy-duty, single package (58536) A-A-52624	
		6850-01-464-9125	1 Gallon Container	GAL.
		6850-01-464-9137	5 Gallon Container	GAL.
		6850-01-464-9152	55 Gallon Container	GAL.
4	C		CLEANING COMPOUND: windshield washer (81348) O-C-1901	
		6850-00-926-2275	1 Pint	PT
5	C		CREAM CLEANER: plastic, liquid (09177) 200-767-4A	
		8520-00-262-7177	1 Pint Container	PT
6	C		DETERGENT: general purpose (74188) ORVUS WA PASTE	
		7930-01-107-6997	1 Gallon Container	GAL.
7	C		DISTILLED WATER:	
		6810-00-682-6867	1 Gallon Container	GAL.
		6810-00-356-4936	5 Gallon Container	GAL.
		6810-00-356-4936	5 Gallon Container	GAL.

Section II. EXPENDABLE/DURABLE SUPPLIES AND MATERIALS LIST (Cont'd)

(1) ITEM NUMBER	(2) LEVEL	(3) NATIONAL STOCK NUMBER	(4) DESCRIPTION	(5) U/M
8	C		DRYCLEANING SOLVENT: (81348) P-D-680, Type II	
		6850-00-110-4498	1 Pint Can	PT
		6850-00-274-5421	5 Gallon Drum	GAL.
		6850-00-285-8011	55 Gallon Drum	GAL.
		6850-00-637-6135	Bulk	GAL.
9	C		ETHANOL/SOLUTION: (66735) F89I244	
		6550-01-315-5333	32 Ounce Bottle	OZ
9.1	C		FUEL: aviation, turbine, all temperature, JP-8	
		9130-01-031-5816	Bulk	GAL.
10	C		FUEL OIL: diesel, regular DF-2 (81348) VV-F-800	
		9140-00-286-5296	55 Gallon Drum	GAL.
		9140-00-286-5294	Bulk	GAL.
11	C		FUEL OIL: diesel, winter, DF-1 (81348) VV-F-800	
		9140-00-286-5288	55 Gallon Drum	GAL.
		9140-00-286-5286	Bulk	GAL.
12	C		FUEL OIL: diesel, arctic DF-A (81348) VV-F-800	
		9140-00-286-5284	55 Gallon Drum	GAL.
		9140-00-286-5283	Bulk	GAL.
13	C		GREASE: automotive and artillery (81349) MIL-G-10924	
		9150-01-197-7693	14 Ounce Cartridge	OZ
		9150-01-197-7690	1-3/4 Pound Can	LB
		9150-01-197-7689	6-1/2 Pound Can	LB

Section II. EXPENDABLE/DURABLE SUPPLIES AND MATERIALS LIST (Cont'd)

(1) ITEM NUMBER	(2) LEVEL	(3) NATIONAL STOCK NUMBER	(4) DESCRIPTION	(5) U/M
14	C		GREASE: lithium base w/ molybdenum disulfide (60218) LS 2267	
		9150-01-015-1542	14-1/2 Ounce Cartridge	OZ
15	C		GREASE: wire-rope, EX (81349) MIL-G-18458	
		9150-00-530-6814	35 Pound Can	LB
16	C		HAND CLEANER (19410) PAXSOLV16	
		8520-00-082-2146	1 Pound Container	LB
17	C		HYDRAULIC FLUID: transmission (24617) Dexron® VI	
		9150-01-353-4799	1 Quart (plastic) Can	QT
		9150-01-114-9968	55 Gallon Drum	GAL.
18	C		ISOPROPYL ALCOHOL: (97403) 13222E0694	
		6810-01-075-5546	4 Ounce Bottle	OZ
19	C		LUBRICANT: interlock (96980) Zipperease	
		9150-00-999-7548	24 Stick Box	OZ
20	C		LUBRICANT: solid film (F0301) P0329	OZ
		9150-01-064-6511	11 ounce can, aerosol	
21	C		LUBRICATING OIL: gear, multipurpose, GO 75 (81349) MIL-L-2105	
		9150-01-035-5390	1 Quart Can	QT
		9150-01-035-5391	5 Gallon Drum	GAL.

Section II. EXPENDABLE/DURABLE SUPPLIES AND MATERIALS LIST (Cont'd)

(1) ITEM NUMBER	(2) LEVEL	(3) NATIONAL STOCK NUMBER	(4) DESCRIPTION	(5) U/M
22	C		LUBRICATING OIL: gear, multipurpose, GO 80/90 (81349) MIL-L-2105	
		9150-01-035-5392	1 Quart Can	QT
		9150-01-035-5393	5 Gallon Drum	GAL.
`23	C		LUBRICATING OIL: general purpose, preservative, PL-S (81348) VV-L-800	
		9150-00-231-6689	1 Quart Can	QT
24	C		LUBRICATING OIL: internal combustion engine, arctic, OEA (81349) MIL-L-46167	
		9150-00-402-4478	1 Quart Can	QT
		9150-00-402-2372	5 Gallon Drum	GAL.
		9150-00-491-7197	55 Gallon Drum	GAL.
25	C		LUBRICATING OIL: internal combustion engine, tactical service, OE/HDO 10 (81349) MIL-L-2104	
		9150-00-189-6727	1 Quart Can	QT
		9150-00-186-6668	5 Gallon Drum	GAL.
		9150-00-191-2772	55 Gallon Drum	GAL.
26	C		LUBRICATING OIL: internal combustion engine, tactical service, OE/HDO 30 (81349) MIL-L-2104	
		9150-00-186-6681	1 Quart Can	QT
		9150-00-188-9858	5 Gallon Drum	GAL.
		9150-00-189-6729	55 Gallon Drum	GAL.

Section II. EXPENDABLE/DURABLE SUPPLIES AND MATERIALS LIST (Cont'd)

(1) ITEM NUMBER	(2) LEVEL	(3) NATIONAL STOCK NUMBER	(4) DESCRIPTION	(5) U/M
27	C		LUBRICATING OIL: internal combustion engine, tactical service, OE/HDO 15/40 (81349) MIL-L-2104	
		9150-01-152-4117	1 Quart Can	QT
		9150-01-152-4118	5 Gallon Drum	GAL.
		9150-01-152-4119	55 Gallon Drum	GAL.
28	C		RAG: wiping, cotton and cotton-synthetic (58536) A-A-531	
		7920-00-205-1711	50 Pound Bale	LB

APPENDIX E
STOWAGE AND SIGN GUIDE

E-1. SCOPE

This appendix shows the location for stowage of equipment and material required to be carried on the vehicles.

E-2. GENERAL

The equipment stowage locator is designed to help inventory items required for safe and efficient operation. This equipment locator is representative of BII and applicable AAL stowage on the vehicles. Refer to appendix F for specific deviations from this equipment locator and to para. 2-2 for additional information concerning stowage of equipment.

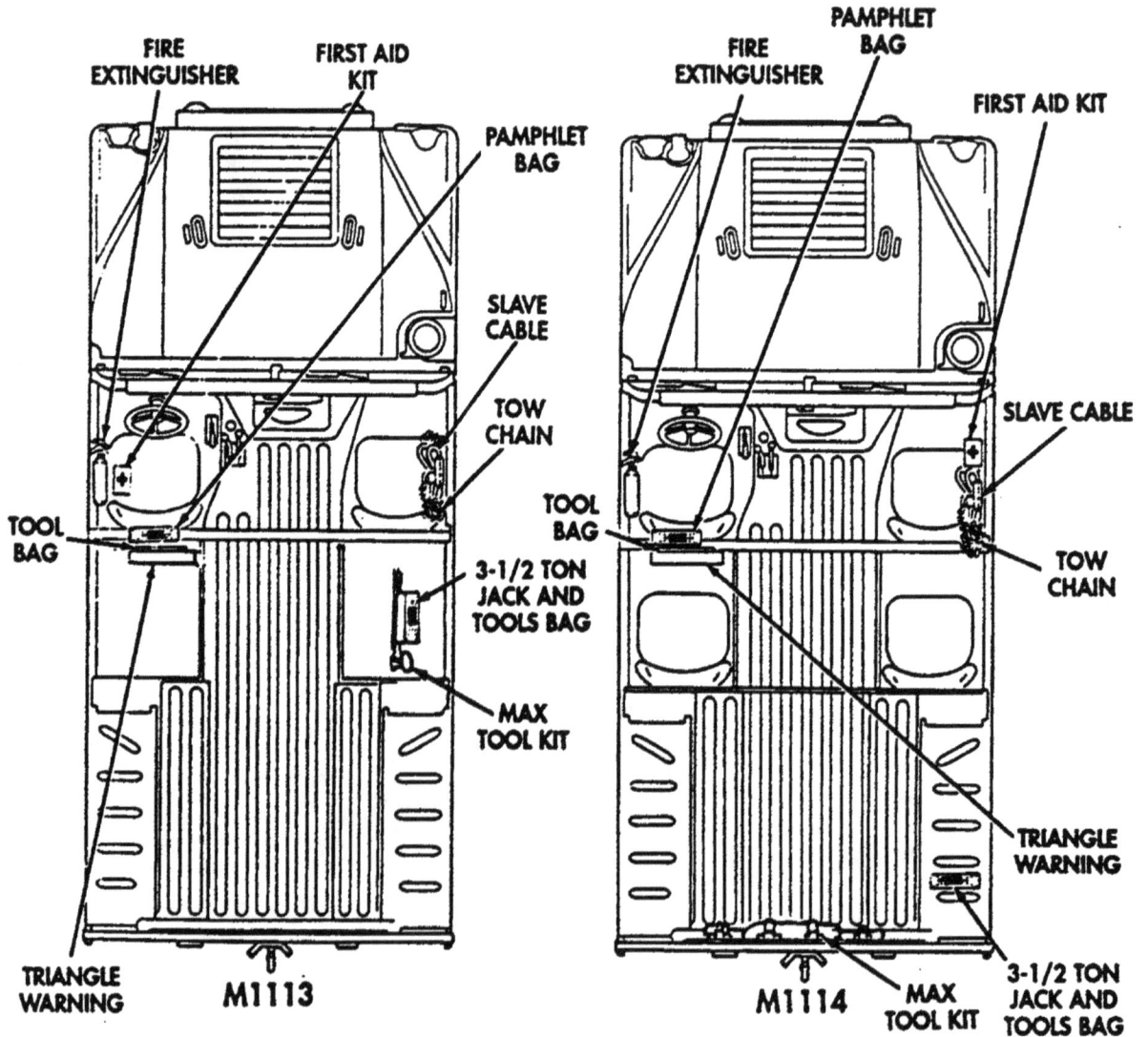

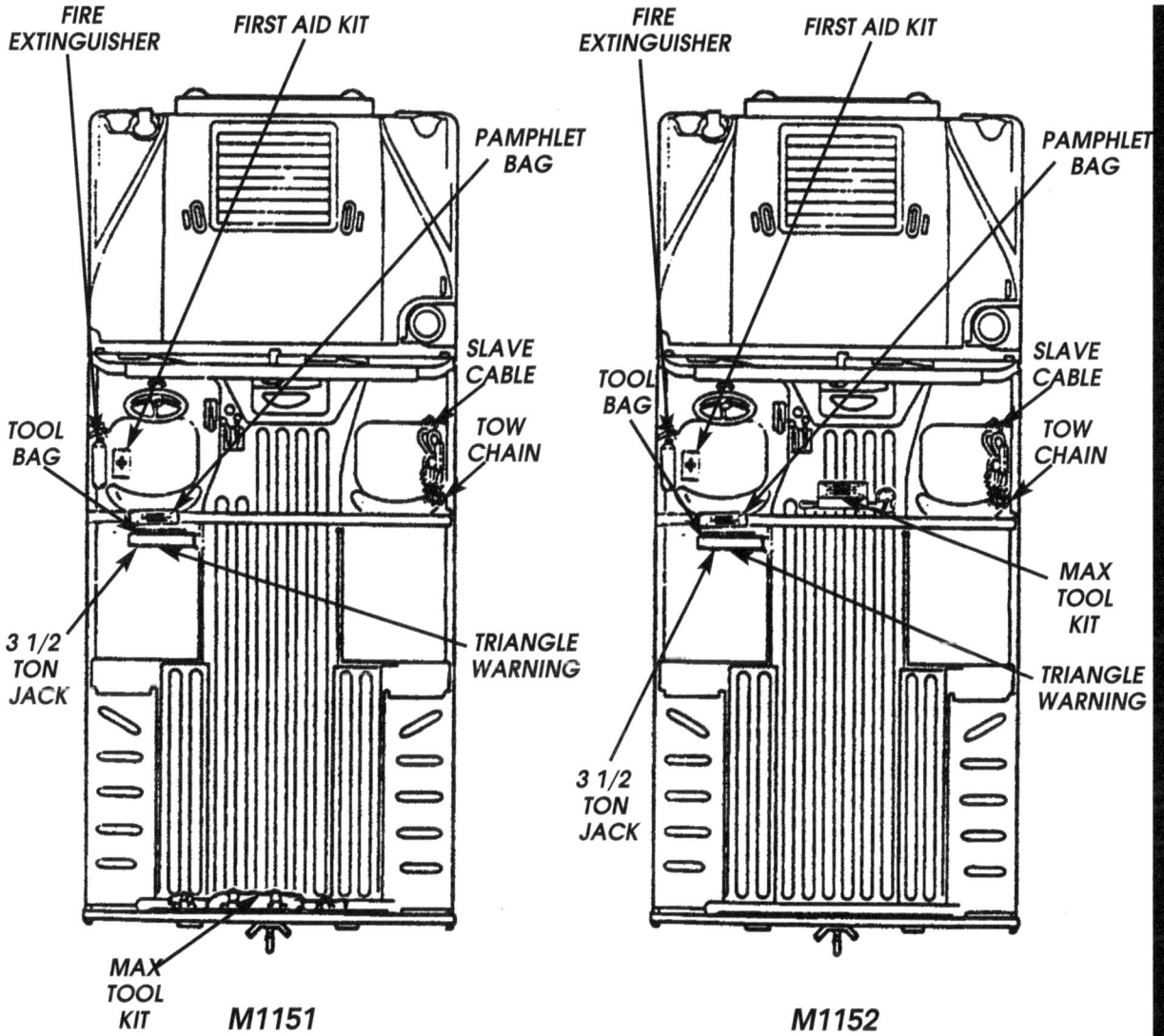

FIRE EXTINGUISHER
FIRST AID KIT
PAMPHLET BAG
SLAVE CABLE
TOOL BAG
TOW CHAIN
3 1/2 TON JACK
TRIANGLE WARNING
MAX TOOL KIT
M1151

FIRE EXTINGUISHER
FIRST AID KIT
PAMPHLET BAG
SLAVE CABLE
TOOL BAG
TOW CHAIN
MAX TOOL KIT
TRIANGLE WARNING
3 1/2 TON JACK
M1152

E-3. SIGN GUIDE

LOCATION OF CREW AREA AND EXTERIOR DECALS AND DATA PLATES

KEY	ITEM	VEHICLE APPLICATION
1	Decal, caution, windshield cleaning	M1114/M1151
2	Plate, identification, ignition	All
3	Decal, instruction, vehicle, break-in service	M1114
4	Plate, instruction, steering wheel lock	All
5	Decal, speedometer	All
6	Decal, instruction, vehicle, break-in service	M1113, M1151, M1152
7	Decal, warning, hand throttle	All
8	Plate, identification, heater, defroster, temperature, fan	All
9	Decal, identification, M16A2/M203 rifle, front	M1113, M1151
10	Decal, floor heat control	M1114
11	Decal, A/C control	M1114
12	Plate, instruction, heater air control	All
13	Decal, procedures, tiedown	All
14	Decal, service and data manuals	All
15	Plate, instruction, seatbelt retractor (front)	All
16	Plate, instruction, operating	All
17	Plate, instruction, driver's seat adjustment	M1113, M1151, M1152
18	Plate, identification, vehicle	All
19	Plate, instruction, driver's seat adjustment	M1114
20	Decal, identification, fire extinguisher	All
21	Plate, information, deep water fording	Vehicles so equipped
22	Decal, bar code	All
23	Plate, identification, vehicle	M1114 (under hood)
24	Plate, sling and tiedown/weights and dimensions data	M1113, M1151, M1152
25	Plate, instruction, battery cable connections	M1113, M1151, M1152
26	Decal, 24V	All
27	Plate, identification, slave receptacle	All
28	Plate, instruction, battery cable connections	M1114
29	Plate, sling and tiedown/weights and dimensions data	M1114

E-3. SIGN GUIDE (Cont'd)

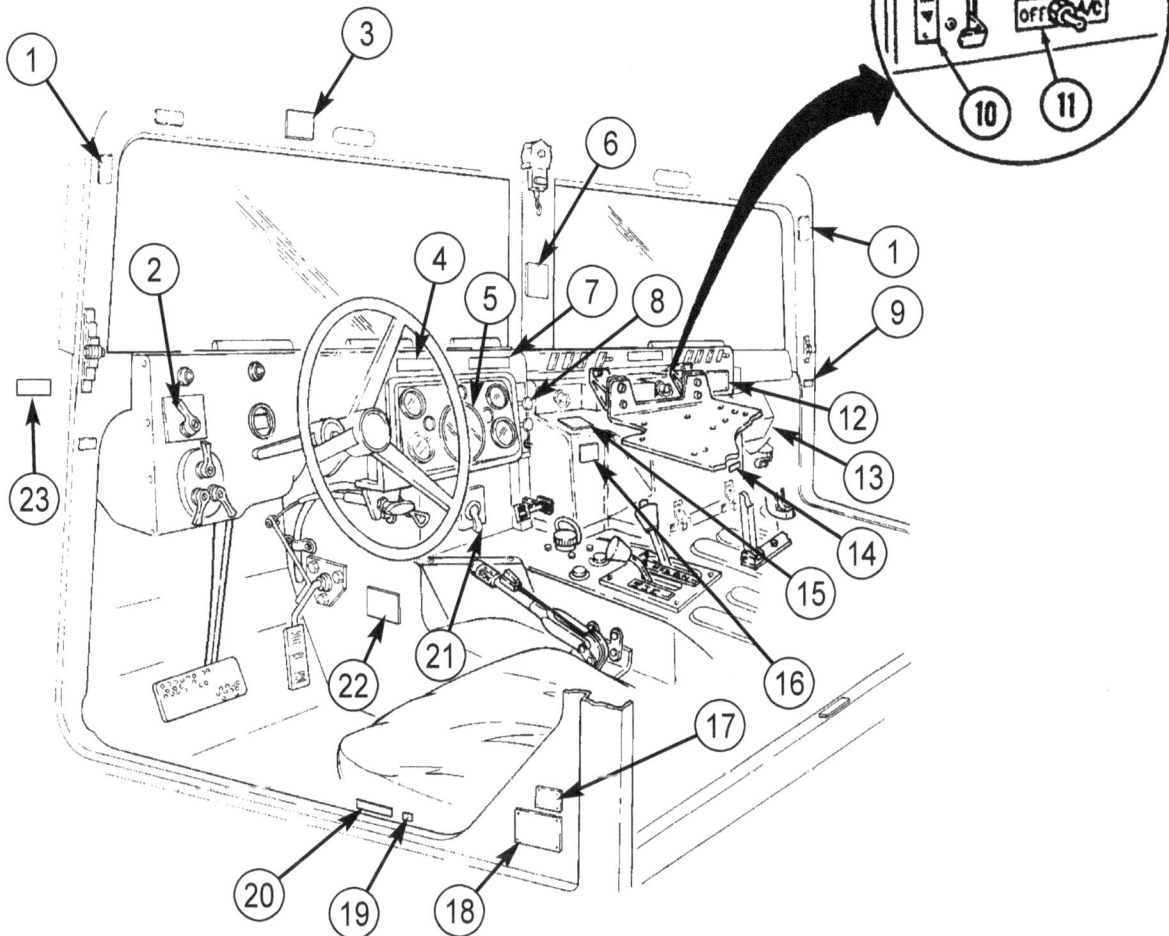

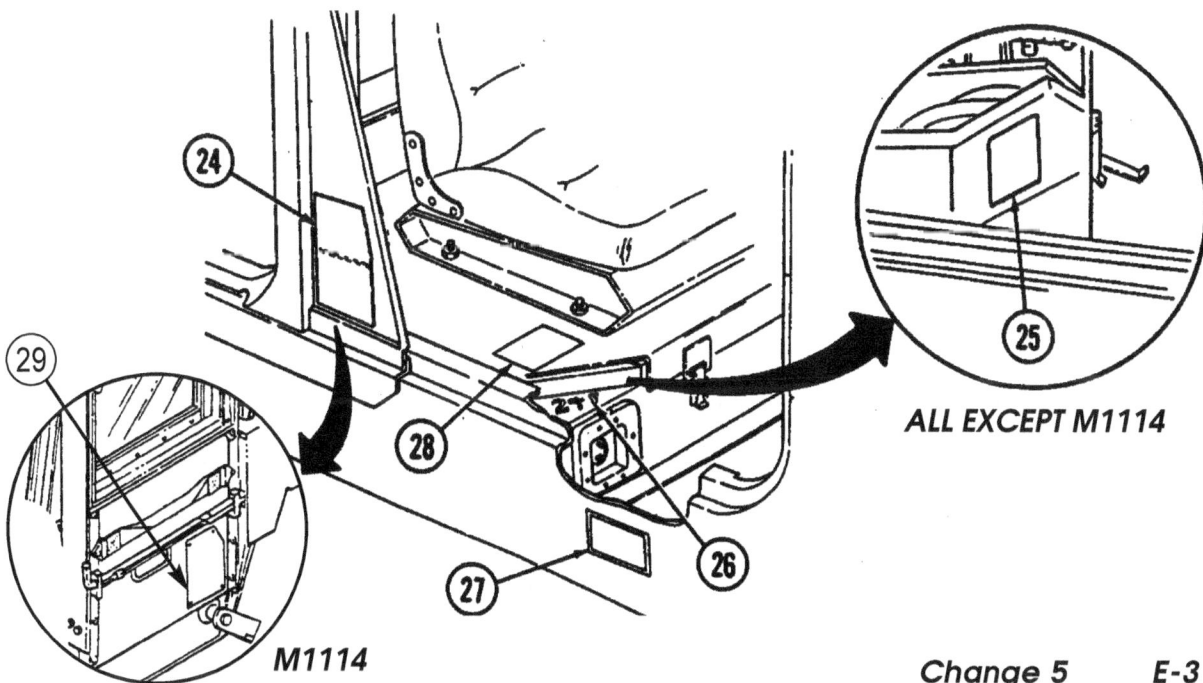

ALL EXCEPT M1114

M1114

E-3. SIGN GUIDE (Cont'd)

LOCATION OF CREW AREA AND EXTERIOR DECALS AND DATA PLATES		
KEY	ITEM	VEHICLE APPLICATION
1	Decal, warning, fan blade	All
2	Decal, no step	All
3	Decal, sling	All
4	Decal, no step	All
5	Decal, warning, surge tank cap	All
6	Decal, no step	All
7	Decal, identification, star	All
8	Decal, warning, winch	M1114 (when equipped)
9	Decal, identification, sling and tiedown	All
10	Decal, identification, vehicle number	All
11	Decal, identification, tiedown	All
12	Decal, warning, winch	M1113 (when equipped)
13	Decal, identification, USMC registration number	USMC vehicles only
14	Decal, instruction sling guide	M1151
15	Decal, tire pressure, rear	All
16	Decal, tire pressure, front	All

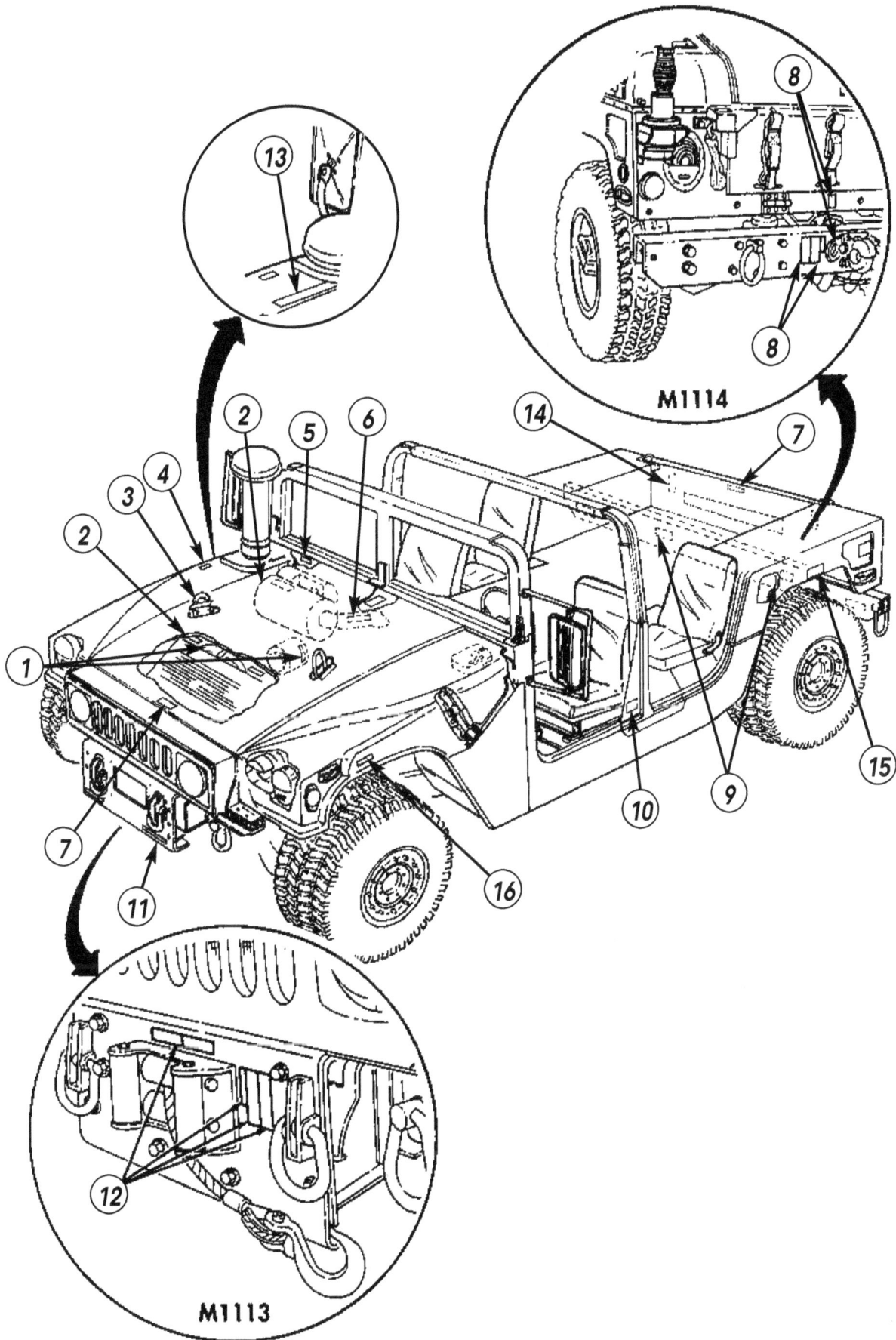

M1114

M1113

E-3. SIGN GUIDE (Cont'd)

LOCATION OF CREW AREA AND EXTERIOR DECALS AND DATA PLATES

KEY	ITEM	VEHICLE APPLICATION
1	Decal, fire control system	M1167
2	Decal, target acquisition system	M1167
3	Decal, vehicle mounted charger	M1167
4	Decal, battery box	M1167
5	Decal, traversing unit	M1167
6	Decal, missile rack configuration	M1167

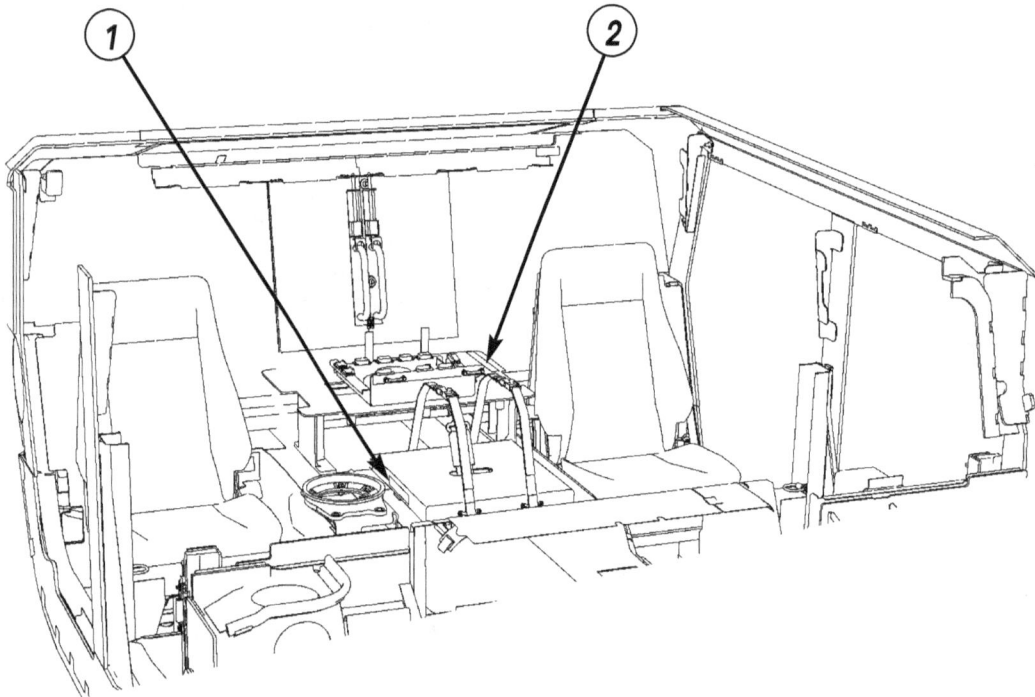

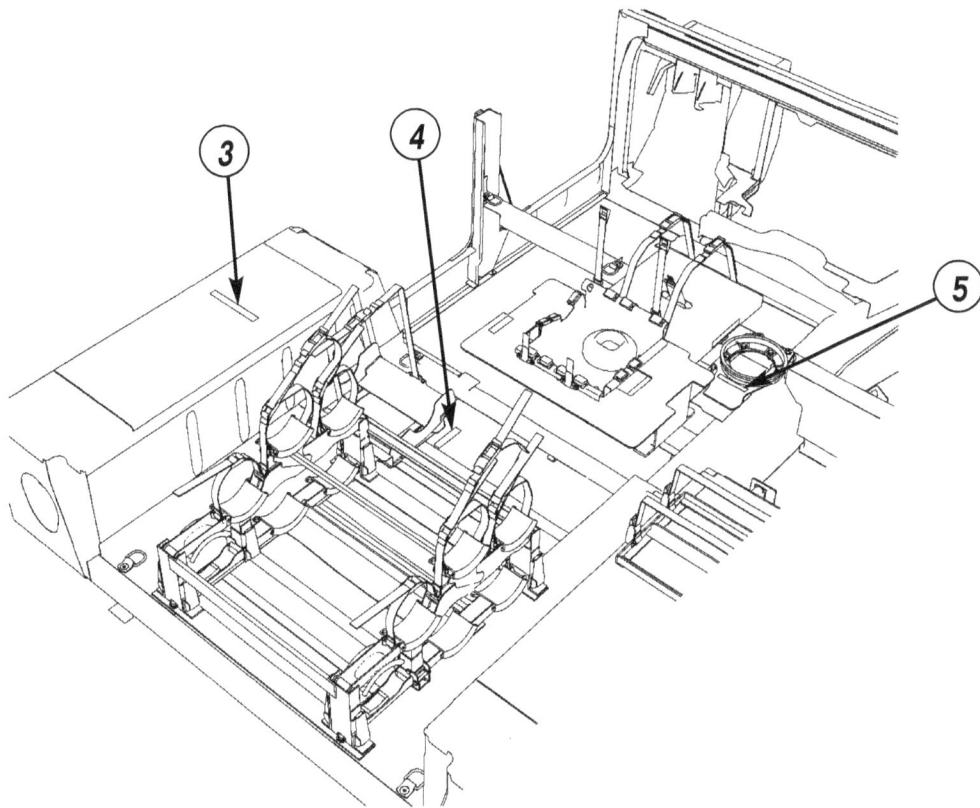

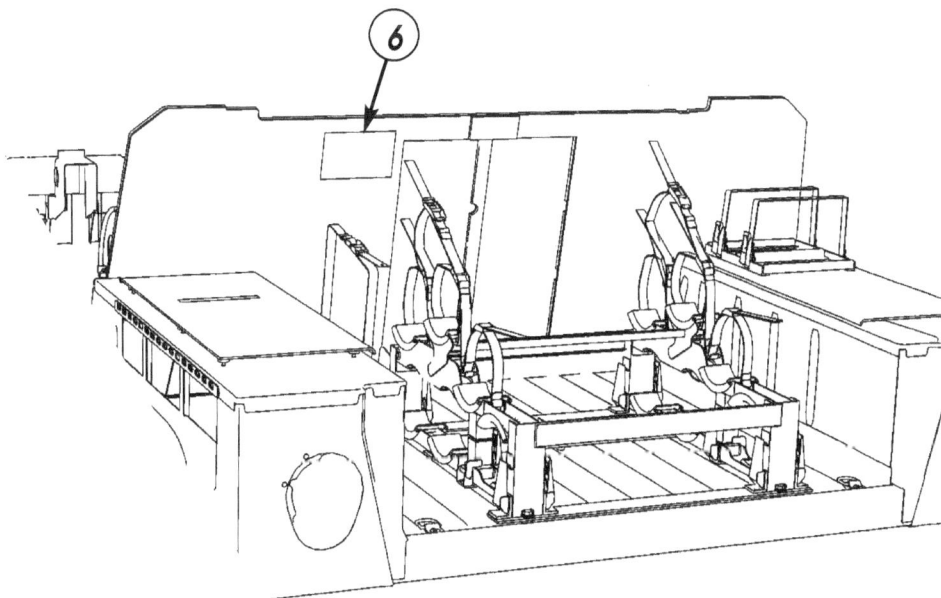

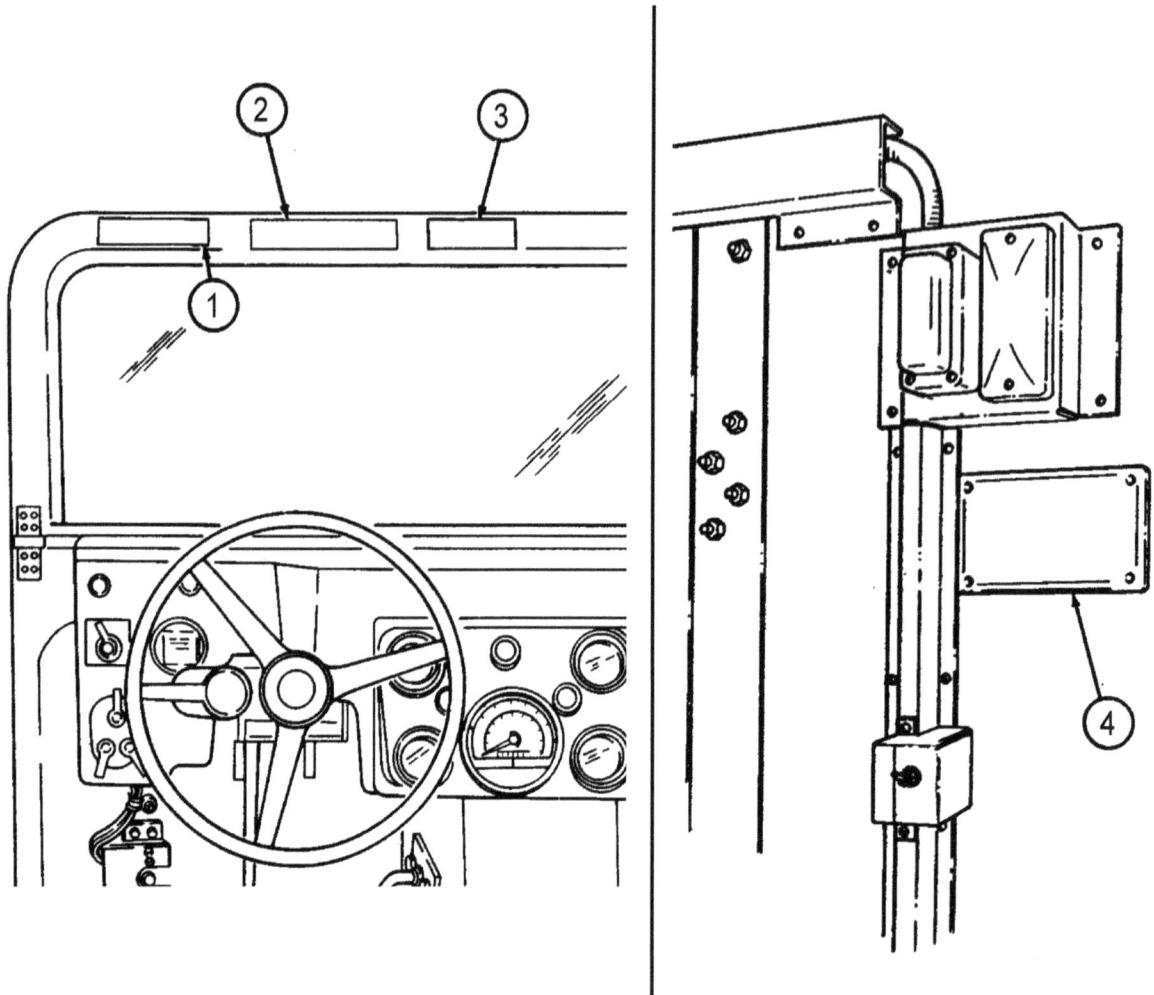

APPENDIX F
ON-VEHICLE EQUIPMENT LOADING PLANS

F-1. SCOPE

This appendix shows the on-vehicle equipment loading plan for ECV vehicles.

Proponent: Commandant
US Army Infantry School
Fort Benning, GA 31905

F-2. M1114 UP-ARMORED CARRIERS W/SUPPLEMENTAL ARMOR LOAD PLAN (MK19/40MM)

This is a standard load plan for the M1114 carriers w/supplemental armor. It includes selected items of personnel and unit equipment issued to most units equipped with this vehicle. Equipment not shown in this document may be loaded in accordance with local command policy.

M1114 UP-ARMORED CARRIERS W/SUPPLEMENTAL ARMOR STOWAGE PLAN (MK19/40MM)

NO.	ITEM
\multicolumn{2}{c}{LOAD PLAN}
1	M16A2/M203 Rifle
2	Box Crewman (VIC 1)
3	Binoculars, M17/22
4	Amplifier, 1780 VIC 1
5	Radio (121.3), (VRC-91)
6	M16A2/M203 Rifle
7	Gunner's Platform
8	First Aid Kit
9	Vehicle Power Conditioner
10	Night Vision Goggles, AN/PVS 7B
11	Laser Range Finder, AN/GVS 5 Case
12	Panel, VS17
13	Radiac Meter, AN/VDR
14	Ammunition, 40 mm, MK19 (240 RDS)
15	Chain Saw
16	Ammunition/Tray, M16A2 (1680 RDS)
17	Hand Grenade (24 ea)
18	Ammunition, 40 mm, M203 (72 RDS)
19	3-1/2-Ton Jack
20	Night Vision Sight, AN/PVS 4
21	Radio W/Accessories, Back Pack
22	Tripod, AN/VAS 11
23	Pintle Adapter
24	Machine Gun, M60
25	Max Tool Kit
26	Batteries, AN/VAS 11
27	CE-11 and Spools, TA 312/TA-1
28	Spare Barrel, M60
29	Camouflage Screen and Support System
30	Demolition Kit (2 bags)
31	Claymore Mine, M18A1 (3 ea)
32	Night Vision Sight, AN/TVS 5
33	Water Can
34	Ammunition, 7.62, M60 (800 RDS)
35	Deleted
36	Storage Case, AN/TAS 46

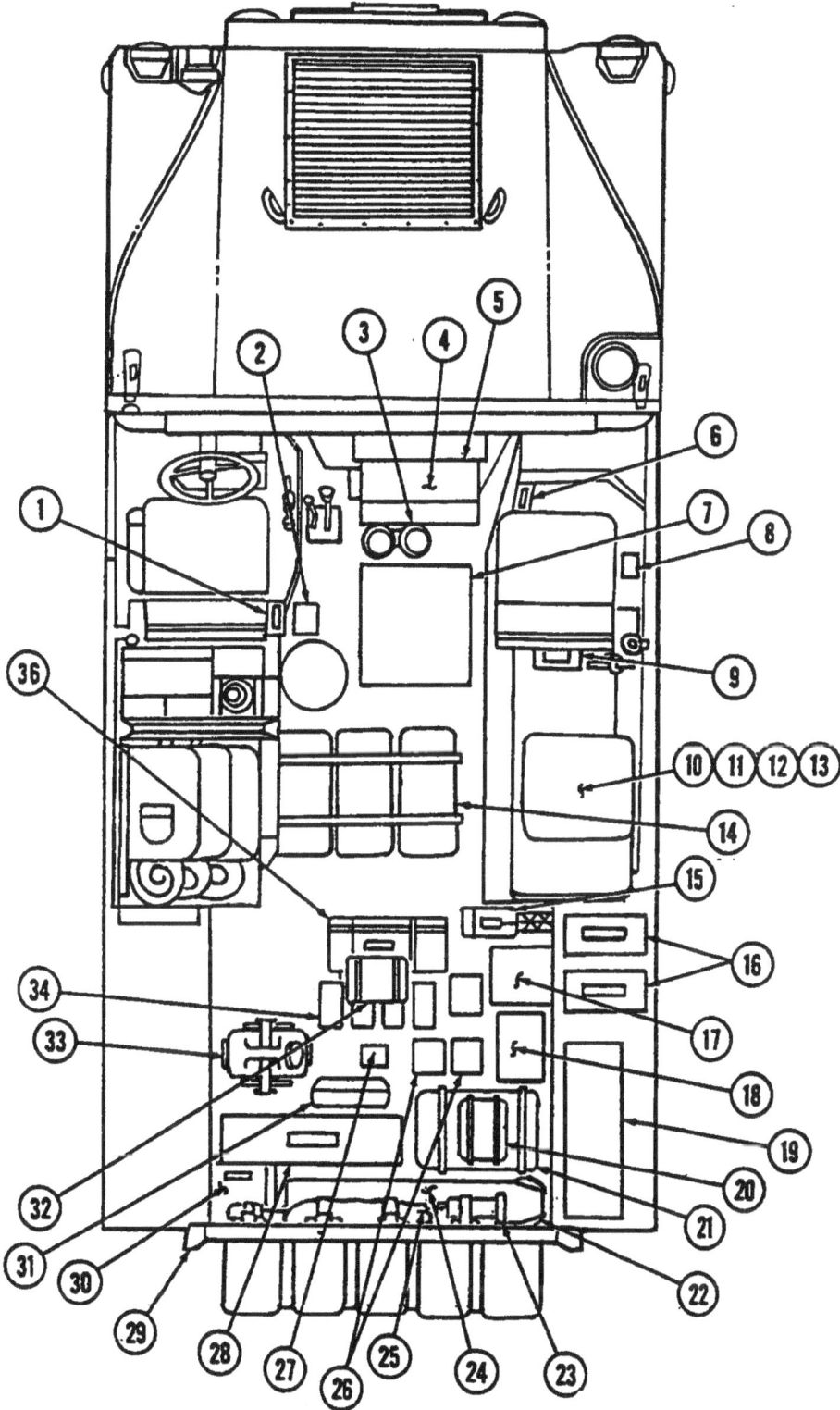

M1114 UP-ARMORED CARRIERS W/SUPPLEMENTAL
ARMOR STOWAGE PLAN (MK19/40MM) (Cont'd)

	LOAD PLAN
NO.	ITEM
37	Chemical Detector, M8A1
38	Radiac Charger, PQ1578
39	Total Dose Meter, IM93
40	Radiac Meter, IM 174
41	ALICE Packs (3 ea)
42	Machine Gun Toolbox
43	Wire, Antenna
44	MRE's (2 cases)
45	ICE Storage
46	Warning Triangle and BII Tool Bag
47	Storage Ring, VAS 11
48	Fire Extinguisher
49	Decontaminating Bottle, M-13
50	MK19 W/Ammunition (48 RDS)
51	Tripod, AN/VAS 11
52	Laser Range Finder, AN/VAS 5
53	Duffle Bags (3 ea)
54	Antitank Missile (AT-4), (under cargo hatch)
55	Hatch Stowage Net
56	Switch and Cable, Remote, Winch
57	Stowage Compartment Net

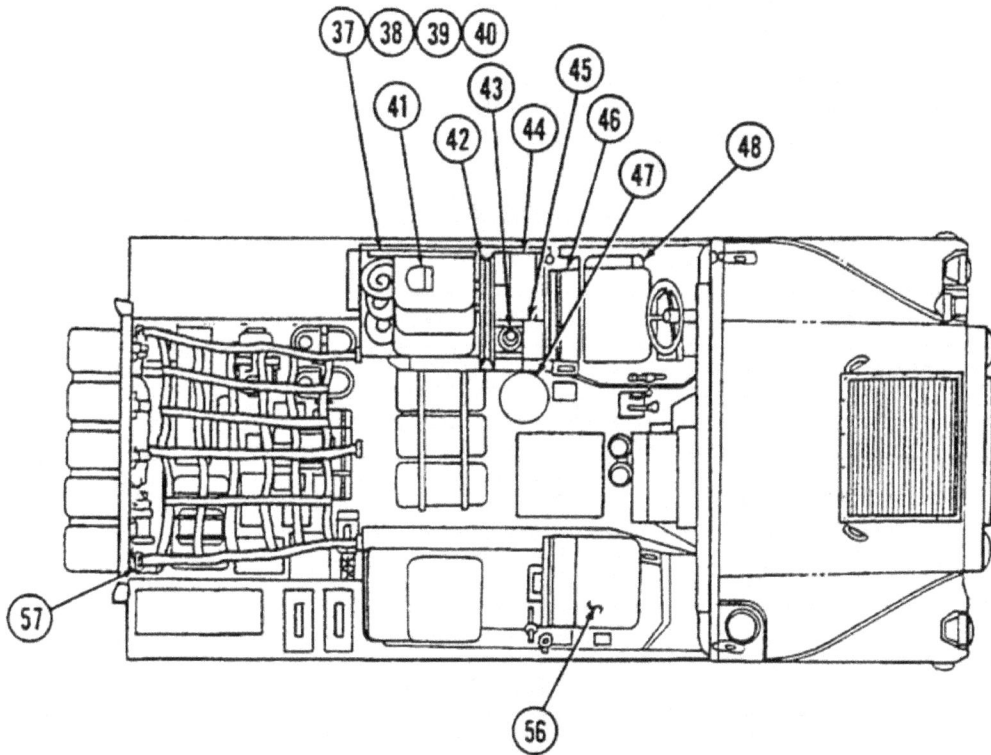

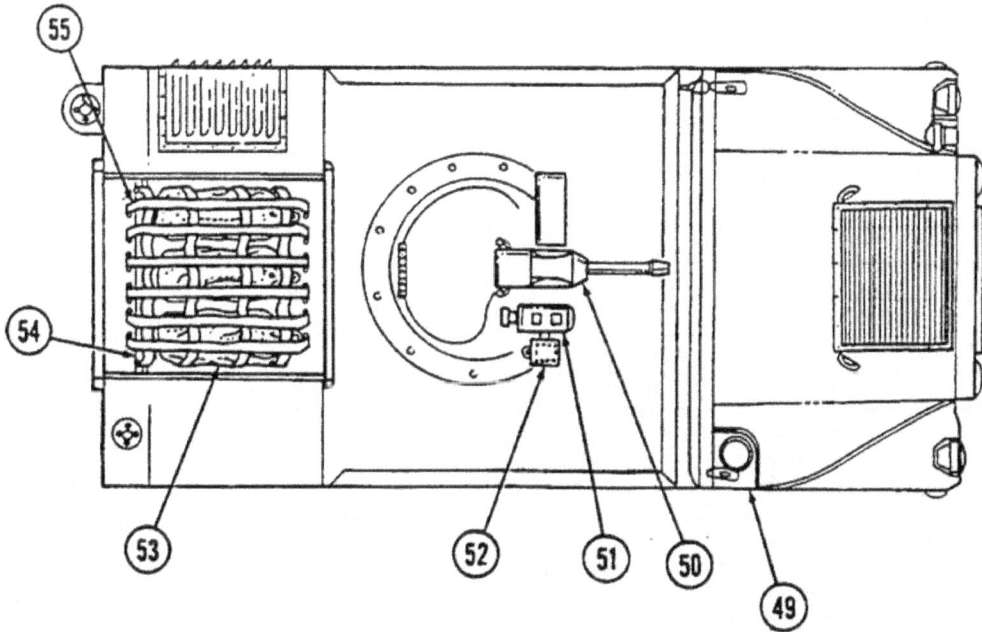

F-3. M1114 UP-ARMORED CARRIERS W/SUPPLEMENTAL ARMOR LOAD PLAN (M2/.50 CAL)

This is a standard load plan for the M1114 carriers w/supplemental armor. It includes selected items of personnel and unit equipment issued to most units equipped with this vehicle. Equipment not shown in this document may be loaded in accordance with local command policy.

M1114 UP-ARMORED CARRIERS W/SUPPLEMENTAL ARMOR STOWAGE PLAN (M2/.50 CAL)

LOAD PLAN	
NO.	ITEM
1	M16A2/M203 Rifle
2	Box, Crewman (VIC 1)
3	Binoculars, M17/22
4	Amplifier, 1780 VIC 1
5	Radio (121.3), (VRC-91)
6	M16A2/M203 Rifle
7	Gunner's Platform
8	First Aid Kit
9	Night Vision Goggles, AN/PVS 7B
10	Laser Range Finder, AN/GVS 5, Case
11	Panel, VS17
12	Radiac Meter, AN/VDR
13	Chain Saw
14	Hand Grenade (24 ea)
15	Mine Detector, AN/PSS 11
16	Ammunition, .40 mm, M203 (72 RDS)
17	Radio W/Accessories, Back Pack
18	Night Vision Sight, AN/PVS 4
19	Pintle Adapter
20	Batteries, AN/TAS 4
21	CE-II and Spools, TA-1/TA 312
22	Machine Gun, M60
23	Max Tool Kit
24	Camouflage Screen and Support System
25	Demolition Kit (2 bags)
26	Space Barrel, M60/M2
27	Claymore Mine, M18A1 (3 ea)
28	Water Can
29	Machine Gun Ammunition, M60 (800 RDS)
30	Deleted
31	Night Vision Sight, AN/TVS 5
32	Storage Case, AN/TVS 4/6

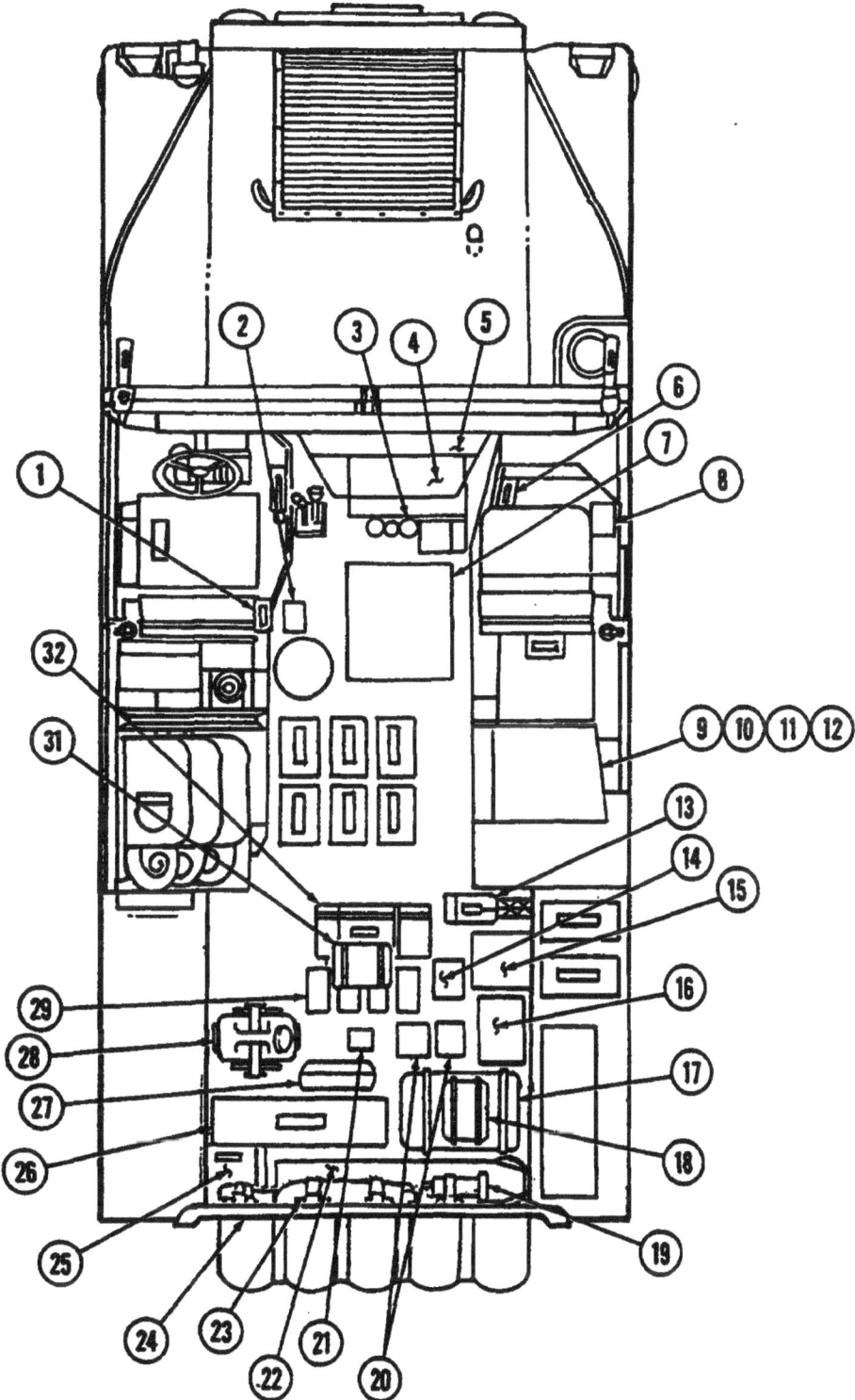

M1114 UP-ARMORED CARRIERS W/SUPPLEMENTAL ARMOR STOWAGE PLAN (M2/.50 CAL) (Cont'd)

LOAD PLAN	
NO.	**ITEM**
33	Chemical Detector, M8A1
34	Radiac Charger, PQ 1578
35	Total Dose Meter, IM93
36	Radiac Meter, IM 174
37	ALICE Packs (3 ea)
38	Machine Gun Toolbox
39	Wire, Antenna
40	MRE's (2 Cases)
41	ICE Storage
42	Warning Triangle and BII Tool Bag
43	Storage Ring, VAS 11
44	Fire Extinguisher
45	Decontaminating Bottle, M-13
46	M2 .50 Cal, Machine Gun
47	Tripod, AN/VAS 11
48	Long Range Finder, AN/GVS 5
49	Duffle Bags (3 ea)
50	Antitank Missile (AT-4), (under cargo hatch)
51	Hatch Stowage Net
52	Switch and Cable, Remote, Winch
53	Vehicle Power Conditioner
54	Ammunition, M2, .50 Cal. (700 RDS)
55	Ammunition/Tray, 556 M16A2 (1680 RDS)
56	3-1/2-Ton Jack
57	Stowage Compartment Net

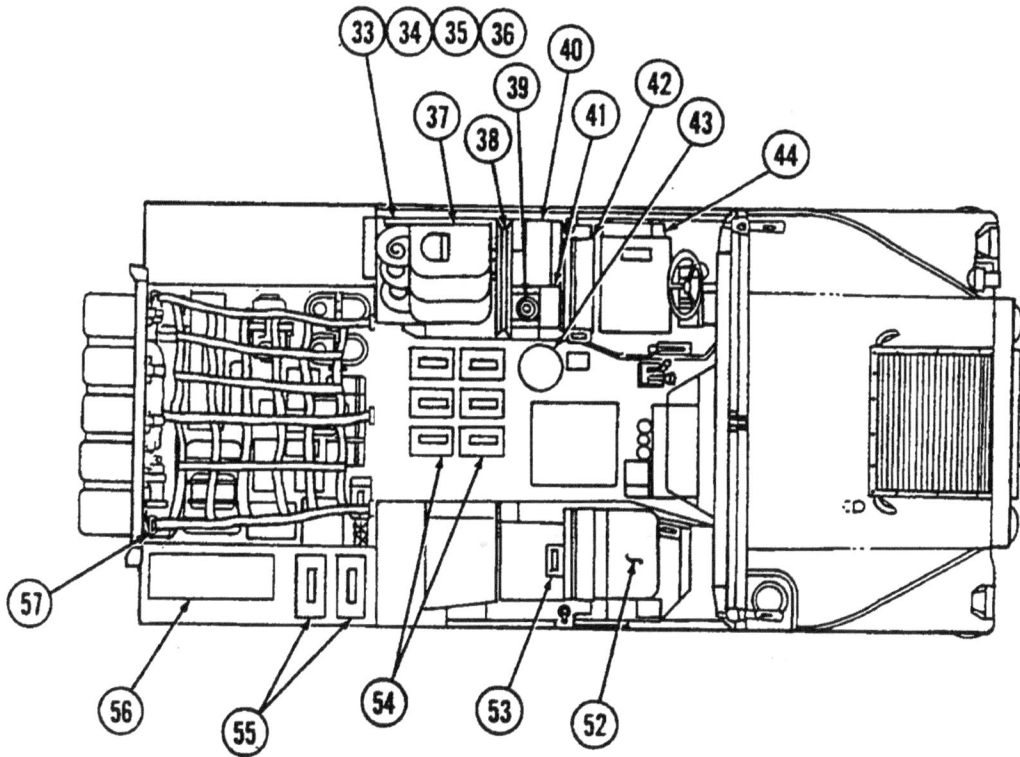

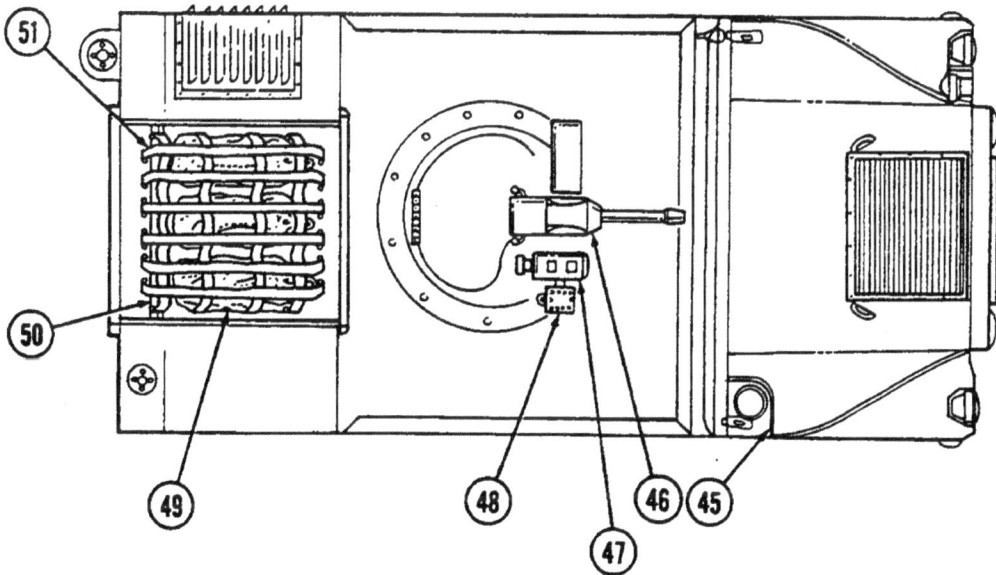

F-4. M1113 LOAD PLAN

This is a standard load plan for the M1113 vehicle. It is designed to supplement the stowage and sign guide contained in appendix E. It includes selected items of personnel and unit equipment issued to most units within the Army for this vehicle. Equipment not shown in this document may be loaded in accordance with local command policy.

M1113 STOWAGE PLAN

LOAD PLAN	
NO.	ITEM
1	M16A2/M203 Rifle
2	M16A2/M203 Rifle
3	Radio, AN/GRC-121.3
4	Sincgars (AV/VRC-89, AN/VRC-91 or AN/VRC-92)
5	Chain, Tow
6	Cable, NATO-Slave
7	Switch and Cable, Remote, Winch
8	Jack and Tools Bag, 3-1/2-Ton
9	Max Tool Kit
10	Decontaminating Bottle, M-13
11	Warning Triangle
12	Tool Kit
13	Fire Extinguisher
14	First Aid Kit (under driver seat)

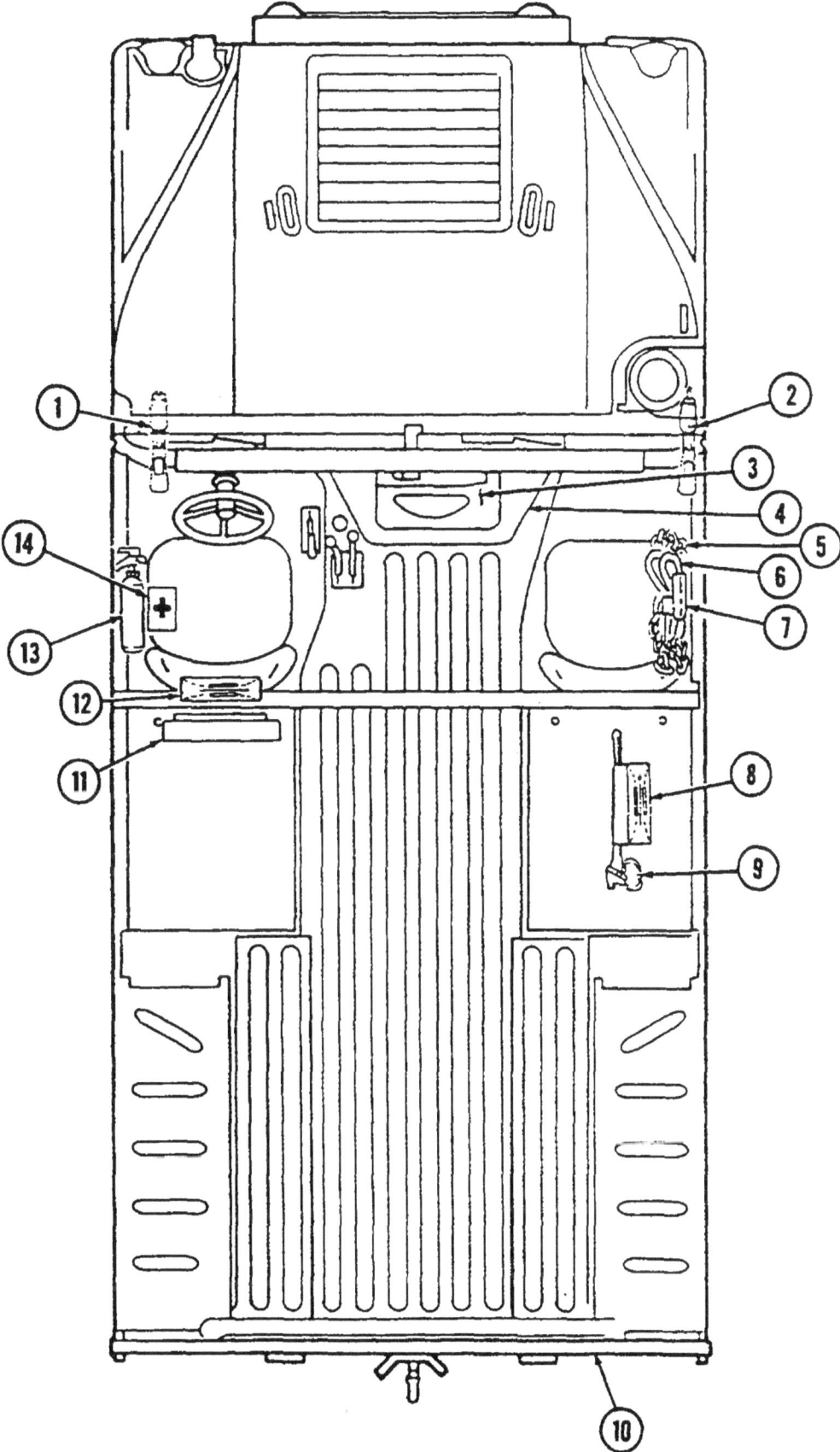

F-5. ARMAMENT CARRIER LOAD PLAN

This is a standard load plan for the expanded capacity armament carriers (M1151). It is designed to supplement the stowage and sign guide contained in Appendix E. It includes selected items of personnel and unit equipment issued to most units within the Army equipped with this vehicle. Equipment not shown in this document may be loaded in accordance with local command policy. For Marine Corps Load Plans, refer to TI 11240-24/42.

ARMAMENT CARRIER STOWAGE PLAN

LOAD PLAN	
NO.	ITEM
1	M16A1/M203 Rifle
2	Rifle M16A1
3	Tool Box
4	Pedestal
5	Ammunition Tray (caliber 5.56 mm)
6	Radio Antenna (on cargo shell)
7	Max Tool Kit
8	Water Can Tray (2)
9	Water Can Tray
10	Ammunition Tray (caliber 5.56 mm)
11	40 mm/50 cal/Ammunition Tray/A/C Evaporator
12	Jack, Scissors
13	Fire Extinguisher
14	AN/GRC-160 Radio

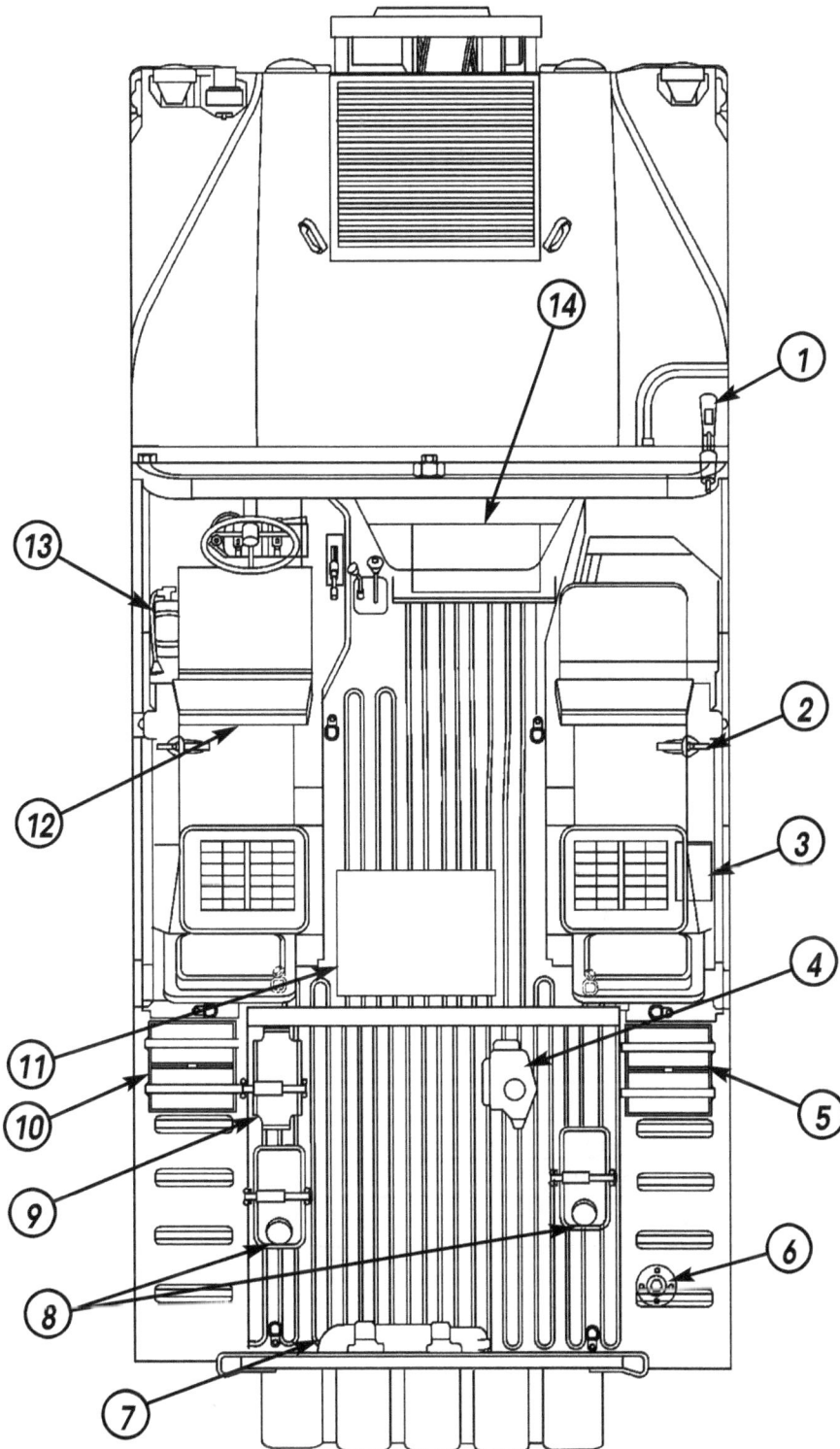

F-6. HEAVY VARIANT STOWAGE PLAN

This is the standard load plan for the M1152, and variants equipped with a howitzer prime mover or towed vulcan kit. It is designed to supplement the stowage and sign guide contained in Appendix E. It includes selected items of personnel and unit equipment issued to most units within the Army equipped with this vehicle. Equipment not shown in this document may be loaded in accordance with local command policy.

HEAVY VARIANT STOWAGE PLAN

LOAD PLAN	
NO.	ITEM
1	Camouflage Screen and Support System
2	Fuel (Diesel) Cans (2)
3	Remote
4	M122 Tripod
5	Telephone
6	DR8 Cable Reels (2)
7	Paralleloscope and Aiming Posts
8	Jack Strut
9	Paralleloscope Spikes (2)
10	Ammo (22 Rounds)
11	M60 Machine Gun
12	Sight Box #2
13	Sight Box #1
14	GDU Battery
15	GDU Box
16	Section Chest
17	Water Cans (2)
18	Spade

INTERNAL TOP VIEW

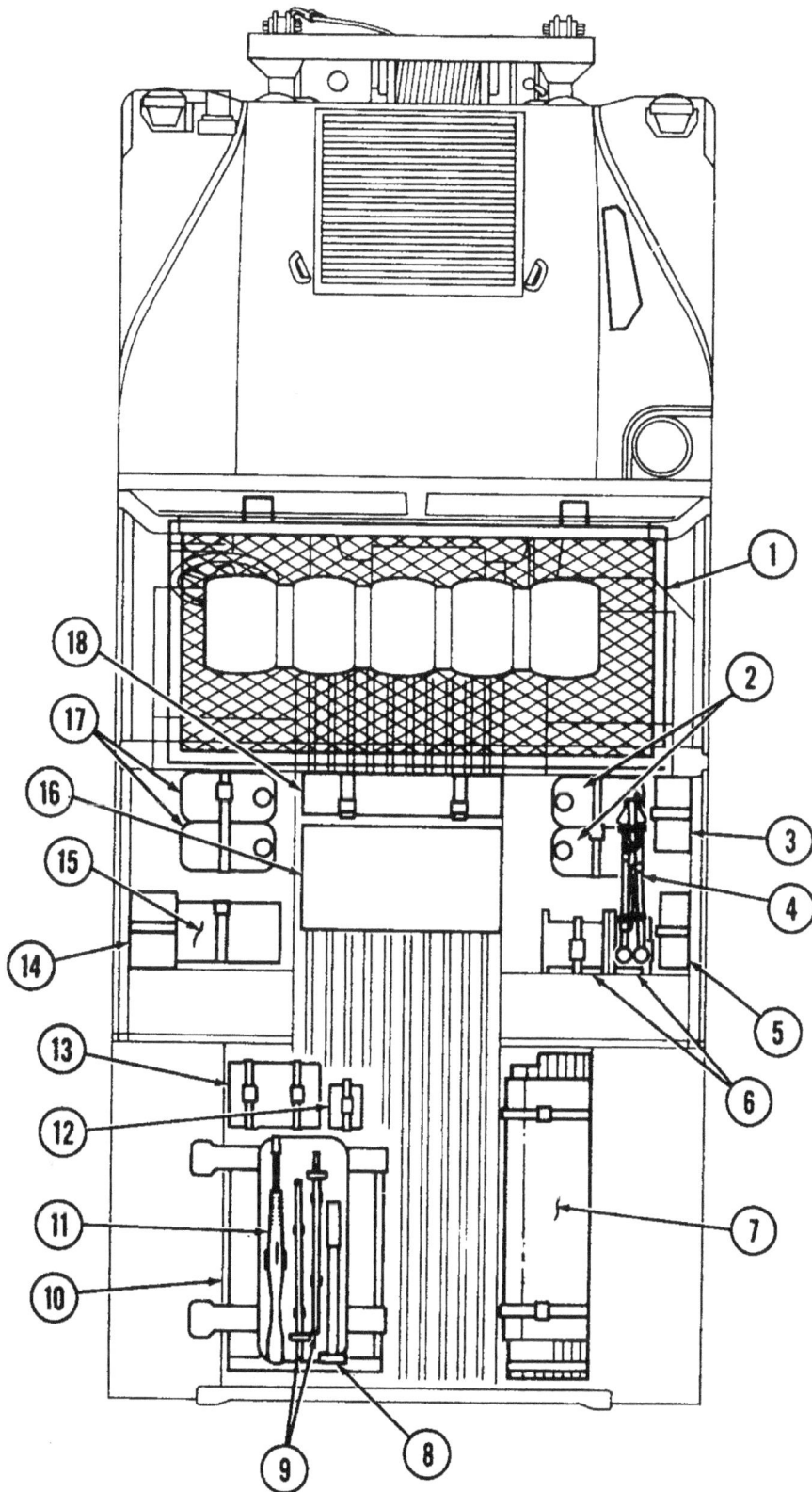

APPENDIX G
LUBRICATION INSTRUCTIONS

G-1. SCOPE

This appendix gives lubrication requirements for the ECV vehicles which are the
responsibility of the operator/crew.

G-2. GENERAL LUBRICATION REQUIREMENTS

a. Maintaining Lubricant Levels. Lubricant levels must be checked as
specified in the PMCS (chapter 2, section 2) and table G-1, Lubrication. Steps must
be taken to replenish and maintain lubricant levels.

WARNING

- Drycleaning solvent P-D-680 is TOXIC and flammable.Wear
  protective goggles and gloves, use only in well-ventilated area, avoid
  contact with skin, eyes, and clothes, and do not breathe vapors.
  Keep away from heat and flame. Never smoke when
  using solvent.The flashpoint for type I drycleaning solvent is 100°F
  (38°C), and for type II, is 138°F (59°C). Failure to comply may
  result in injury or death to personnel.
- If personnel become dizzy while using cleaning solvent,
  immediately get fresh air and medical help. If solvent contacts skin
  or clothes, flush with cold water. If solvent contacts eyes,
  immediately flush eyes with water and get medical attention.

b. Cleaning Fittings Before Lubrication. Clean parts with drycleaning
solvent P-D-680 or equivalent. Dry before lubricating.

NOTE

Dotted arrow points indicate lubrication on both sides of the
equipment.

c. Lubrication After Fording. Following fording operation, lubricate all
fittings below fording depth and check submerged gear boxes for presence of water.

d. Lubrication After High-Pressure Washing. After a thorough washing,
lubricate all grease fittings and oil can points outside and underneath vehicle.

e. Localized Views. A reference to the appropriate localized view is given
after most lubrication entries. Localized views begin on page G-11.

G-3. LUBRICATION INTERVALS

a. Service Interval Under Normal Conditions. Service intervals listed are for normal operation in moderate temperatures, humidity, and atmospheric conditions. Hard-time intervals may be shortened if your lubricants are contaminated or if you are operating the equipment under adverse conditions, including longer-than-usual operating hours. Hard-time intervals may be extended during periods of low activity, though adequate preservation precautions must be taken. Perform semiannual service intervals every six months, or 3,000 miles (4,827 km), whichever comes first.

b. Service Interval Under Unusual Conditions. Increase frequency of lubricating service when operating under abnormal conditions such as high or low temperatures, prolonged high-speed driving, or extended cross-country operations. Such operation can diminish lubricant's protective qualities. More frequent lubricating service intervals are necessary to maintain vehicle readiness when operating under abnormal conditions.

c. Hard-Time Intervals. Intervals shown in this appendix are based on mileage and calendar times. An example of mileage and calendar interval is: 3/S, in which 3 stands for 3,000 mi (4,827 km), and S stands for semiannually (every six months). The lubrication of the vehicle is to be performed at whichever interval occurs first. For equipment under manufacturer's warranty, hard-time oil service intervals shall be followed.

d. Deleted

G-4. LUBRICATION FOR OPERATION UNDER EXTREME TEMPERATURES

a. Changes in Lubricant Grades. Lubricant grades change with weather conditions. Refer to lubrication table for lubricant grade changes.

b. Arctic Conditions. Refer to FM 9-207, Operation and Maintenance of Ordnance Materiel in Cold Weather (0°F to -65°F) (-18°C to -54°C), or the lubrication table.

G-5. CORROSION CONTROL

Refer to para. 2-4 for appropriate corrosion control procedures.

Table G-1. Lubrication.

USAGE	FLUID/LUBRICANT	CAPACITIES	EXPECTED TEMPERATURES
Engine Oil (MIL-L-2104) (MIL-L-46167)	OE/HDO 30 OE/HDO 10 OEA-30 OE/HDO 15W-40	Crankcase: w/o filter 7 qt (6.6 L) w/ filter 8 qt (7.6 L) Dry System 10 qt (9.5 L)	Above +15°F (-9°C) +40° to -15°F (+4° to -26°C) +40° to -65°F (+4° to -54°C) +120° to -55°F (+49° to -48°C) Above 0°F (-18°C)
Engine Coolant	Ethylene Glycol and Water 1/2 Ethylene Glycol, 1/2 Water 1/2 Ethylene Glycol, 1/2 Water 3/5 Ethylene Glycol, 2/5 Water	Radiator: 5 qt (4.7 L) Complete System: 26 qt (24.6 L)	Above +15°F (-9°C) +40° to -15°F (+4° to -26°C) +40° to -65°F (+4° to -54°C)
Transmission	Dexron® VI only	Dry 13.5 qt (12.8 L) Drain & Refill 7.7 qt (7.3 L)	All Temperatures
Transfer Case	Dexron ® VI	Drain & Refill 3.35 qt (3.17 L)	All Temperatures
Differentials	GO 80/90 GO 75	Dry 2 qt (1.89 L) "	All Temperatures +40° to -65°F (+4° to -54°C)
Differential with diff. cooler	GO 80/90 GO 75	Dry 2.68 qt (2.53 L) "	Above +15°F (-9°C) +40° to -65°F (+4° to -54°C)
Geared Hubs	GO 80/90 GO 75	Dry 1 pt (0.47 L) "	All Temperatures +40° to -65°F (+4° to -54°C)
Geared Fan Drive	GO 80/90 GO 75	Dry 1.2 pt (0.51 L) "	All Temperatures +40° to -65°F (+4° to -54°C)

Table G-1. Lubrication (Cont'd).

USAGE	FLUID/LUBRICANT	CAPACITIES	EXPECTED TEMPERATURES
Steering System	Dexron ® VI	Refer to table 1-2.1, Steering System Capacities	All Temperatures
Upper and Lower Ball Joints, Tie Rod Ends, Pitman Arm, Propeller Shafts, etc.	GAA (MIL-G-10924)	As Required	All Temperatures
Hinges, Cables, and Linkages	OE/HDO	As Required	All Temperatures

OE/HDO 15/40 (Grade 15W-40) lubricant may be used when expected temperatures are above +5°F (-15°C). If OEA-30 lubricant is required to meet the temperature ranges prescribed in the lubrication table, then the OEA-30 lubricant is to be used in place of OE/HDO 10 lubricant for all temperature ranges. If operating conditions are severe or abnormal, service chassis lubrication points at 1,000 mi (1,609 km).

Door Latches and Wire Handle Lock	PL-S (VV-L-800)	As Required	All Temperatures
GREASE, WIRE ROPE (MIL-G-18458)		ALL TEMPERATURES	

FUEL REQUIREMENTS — TEMPERATURE LIMITS (VV-F-800)

Grade DF2 Fuel	For use above +10°F (-12°C)*
Grade DF1**	For use below +10°F (-12°C) to above -20°F (-29°C)
Grade DFA	For use below -20°F (-29°C)
Grade JP-8	All Temperatures

* Temperature limits may vary, dependent on the cloud point of the actual DF2 fuel being supplied in the geographical area.

** DF1 is not normally procured in CONUS or OCONUS. Refineries will blend DF2 with kerosene to meet temperature requirements of DF1.

LUBRICATION POINTS

INTERVAL • LUBRICANT

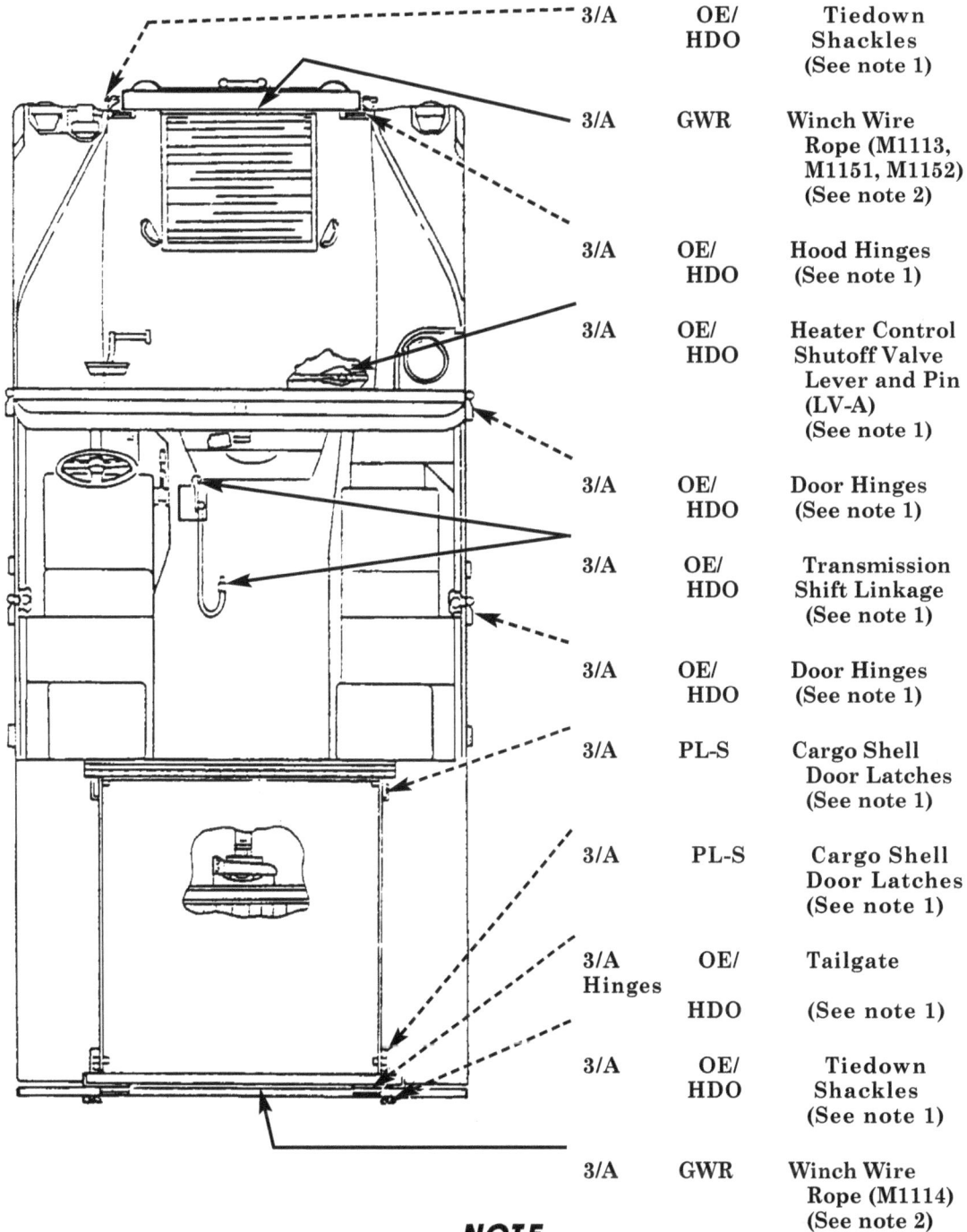

3/A	OE/HDO	Tiedown Shackles (See note 1)
3/A	GWR	Winch Wire Rope (M1113, M1151, M1152) (See note 2)
3/A	OE/HDO	Hood Hinges (See note 1)
3/A	OE/HDO	Heater Control Shutoff Valve Lever and Pin (LV-A) (See note 1)
3/A	OE/HDO	Door Hinges (See note 1)
3/A	OE/HDO	Transmission Shift Linkage (See note 1)
3/A	OE/HDO	Door Hinges (See note 1)
3/A	PL-S	Cargo Shell Door Latches (See note 1)
3/A	PL-S	Cargo Shell Door Latches (See note 1)
3/A Hinges	OE/HDO	Tailgate (See note 1)
3/A	OE/HDO	Tiedown Shackles (See note 1)
3/A	GWR	Winch Wire Rope (M1114) (See note 2)

NOTE

A REFERENCE TO THE LOCALIZED VIEW (LV) IS PROVIDED AFTER LUBRICATION POINT ENTRY, WHERE APPLICABLE.

LUBRICATION POINTS (Cont'd)

LUBRICANT • INTERVAL

■ Accelerator Linkage (LV-C) (See note 1)   OE/HDO   3/A

■ Heater Control Knob and Plunger (LV-B) (See note 1)   OE/HDO   3/A

■ Transfer Case Shift Linkage (See note 1)   OE/HDO   3/A

■ Door Handles (See note 1)   OE/HDO   3/A

■ Cargo Shell Door Rear Wire Handle Lock (See note 1)   PL-S   3/A

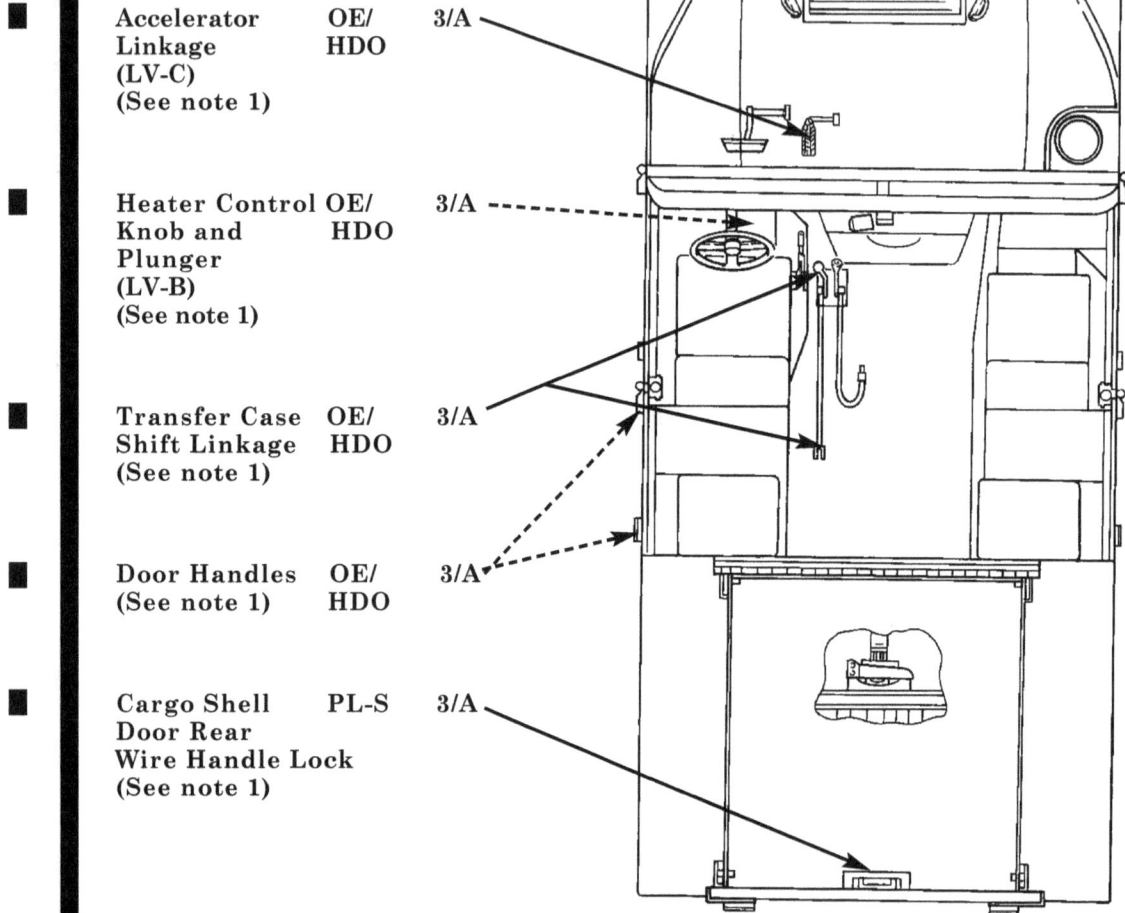

NOTE
A REFERENCE TO THE LOCALIZED VIEW (LV) IS
PROVIDED AFTER LUBRICATION POINT ENTRY,
WHERE APPLICABLE.

LUBRICATION POINTS (Cont'd)

INTERVAL • LUBRICANT

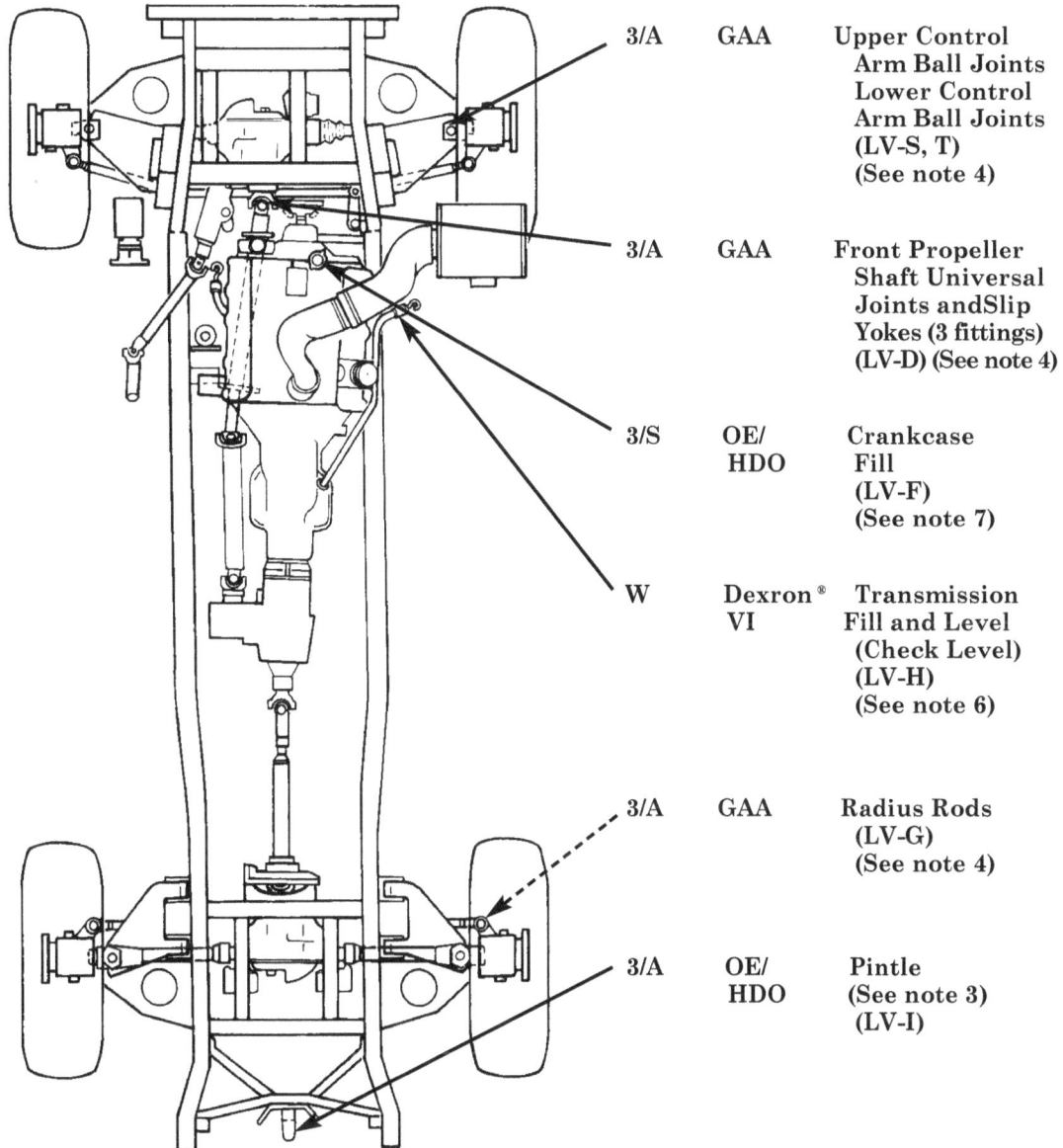

3/A	GAA	Upper Control Arm Ball Joints Lower Control Arm Ball Joints (LV-S, T) (See note 4)
3/A	GAA	Front Propeller Shaft Universal Joints and Slip Yokes (3 fittings) (LV-D) (See note 4)
3/S	OE/ HDO	Crankcase Fill (LV-F) (See note 7)
W	Dexron® VI	Transmission Fill and Level (Check Level) (LV-H) (See note 6)
3/A	GAA	Radius Rods (LV-G) (See note 4)
3/A	OE/ HDO	Pintle (See note 3) (LV-I)

NOTE

A REFERENCE TO THE LOCALIZED VIEW (LV) IS
PROVIDED AFTER LUBRICATION POINT ENTRY,
WHERE APPLICABLE

LUBRICATION POINTS (Cont'd)

LUBRICANT • INTERVAL

■
■

Power
Steering
Reservoir
(P/N RCSK 18330)
Fill and Level
(Check level)
(LV-J)
(See note 5)

Dexron® VI

M

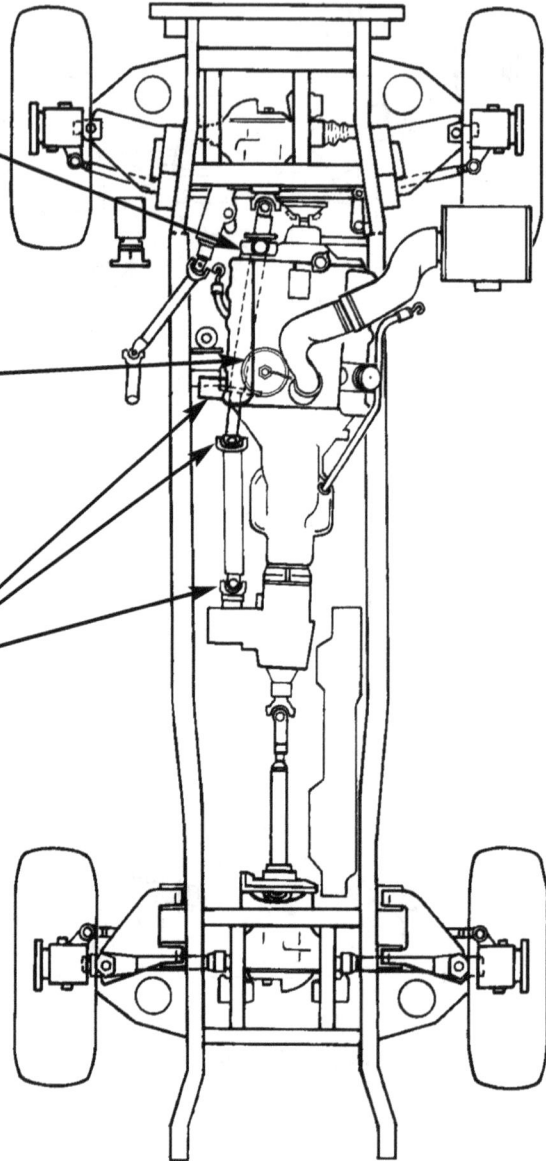

Power
Steering
Reservoir
(P/N 94252A)
Fill and Level
(Check level)
(LV-J1)
(See note 5.1)

Dexron
VI   M

Front
Propeller
Shaft
Universal
Joints and
Slip Yokes
(3 fittings)
(LV-D, E)
(See note 4)

GAA
HDO   3/A

NOTE
A REFERENCE TO THE LOCALIZED VIEW (LV) IS
PROVIDED AFTER LUBRICATION POINT.

LUBRICATION POINTS (Cont'd)

INTERVAL • LUBRICANT

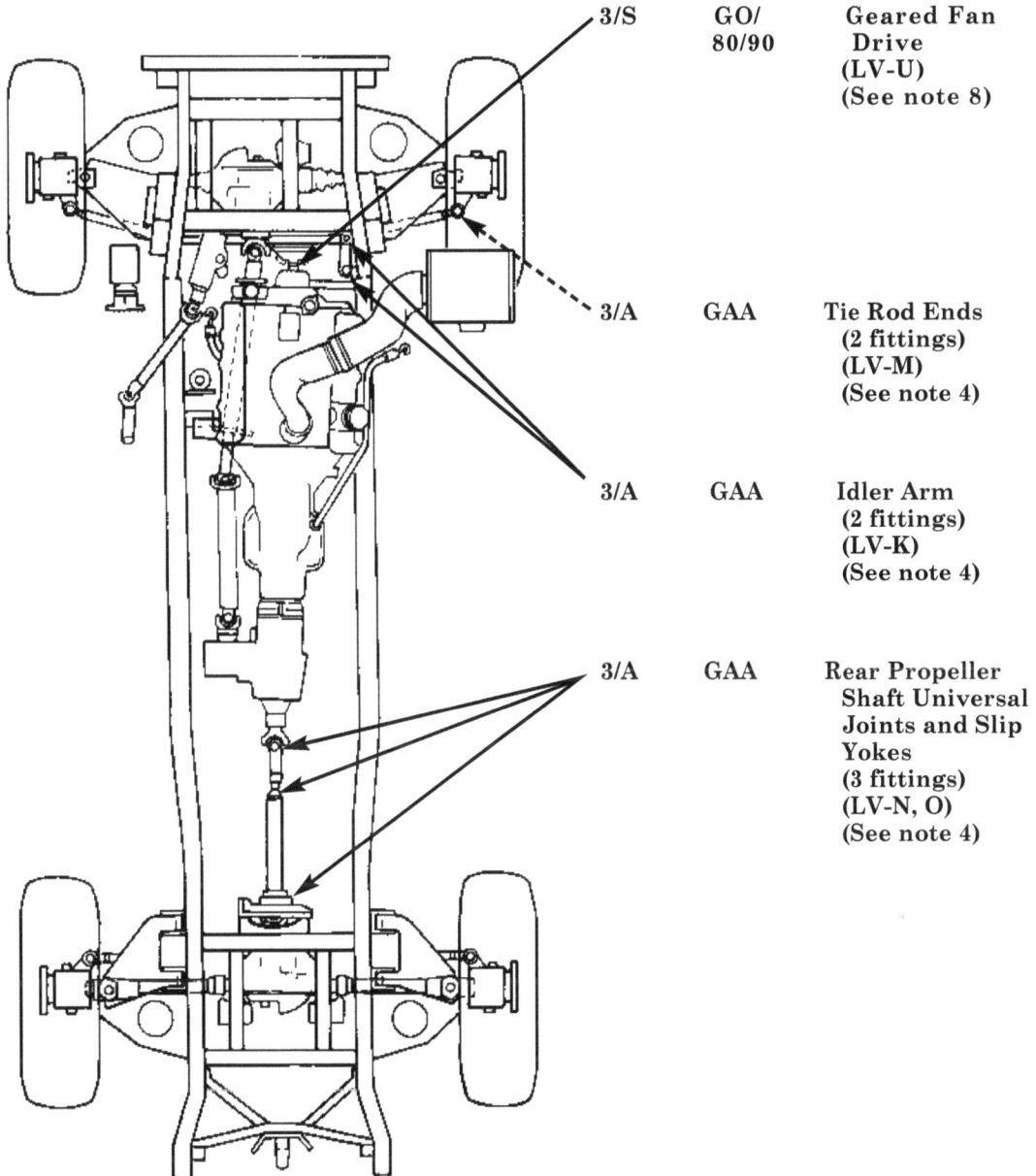

3/S	GO/ 80/90	Geared Fan Drive (LV-U) (See note 8)
3/A	GAA	Tie Rod Ends (2 fittings) (LV-M) (See note 4)
3/A	GAA	Idler Arm (2 fittings) (LV-K) (See note 4)
3/A	GAA	Rear Propeller Shaft Universal Joints and Slip Yokes (3 fittings) (LV-N, O) (See note 4)

NOTE

A REFERENCE TO THE LOCALIZED VIEW (LV) IS PROVIDED AFTER LUBRICATION POINT ENTRY, WHERE APPLICABLE.

LUBRICATION POINTS (Cont'd)

LUBRICANT • INTERVAL

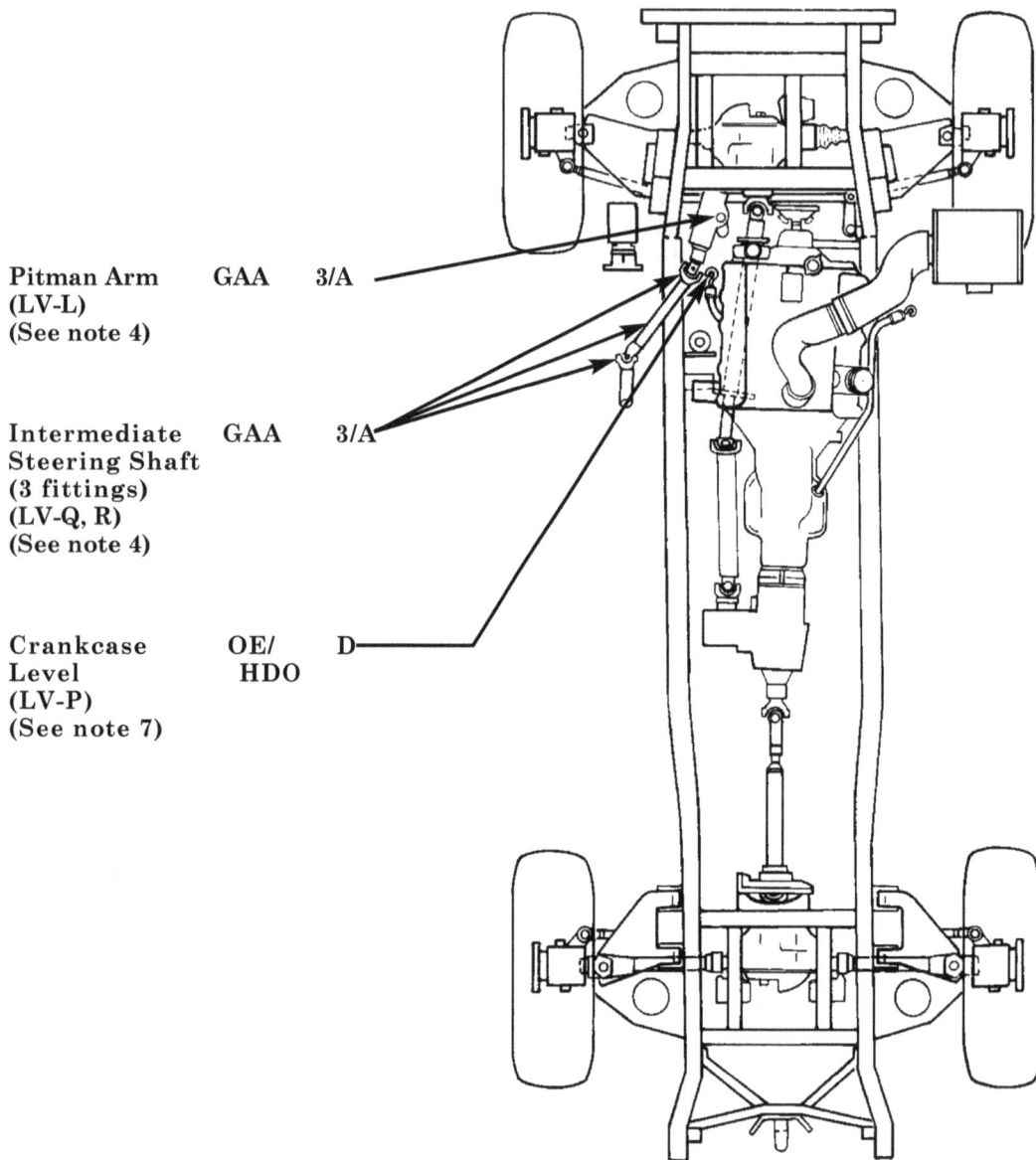

Pitman Arm   GAA   3/A
(LV-L)
(See note 4)

Intermediate   GAA   3/A
Steering Shaft
(3 fittings)
(LV-Q, R)
(See note 4)

Crankcase   OE/   D
Level   HDO
(LV-P)
(See note 7)

NOTE
A REFERENCE TO THE LOCALIZED VIEW (LV) IS
PROVIDED AFTER LUBRICATION POINT ENTRY,
WHERE APPLICABLE.

LOCALIZED LUBRICATION POINTS

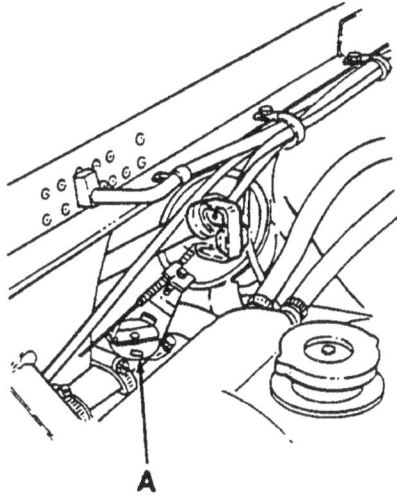

HEATER CONTROL SHUTOFF
VALVE LEVER AND PIN

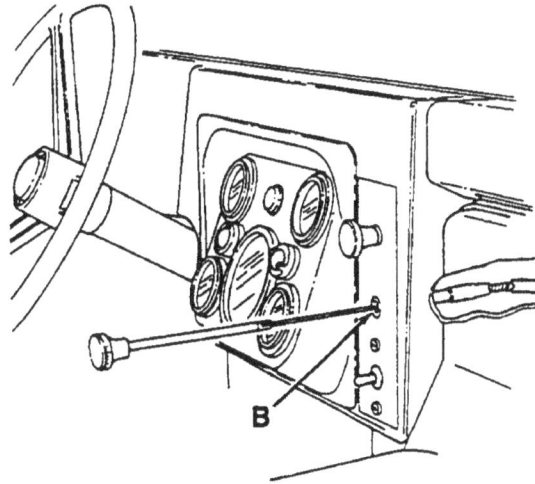

HEATER CONTROL KNOB AND
PLUNGER

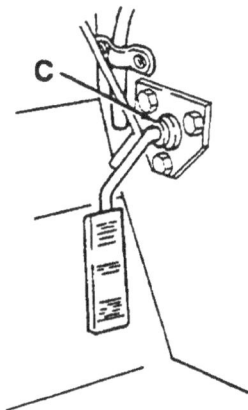

ACCELERATOR LINKAGE

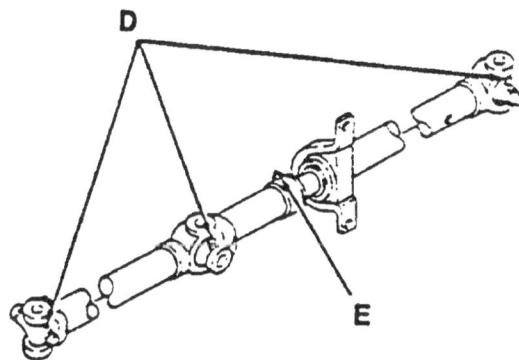

FRONT PROPELLER SHAFT UNIVERSAL
JOINT AND SLIP YOKES

LOCALIZED LUBRICATION POINTS (Cont'd)

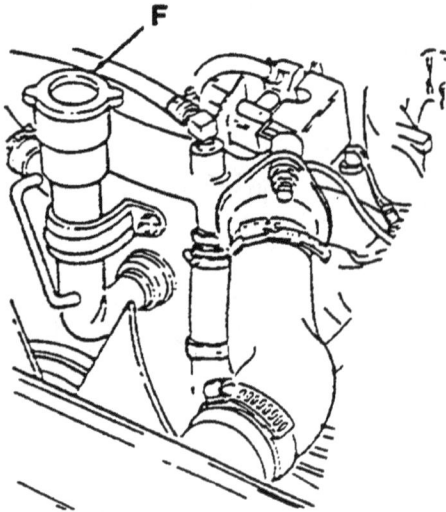

CRANKCASE FILL

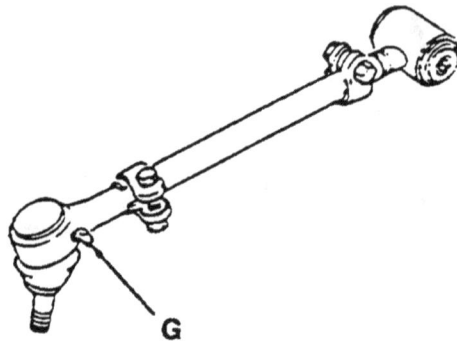

RADIUS ROD

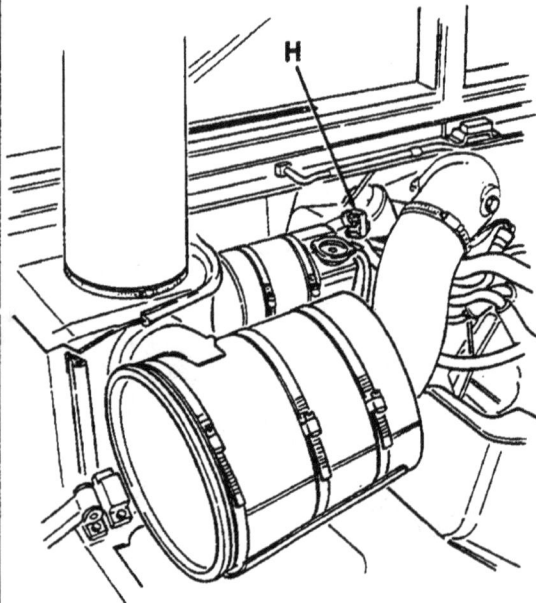

TRANSMISSION FILL
AND LEVEL

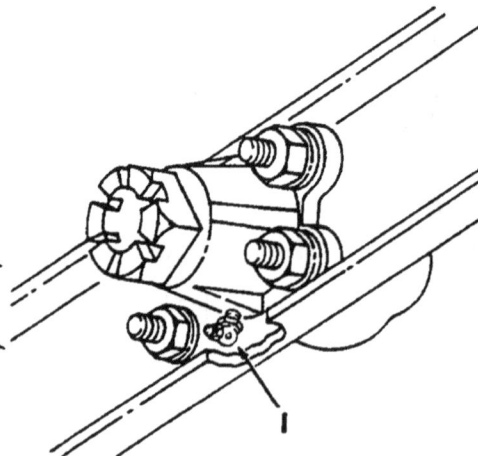

PINTLE

LOCALIZED LUBRICATION POINTS (Cont'd)

POWER STEERING RESERVOIR FILL
AND LEVEL (P/N RCSK 18330)

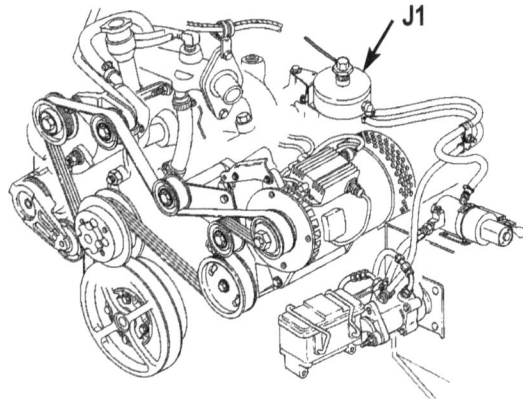

POWER STEERING RESERVOIR FILL
AND LEVEL (P/N 94252A)

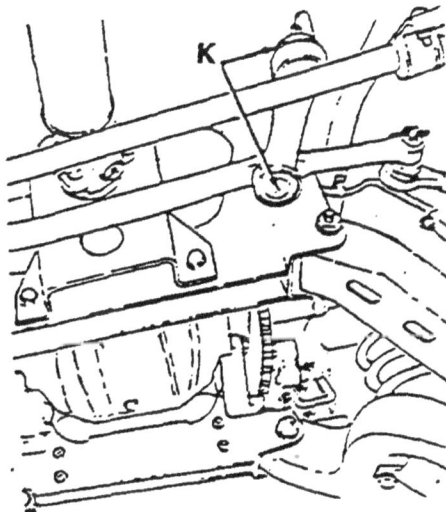

IDLER ARM

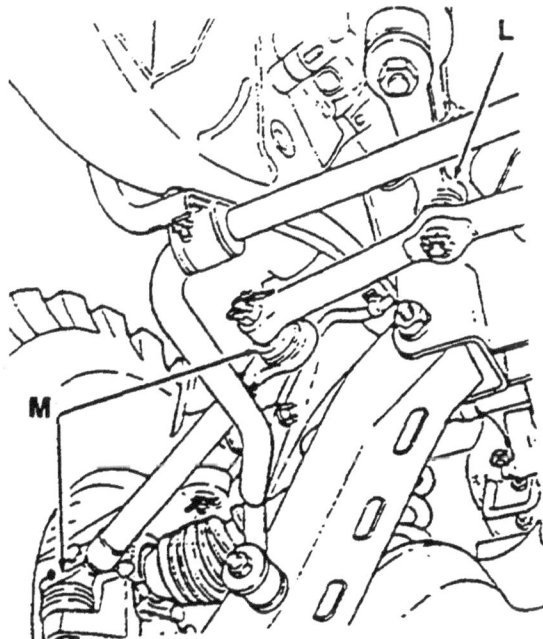

TIE ROD ENDS AND PITMAN ARM

LOCALIZED LUBRICATION POINTS (Cont'd)

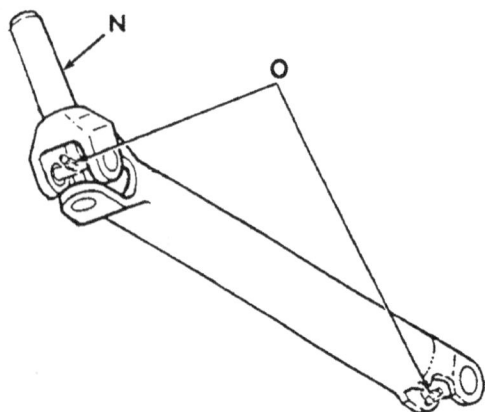

REAR PROPELLER SHAFT UNIVERSAL JOINTS
AND SLIP YOKES

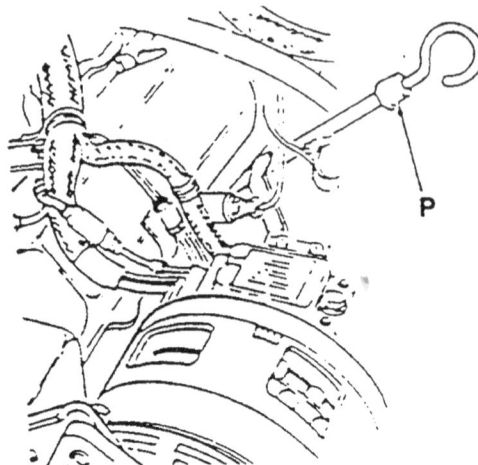

CRANKCASE LEVEL

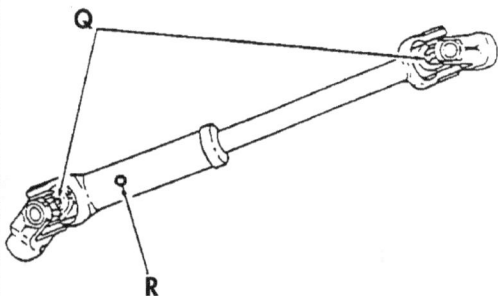

INTERMEDIATE STEERING SHAFT

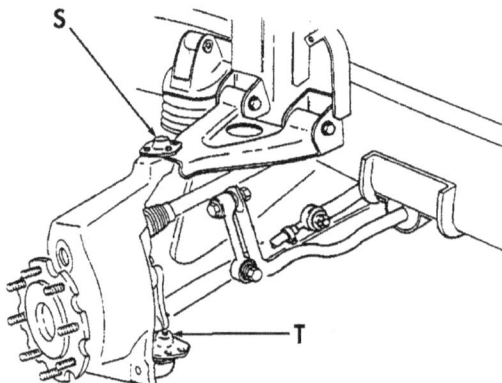

LOWER CONTROL ARM BALL JOINTS
AND UPPER CONTROL ARM BALL
JOINTS

LOCALIZED LUBRICATION POINTS (Cont'd)

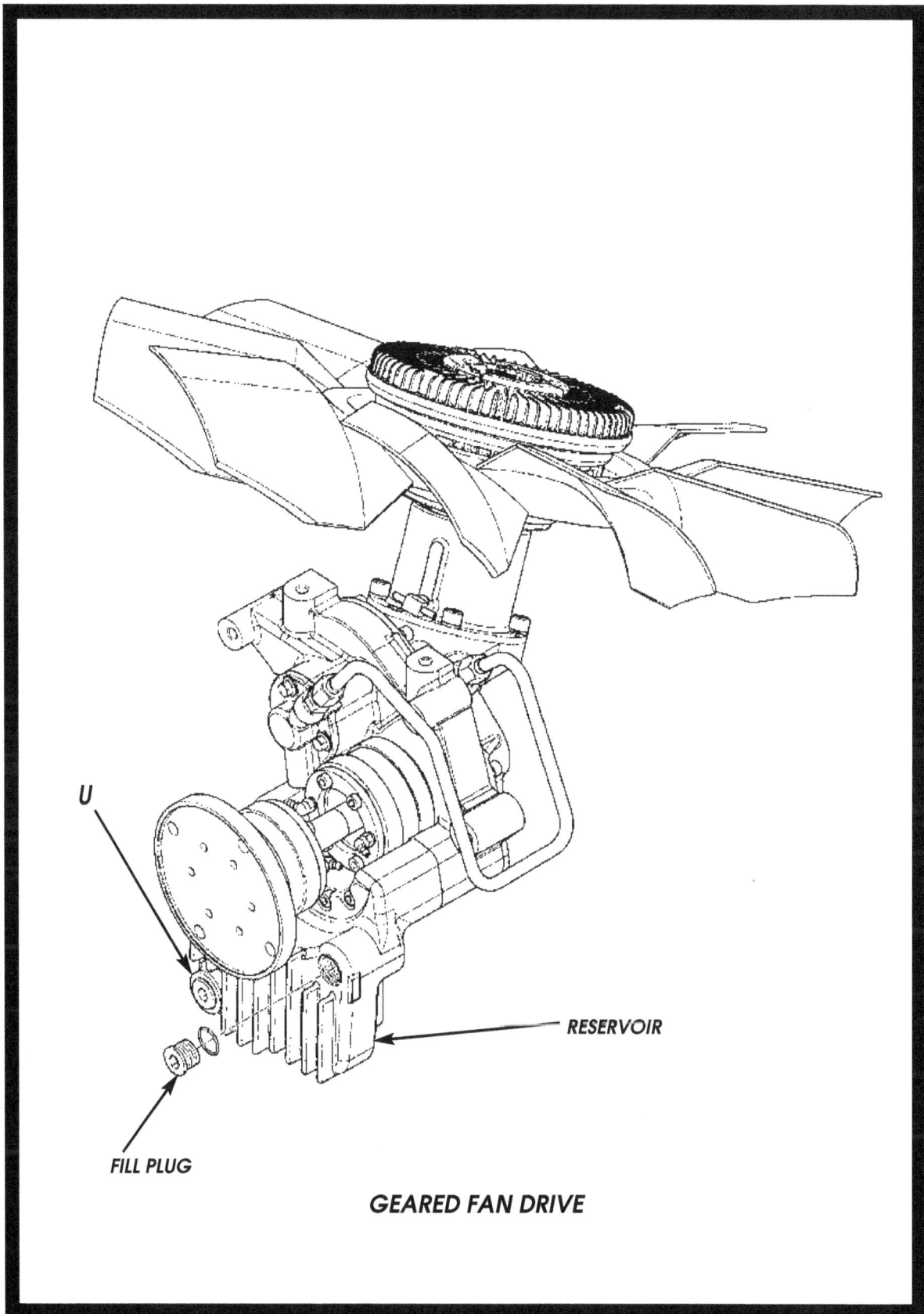

U

FILL PLUG

RESERVOIR

GEARED FAN DRIVE

NOTES

1. Oil Can Points.

Lubricate all oil can points every 3,000 mi (4,827 km), or semi-annually, whichever occurs first. Use seasonal grade OE on hood hinges, tailgate hinges, door hinges, door handles, transfer case shift linkage, accelerator linkage, transmission shift linkage, heater control shutoff valve lever and pin, heater control knob, and tiedown shackles and plunger. Lubricate cargo shell door paddle lock, rear wire handle lock, and cargo shell door latches with PL-S.

2. Winch Wire Rope.

WARNING

Wear leather gloves when handling winch wire rope. Do not handle wire rope with bare hands. Broken wires may cause injury.

NOTE

Front and rear winch lubrication is identical.

After each use, clean and lubricate winch wire rope with new OE/HDO 30. Clean entire wire rope with wire brush. Perform winch wire rope cleaning and lubrication every 3,000 mi (4,827 km), or annually, whichever occurs first, when wire rope is not used.

3. Pintle.

Every 3,000 mi (4,827 km), or annually, whichever occurs first, clean pintle with wire brush and lubricate rear plate fitting with seasonal grade OE.

4. Steering System.

CAUTION

Do not overlubricate tie rod ends and upper and lower control arm ball joints. One or two shots is adequate. Excessive lubrication will result in the boot rupturing. Observe the boot during lubrication: a seeping condition indicates adequate lubrication; expansion of the boot indicates over lubrication.

Lubricate front and rear propeller shaft, steering column, U-joints, slip yokes, tie rods, upper and lower control arm ball joints, radius rods, pitman arm, intermediate steering shafts, and idler arm with GAA/HDO every 3,000 mi (4,827 km), or annually, whichever comes first. Insert enough grease in each U-joint to purge air and impurities of all four cross-bearings. If one or more cross bearings do not purge, then shake, rotate and/or hit shaft with a rubber mallet to help purge cross-bearings. Repair or replace bearing if necessary, notify your supervisor

5. Power Steering Reservoir (P/N RCSK 18330).

CAUTION

Use Dexron® VI for filling power steering reservoir. Failure to use Dexron® VI may cause damage to power steering system.

Check the fluid level in the power steering reservoir monthly and adjust level as necessary. If fluid is hot, level should be between HOT and COLD marks on the cap indicator. If cool, level should be between ADD and COLD marks. In either condition, level must be above ADD mark. Fluid does not require periodic changing.

NOTES

5.1. Power Steering Reservoir (P/N 94252A).

CAUTION

Use Dexron® VI for filling power steering reservoir.Failure to use Dexron® VI may cause damage to power steering system.

Check the fluid level in the power steering reservoir monthly and adjust level as necessary. If fluid is HOT, level should be at top of sightglass. If fluid is COLD, fluid should be in center of sightglass. If fluid is at bottom of sightglass ADD fluid as required, till in center of sightglass. Fluid does not require periodic changing.

6. Transmission.

CAUTION

• Do not overfill transmission. The fluid level rises as the fluid temperature increases. Therefore, do not check level before the transmission has reached normal operating temperature. The safe operating level is within the crosshatch marks on the dipstick.Overfilling may result in damage to transmission.

• Use Dexron ® VI for filling transmission. Failure to use Dexron® VI may cause damage to transmission.

Check and fill transmission to proper level weekly. Operate transmission through all operating ranges to fill cavities and fluid passages. With vehicle positioned on level ground, allow engine to idle, shift transmission to P, and set parking brake. Check fluid level on dipstick. It should register within the crosshatch marks under the conditions stated above. Have fluid changed every 12,000 mi (19,308 km), or biennially, whichever occurs first.

7. Crankcase Oil Level.

CAUTION

There are two marks on the dipstick: FULL and ADD 1 QT. The quantity of oil required to raise the oil level from ADD 1 QT mark to FULL mark is 1 qt (0.9 L). Do not overfill crankcase. Overfilling may result in damage to engine.

NOTE

• If oil level is above FULL, it may be due to oil cooler drainback. Operate the engine for one minute, shut down, wait one minute, then recheck oil level.

• Oil is added to crankcase through fill tube which is located on top of engine.

Check crankcase oil level daily. Start engine and visually check for oil leaks at drainplug and oil filter. Stop engine and wait approximately one minute for oil to drain back into oil pan, then recheck oil level with dipstick. On vehicles equipped for deep water fording, the dipstick has a seal which fits into the opening of the dipstick tube. The dipstick handle must be turned counter-clockwise to be released before dipstick is withdrawn. Turn handle clockwise to seat after installing dipstick. Have oil changed every 3,000 mi (4,827 km), or semiannually, whichever occurs first.

8. Geared Fan Drive.

WARNING

If vehicle has been operating, use extreme care to avoid being burned when removing geared fan drive fill plug. Use rags or heavy gloves to protect hands.

Check and fill geared hub reservoir every 3,000 mi (4,827 km) or semi-annually, whichever comes first. Fluid should be even with bottom of fill plug hole.

INDEX

INDEX (Cont'd)

INDEX (Cont'd)

INDEX (Cont'd)

INDEX (Cont'd)

INDEX (Cont'd)

INDEX (Cont'd)

INDEX (Cont'd)

THE METRIC SYSTEM AND EQUIVALENTS

LINEAR MEASURE
1 Centimeter = 10 Millimeters = 0.01 Meters =
  0.3937 Inches
1 Meter = 100 Centimeters = 1,000 Millimeters =
  39.37 Inches
1 Kilometer = 1,000 Meters = 0.621 Miles

SQUARE MEASURE
1 Sq Centimeter = 100 Sq Millimeters = 0.155 Sq Inches
1 Sq Meter = 10,000 Sq Centimeters = 10.76 Sq Feet
1 Sq Kilometer = 1,000,000 Sq Meters = 0.386 Sq Miles

CUBIC MEASURE
1 Cu Centimeter = 1,000 Cu Millimeters = 0.06 Cu Inches
1 Cu Meter = 1,000,000 Cu Centimeters = 35.31 Cu Feet

LIQUID MEASURE
1 Milliliter = 0.001 Liters = 0.0338 Fluid Ounces
1 Liter = 1,000 Milliliters = 33.82 Fluid Ounces

TEMPERATURE
Degrees Fahrenheit (F) = °C · 9 ÷ 5 + 32
Degrees Celsius (C) = °F - 32 · 5 ÷ 9
212° Fahrenheit is equivalent to 100° Celsius
89.96° Fahrenheit is equivalent to 32.2° Celsius
32° Fahrenheit is equivalent to 0° Celsius

WEIGHTS
1 Gram = 0.001 Kilograms = 1,000 Milligrams =
  0.035 Ounces
1 Kilogram = 1,000 Grams = 2.2 Lb
1 Metric Ton = 1,000 Kilograms = 1 Megagram =
  1.1 Short Tons

APPROXIMATE CONVERSION FACTORS

TO CHANGE	TO	MULTIPLY BY
Inches	Millimeters	25.400
Inches	Centimeters	2.540
Feet	Meters	0.305
Yards	Meters	0.914
Miles	Kilometers	1.609
Square Inches	Square Centimeters	6.451
Square Feet	Square Meters	0.093
Square Yards	Square Meters	0.836
Square Miles	Square Kilometers	2.590
Acres	Square Hectometers	0.405
Cubic Feet	Cubic Meters	0.028
Cubic Yards	Cubic Meters	0.765
Fluid Ounces	Milliliters	29.573
Pints	Liters	0.473
Quarts	Liters	0.946
Gallons	Liters	3.785
Ounces	Grams	28.349
Pounds	Kilograms	0.4536
Short Tons	Metric Tons	0.907
Pound-Feet	Newton-Meters	1.356
Pounds Per Square Inch	Kilopascals	6.895
Miles Per Gallon	Kilometers Per Liter	0.425
Miles Per Hour	Kilometers Per Hour	1.609

TO CHANGE	TO	MULTIPLY BY
Millimeters	Inches	0.03937
Centimeters	Inches	0.3937
Meters	Feet	3.280
Meters	Yards	1.094
Kilometers	Miles	0.621
Square Centimeters	Square Inches	0.155
Square Meters	Square Feet	10.764
Square Meters	Square Yards	1.196
Square Kilometers	Square Miles	0.386
Square Hectometers	Acres	2.471
Cubic Meters	Cubic Feet	35.315
Cubic Meters	Cubic Yards	1.308
Milliliters	Fluid Ounces	0.034
Liters	Pints	2.113
Liters	Quarts	1.057
Liters	Gallons	0.264
Grams	Ounces	0.035
Kilograms	Pounds	2.2046
Metric Tons	Short Tons	1.102
Newton-Meters	Pound-Feet	0.738
Kilopascals	Pounds Per Square Inch	0.145
Kilometers Per Liter	Miles Per Gallon	2.354
Kilometers Per Hour	Miles Per Hour	0.621

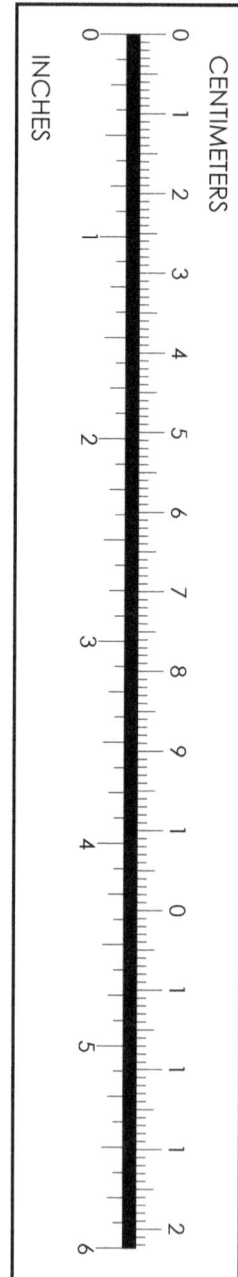

PIN 075913-000